Studies in Computational Intelligence

Volume 940

The series "Studies in Computational Intelligence" (SCI) publishes new developments and advances in the various areas of computational intelligence—quickly and with a high quality. The intent is to cover the theory, applications, and design methods of computational intelligence, as embedded in the fields of engineering, computer science, physics and life sciences, as well as the methodologies behind them. The series contains monographs, lecture notes and edited volumes in computational intelligence spanning the areas of neural networks, connectionist systems, genetic algorithms, evolutionary computation, artificial intelligence, cellular automata, self-organizing systems, soft computing, fuzzy systems, and hybrid intelligent systems. Of particular value to both the contributors and the readership are the short publication timeframe and the world-wide distribution, which enable both wide and rapid dissemination of research output.

Indexed by SCOPUS, DBLP, WTI Frankfurt eG, zbMATH, SCImago.

All books published in the series are submitted for consideration in Web of Science.

More information about this series at http://www.springer.com/series/7092

Oscar Castillo · Patricia Melin
Editors

Fuzzy Logic Hybrid Extensions of Neural and Optimization Algorithms: Theory and Applications

Springer

Editors
Oscar Castillo
Division of Graduate Studies and Research
Tijuana Institute of Technology
Tijuana, Mexico

Patricia Melin
Division of Graduate Studies and Research
Tijuana Institute of Technology
Tijuana, Mexico

ISSN 1860-949X ISSN 1860-9503 (electronic)
Studies in Computational Intelligence
ISBN 978-3-030-68775-5 ISBN 978-3-030-68776-2 (eBook)
https://doi.org/10.1007/978-3-030-68776-2

This Springer imprint is published by the registered company Springer Nature Switzerland AG
The registered company address is: Gewerbestrasse 11, 6330 Cham, Switzerland

Preface

We describe in this book, recent developments on fuzzy logic, neural networks and meta-heuristic optimization algorithms, as well as their hybrid combinations, and their application in areas such as intelligent control and robotics, pattern recognition, medical diagnosis, time series prediction and optimization of complex problems. There are papers with the main theme of type-1 and type-2 fuzzy logic, which basically consists of papers that propose new concepts and algorithms based on type-1 and type-2 fuzzy logic and their applications. There are also papers that present theory and practice of meta-heuristics in different areas of application. There are interesting papers on diverse applications of fuzzy logic, neural networks and hybrid intelligent systems in medical applications. In addition, we can find papers describing applications of fuzzy logic, neural networks and meta-heuristics in robotics problems. Another set of papers are presenting theory and practice of neural networks in different areas of application, including convolutional and deep learning neural networks. There are also a group of papers that present theory and practice of optimization and evolutionary algorithms in different areas of application. Finally, we can find a set of papers describing applications of fuzzy logic, neural networks and meta-heuristics in pattern recognition problems.

In conclusion, the edited book comprises papers on diverse aspects of fuzzy logic, neural networks and nature-inspired optimization meta-heuristics for forming hybrid intelligent systems and their application in areas such as intelligent control and robotics, pattern recognition, time series prediction and optimization of complex problems. There are theoretical aspects as well as application papers.

Tijuana, Mexico
October 2020

Oscar Castillo
Patricia Melin

Contents

Estimation of the Number of Filters in the Convolution Layers of a Convolutional Neural Network Using a Fuzzy Logic System

Yutzil Poma and Patricia Melin

Abstract In this paper, we propose to search for the best number of filters in the convolution layer of a convolutional neural network, we used a fuzzy logic system to find the most suitable parameters for the proposed case study. In addition to this we make use of the Fuzzy Gravitational Search Algorithm method to find the parameters of the fuzzy system memberships.

Keywords Optimization · Convolutional neural networks · Fuzzy gravitational search algorithm · Fuzzy system · Fuzzy logic

1 Introduction

The convolutional neural networks or also commonly called CNN, are used for image classification and recognition mainly, the first layers can detect lines, curves and specialize until they reach deeper layers that recognize complex shapes such as a face or the silhouette of an animal or anything object. In 1958 Scientist Frank Rosenblatt, inspired by the work of Warren McCulloch and Walter Pitts created the Perceptron, the unit from where artificial neural networks would be born and would be enhanced (Rosenblatt 1958), Convolutional Neural Networks are multilayered networks that draw their inspiration from the visual cortex of animals. The first CNN was created by Yann LeCun and was focused on handwriting recognition. The architecture consisted of several layers that implemented feature extraction and then classify (Le Cun Jackel 1990). Some works when the neural network were used like in Ashraf et al. (2020), is a new method of image representation was incorporated, which the algorithm is trained to classify medical images through deep learning techniques. It makes use of the application of a deep convolution red neural method which has previously been trained with the focus adjusted to the last three layers of deep red neural, or used in Karambakhsh (2020) they proposed a similar method for 3D segmentation using the Octree coding method and comparing their results to the

Y. Poma · P. Melin (✉)
Tijuana Institute of Technology, Tijuana, Mexico
e-mail: pmelin@tectijuana.mx

O. Castillo and P. Melin (eds.), *Fuzzy Logic Hybrid Extensions of Neural and Optimization Algorithms: Theory and Applications*, Studies in Computational Intelligence 940,
https://doi.org/10.1007/978-3-030-68776-2_1

latest advances with method manipulation, others methods where the using the CNN are like (Zangeneh 2020) A non-linear coupled mapping architecture was created, which uses two deep CNNs and achieves a high percentage of recognition when the image has a low resolution or in Poma (2020) where they using a convolutional neural network in conjunction with the Fuzzy gravitational Search Algorithm for optimizer the parameter Bsize that is a parameter of a the network.

The fuzzy logic has been of great help and importance in the study and research of various methods, which has been applied in various works such as Bernal (2020) where the fuzzy approach was tested with a set of benchmark math functions and in conjunction with a fuzzy controller of the water tank problem to measure performance or in the (Carvajal 2020) which presents the design of a hardware system for an autonomous mobile robot and the development of a fuzzy logic controller to control the movement of a robot to follow a path, or the work when they use the gray wolf optimization algorithm (GWO), where it proposes to solve fuzzy problems for autonomous mobile robots (Hernández 2020), also in the applications of control using fuzzy logic type1 and type 2 like the results of the fuzzy controller optimization are compared using the dynamic adjustment of the Type-1 (T1) parameters and the fuzzy logic of the Type-2 Interval (T2) with the Firefly Algorithm (FA), this algorithm has been used to optimization of the parameters of the controller membership functions (Lagunes 2020) and the others works as in which a method is used to dynamically adjust the parameters in metaheuristics based on the fuzzy logic of the type 2 interval (Olivas et al. 2019).

The convolutional neural network are used in In conjunction with many methods, one of these is fuzzy systems, which in many cases help the network to be more effective in recognizing images as in Choi (2017) is a CNN output optimization method to improve the precision of low precision classes, or in Kh-Madhloom et al. (2020) where CNN is combined with fuzzy logic to more accurately recognize the smile on the human face or the use a fuzzy control system was used in a robot, which previously implemented a classification navigation approach (Born 2018).

As for the method of searching the points of the fuzzy system's membership functions, the FGSA (Sombra 2013) also called Fuzzy Gravitational Search Algorithm was used for this purpose. This method unlike its predecessor, the Gravitational Search Algorithm Method (Rashedi et al. 2009), it uses a fuzzy system to vary the alpha parameter, either increasing or decreasing its value, while in the GSA it is a static parameter. Some works where these methods have been used are (Hatamlou 2011; Verma 2012; Mirjalili 2012).

The main contribution of this work is being able to find the best number of filters for each convolution layer of a convolutional neural network and being able to demonstrate that with the help of Fuzzy system we can vary this parameter in the network, obtaining results with a high percentage of recognition of the images.

This document is sectioned as follows: Sect. 2 shows the basic concepts of CNN and the methods used for the experimentation carried out. In Sect. 3, we can find the proposed method to optimize the neural network. Section 4 shows the results obtained when fuzzy logic is applied in the search for network parameters. Concluding, in Sect. 5 where we can find the conclusions of the experimentation with the selected case study.

2 Literature Review

This section presents the basic concepts necessary to understand the proposed method.

2.1 Convolutional Neural Networks

CNN is a type of Artificial Neural Network with supervised learning that processes its layers imitating the visual cortex of the human eye to identify different characteristics at the entrances that ultimately make it possible to identify objects and "see". For this, CNN contains several specialized hidden layers with a hierarchy: the convolutional layer, pooling layer and classify layer. The main use of convolutional neural networks is the recognition and classification of images (LeCun and Bengio 1998).

2.1.1 Convolution Layer

the convolution layer is the first layer of the network which, together with a filter designed by the programmer or randomly, is in charge of reviewing the image extra-filling its characteristics, thus forming the so-called "characteristics map" which will pass to the next layer of the network, previously this feature extraction goes through an activation function which is generally the ReLU (Rectified linear units) (Nair 2010).

2.1.2 Pooling Layer

This layer, also called grouping layer, is in charge of further reducing the characteristics of the image, it uses an empty mask which goes through the image pixel by pixel and within the set of pixels that cover the masks it will be determined according to the average of all the set of pixels that make up the mask or the maximum of these. This will generate a new map of the main characteristics of the image (Yang 2009).

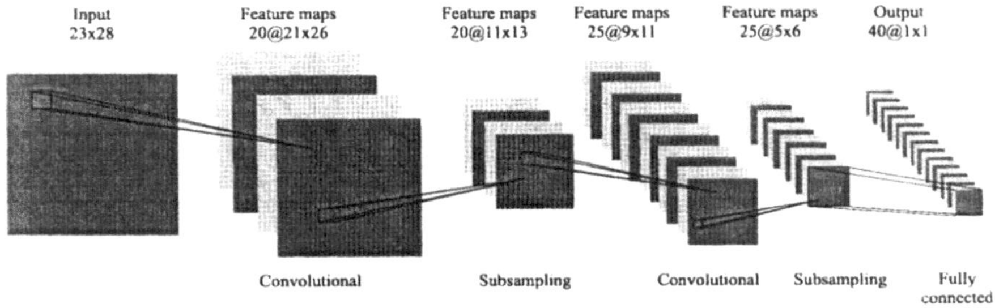

Fig. 1 CNN typical structure

2.1.3 Classifier Layer

Finally, the fully connected layer is used, in which each pixel is a separate neuron like a multilayer perceptron. In this layer the number of classes to recognize is equal to the number of new ones, so this layer fulfills the purpose of classifying and recognizing the images. (Venkatesan 2017).

Figure 1 we can see the typical architecture of a convolutional neural network (Le Cun 1990), where there are layers between loops starting with the convolution layer, moving on to the pooling layer and according to the designer can add more or less layers, ending with a classification layer.

2.2 GSA

In Rashedi et al. (2009) the parent algorithm of the FGSA is introduced, which is based on a probation algorithm, in addition to taking the principles of the law of gravity and Newton's second law. This algorithm has agents which are the "objects" and its performance is measured by the "masses", these objects attract each other thanks to the force of gravity and this causes a global movement between the objects.

The masses use direct communication through the force of gravitation. When an agent with a very heavy mass is slower in its movement.

Therefore it is the best solution. Finally the inertial masses and gravitation is determined thanks to the use of a fitness function.

This algorithm shows that each mass can be a solution and this algorithm properly adjusts the gravitational mass and inertia, while the algorithm is running.

The masses are expected to be tied by the heaviest mass, the latter will be the most optimal solution in the search space.

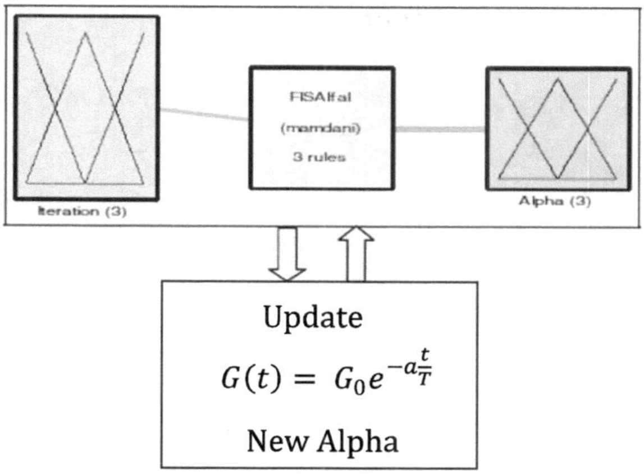

Fig. 2 Fuzzy system for new alpha

2.3 *FGSA*

In this method, agents are called objects which are determined by masses. These attract each other, thanks to the force of gravity which causes a global movement in each of them and maintains direct communication with the masses. Some of the works in which this algorithm has been used have been in González et al. (2015a, b).

While the value of the Alpha parameter varies, different accelerations and gravitations are obtained for each agent and therefore improves its performance. Alpha has been optimized thanks to a fuzzy system in which its functions are triangular and its ranges are determined in a search threshold to find the alpha parameter (Sombra 2013).

It was decided to use the linguistic values and the ranges for each:

Tiny: [−50 0 50], Medium [0 50 100], So High [50 100 150]

The fuzzy system that carries out the modification of the *Alpha* variable has 3 rules which are:

$R = repetition$

- If the R is tiny then α is small.
- If the R is medium and then α is medium.
- If the R is high then α is tall.

In Fig. 2 we can see the fuzzy system used in the method for the obtained the new alpha.

3 Proposed Method

The proposed model the layered architecture for CNN was a s follows:

Conv1 → ReLU → Pool1 → Conv2 → ReLU → Pool2 → Clasif

which is made up of 2 convolution layers and two pooling layers, using between each layer a ReLU(activation function), ending with a clasiffy layer.

A fuzzy system was designed which is shown in Fig. 3, this is made up of a triangular input membership function and two triangular outputs the same membership function, all of which are dynamic.

Each output representing a number of filters for each convolution layer of the neural network convolutional, the output one is the number of filters of convolutional layer 1 and the output 2 is the respectily, at each input and output of the fuzzy system, three membership functions were added, the points of which are given by the FGSA (Input/Error) that represents the error obtained by the neural network and while the output (number of convolution layer filters 1 and 2).

The rules used in the fuzzy system are as follows:

1. If (Error is −REC)then(NumFilters1 is −REC) and (NumFilters2 is −REC)
2. If (Error is 1/2REC) then (NumFilters1 is 1/2REC) and (NumFilters2 is 1/2REC)
3. If (Error is +REC) then (NumFilters1 is +REC) and (NumFilters2 is +REC).

This Fuzzy system is applied in the CNN that the architecture before mentioned.

In the next flowchart in Fig. 4, we can see the steps of how the method is developed in conjunction with the CNN and the FGSA.

Begins with the FGSA which generates the agents, these are the input values of the triangular membership functions, which enter the fuzzy system, the latter has as output the number of filters of the convolution layer 1 (output 1) and the number of filters of the convolution layer 2 (output 2), these values enter the neural network, which applies in their respective convolution layer, once the data is processed, with their respective training epochs (10, 20, 30, 40, 50, 60, 70, 80), the neural network

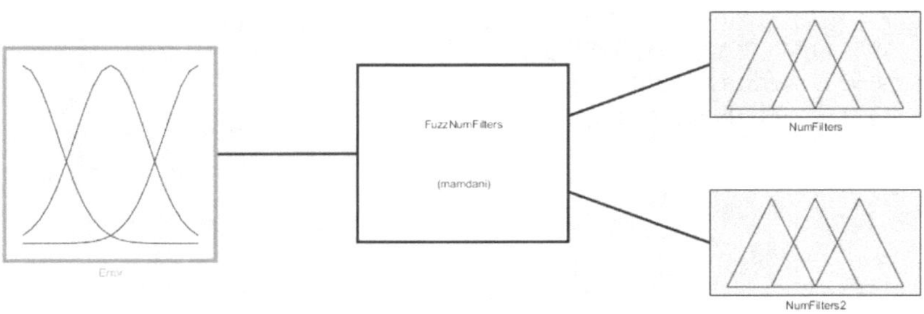

Fig. 3 Proposed fuzzy system

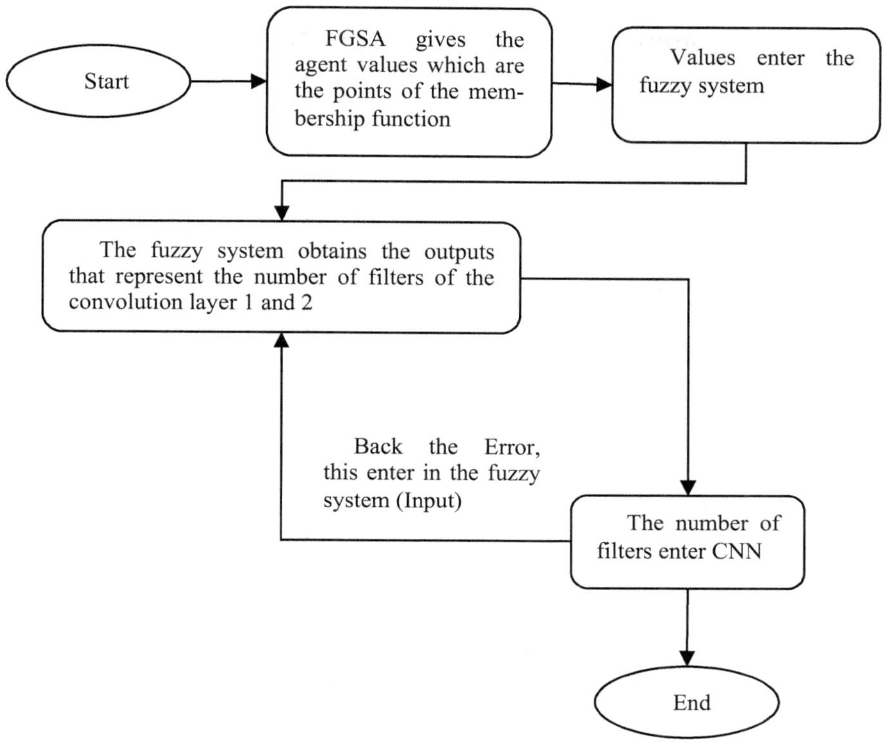

Fig. 4 Flowchart of interaction FGSA with fuzzy system and CNN

returns the classification error, which is entered again to the fuzzy system and this process is carried out until finishing with the stop guideline, in this case the number of iterations (30).

4 Results and Discussion

As a case study, the ORL database was used, which contains 40 different human faces and 10 images for each one of them, making a total of 400 images in total of 112 * 92 pixels each in a .pgm format which in Fig. 5 can have some examples of this database.

30 experiments were performed for each epoch of the neural network, the epochs began in 10, and they were increasing from 10 to 10 until reaching 80 epochs.

In the next tables we can see the results obtained using the method. In Table 1 we shows all the results when we run the experiment 30 times with 10, 20 and 30 epochs, on the other hand in Table 2 we can see the results of epochs 40, 50 and 60 and finally in Table 3 we see the results obtained when the experiment runs with 70

Fig. 5 Images from the ORL database

and 80 epochs.

The results in Table 4, show the increase in the percentage of recognition of the images, clarifying that 60% of the data was used for training and 40% of these were used for testing.

The best solution found for this neural network architecture and using a fuzzy system to find the values of the number of filters for each convolution layer was an average of 92.85% recognition, while the highest value was 94.375% with 80 epochs in both cases and the number of filter con the first convolutional layer is 24 and the second convolutional layer it was 20.

5 Conclusions

Based on the experiments carried out, we can conclude that if a small filter is used, more characteristics of the image can be extracted, and if a diffuse system is used to find the number of filters for each convolution layer of the network, good results are obtained even when CNN is trained with little data compared to when this number of training data increases and other methods of optimization and parameter search are used.

As future work, it is intended to implement other types of membership functions in addition to playing with the points of each one of them to open or delimit the threshold of options between these. In addition, we could consider multiple objective

Table 1 Results the experiments of 10, 20 and 30 epochs

Experiment Number/Epochs	Recognition rate			Number of filter 1			Number of filter 2		
	10	20	30	10	20	30	10	20	30
1	95	90	93.75	22	50	23	23	50	24
2	89.375	92.5	94.375	50	25	24	50	26	24
3	91.875	91.25	91.25	29	25	24	24	26	25
4	90	94.375	92.5	25	25	25	21	24	25
5	89.375	91.875	91.25	22	30	50	26	28	50
6	90.625	91.25	91.25	23	27	25	25	23	27
7	91.875	91.25	91.875	22	24	24	24	24	26
8	89.375	92.5	90	28	23	50	28	26	50
9	90	91.25	90.625	23	22	23	27	25	24
10	90.625	95	93.75	27	26	50	22	24	50
11	91.25	90.625	91.875	23	27	18	24	26	22
12	89.375	91.25	93.75	24	22	25	26	22	22
13	93.75	91.875	93.125	23	24	50	26	24	50
14	90.625	90.625	93.75	25	27	26	28	23	23
15	91.875	94.375	91.25	24	24	50	22	25	50
16	90	91.875	91.25	25	23	30	22	25	24
17	93.125	91.875	92.5	20	25	24	18	24	29
18	91.25	92.5	91.25	24	50	50	23	50	50
19	91.875	91.875	93.75	22	21	22	24	22	29
20	89.375	91.25	93.75	25	23	23	26	27	22
21	92.5	91.25	92.5	26	26	25	25	26	23
22	91.875	92.5	92.5	25	25	50	27	25	50
23	91.25	93.125	92.5	26	24	23	23	29	27
24	90	91.25	92.5	25	27	26	22	26	26
25	90.625	90	93.125	22	24	25	21	20	24
26	89.375	90	91.875	26	23	25	24	22	23
27	91.25	91.875	94.375	23	25	50	20	22	50
28	88.75	91.875	93.75	26	30	25	28	26	27
29	90	92.5	90.625	26	50	24	26	50	24
30	90.625	90.625	92.5	25	28	22	27	28	22
Average	90.89583	91.8125	92.4375						
Standard deviation	1.4466	1.2320	1.2100						

Table 2 Results the experiments of 40, 50 and 60 epochs

Experiment Number/Epochs	Recognition rate			Number of filter 1			Number of filter 2		
	40	50	60	40	50	60	40	50	60
1	91.25	93.75	91.25	23	50	50	25	50	50
2	91.875	92.5	92.5	24	23	23	27	27	24
3	92.5	92.5	93.125	28	50	50	29	50	50
4	93.125	93.75	93.125	24	50	24	24	50	22
5	93.125	93.125	93.75	22	23	29	24	25	28
6	91.25	92.5	90.625	24	50	25	26	50	26
7	90.625	90.625	90.625	21	21	50	22	20	50
8	91.25	93.125	91.875	30	24	21	30	22	23
9	92.5	91.25	92.5	29	27	25	25	29	23
10	93.75	92.5	92.5	25	23	28	22	22	29
11	91.25	93.125	94.375	25	26	50	25	27	50
12	94.375	91.875	93.75	27	26	26	22	27	28
13	92.5	90	94.375	29	23	24	27	25	24
14	91.875	91.875	93.125	22	22	50	21	27	50
15	91.25	91.25	94.375	22	25	50	21	25	50
16	93.125	94.375	93.125	24	20	22	26	22	25
17	92.5	91.875	91.875	50	23	22	50	26	24
18	93.125	91.25	93.125	25	25	27	23	23	26
19	90.625	92.5	92.5	25	25	25	25	25	24
20	93.125	93.125	94.375	26	25	25	28	25	25
21	92.5	93.125	92.5	26	21	50	26	25	50
22	92.5	93.125	93.125	29	25	22	27	27	24
23	93.125	90	92.5	22	28	25	22	25	23
24	90.625	92.5	93.75	22	25	24	24	25	23
25	91.25	90.625	93.125	25	50	21	23	50	24
26	90.625	90.625	91.875	50	26	25	50	23	25
27	92.5	94.375	91.25	22	23	50	24	25	50
28	91.25	93.125	91.875	28	25	25	27	21	25
29	93.125	91.875	91.25	23	21	27	21	23	24
30	93.75	91.25	93.125	50	25	24	50	21	27
Average	92.20833	92.25	92.70833						
Standard deviation	1.060321	1.201651	1.080456						

Table 3 Results of experiments of 70 and 80 epochs

Experiment Number/Epochs	Recognition rate		Number of filter 1		Number of filter 2	
	70	80	70	80	70	80
1	91.875	91.875	25	24	25	23
2	91.875	92.5	27	23	27	24
3	93.75	93.75	24	23	24	23
4	90.625	92.5	24	25	26	24
5	93.125	91.25	25	26	20	26
6	92.5	93.125	22	50	25	50
7	91.875	92.5	23	25	29	26
8	93.125	93.75	22	24	23	22
9	91.875	92.5	23	25	23	28
10	91.875	93.75	27	22	24	23
11	93.75	94.375	23	24	24	20
12	92.5	92.5	22	27	24	26
13	93.125	93.125	25	26	21	24
14	92.5	91.25	25	21	25	25
15	91.25	93.75	22	22	22	25
16	92.5	93.75	24	25	23	25
17	93.125	91.875	23	23	23	25
18	93.125	93.75	23	50	22	50
19	91.875	93.125	24	23	21	24
20	92.5	91.25	25	22	25	22
21	93.125	91.875	25	28	22	29
22	92.5	93.75	22	22	26	28
23	92.5	93.125	24	27	24	27
24	91.25	91.25	27	24	21	23
25	91.25	94.375	26	20	22	21
26	91.25	92.5	22	24	25	22
27	90.625	93.125	25	24	23	24
28	91.875	93.125	23	24	23	22
29	92.5	93.125	23	23	27	22
30	92.5	93.125	50	25	50	27
Average	92.27083	92.85417				
Standard deviation	0.82856	0.923871				

Table 4 Experiments using fuzzy logic in for find the number of filters

Epochs	Experiment number	Recognition rate	Number of filter 1	Number of filter 2	Average	Standard deviation
10	1	95	22	23	90.89	1.44
20	10	95	26	24	91.81	1.23
30	2	94.375	24	24	92.43	1.21
40	12	94.375	27	22	92.20	1.06
50	16	94.375	20	22	92.25	1.20
60	11	94.375	50	50	92.70	1.08
70	3	93.75	24	24	92.27	0.82
80	**11**	**94.375**	**24**	**20**	**92.85**	**0.92**

optimization to improve the results, as in Sánchez and Melin (2014), Sánchez et al. (2017), Sánchez (2017), or other kinds of applications, like in Castillo (1998), Castillo and Melin (2003), Sanchez et al. (2014).

Acknowledgements We thank sour sponsor CONACYT & the **Tijuana Institute of Technology** for the financial support provided with the scholarship number 816488.

References

Ashraf, R., M.A. Habib, M. Akram, M.A. Latif, M.S.A. Malik, M. Awais, S.H. Dar, T. Mahmood, M. Yasir, and Z. Abbas. 2020. Deep convolution neural network for big data medical image classification. *IEEE Access* 8: 105659–105670.

Bernal, E., O. Castillo, J. Soria, F. Valdez. 2020. Fuzzy galactic swarm optimization with dynamic adjustment of parameters based on fuzzy logic. *SN Computer Science* 1(1): 59.

Born, W. and C.J. Lowrance. 2018. Smoother robot control from convolutional neural networks using fuzzy logic. *ICMLA*, 695–700.

Carvajal, O., O. Castillo. 2020. Implementation of a fuzzy controller for an autonomous mobile robot in the PIC18F4550 microcontroller. *Hybrid Intelligent Systems in Control, Pattern Recognition and Medicine*, 315–325.

Castillo, O., and P. Melin. 1998. A new fuzzy-fractal-genetic method for automated mathematical modelling and simulation of robotic dynamic systems. In 1998 IEEE international conference on fuzzy systems (FUZZ-IEEE 1998) Proceedings. Volume 2, 1182–1187.

Castillo, O., and P. Melin. 2003. Intelligent adaptive model-based control of robotic dynamic systems with a hybrid fuzzy-neural approach. *Applied Soft Computing* 3 (4): 363–378.

Choi, H. 2017. CNN output optimization for more balanced classification. *International Journal Fuzzy Logic and Intelligent Systems* 17 (2): 98–106.

González, B., F. Valdez, P. Melin, and G. Prado-Arechiga. 2015a. Fuzzy logic in the gravitational search algorithm enhanced using fuzzy logic with dynamic alpha parameter value adaptation for the optimization of modular neural networks in echocardiogram recognition. *Applied Soft Computing* 37: 245–254.

González, B., F. Valdez, P. Melin, and G. Prado-Arechiga. 2015b. Fuzzy logic in the gravitational search algorithm for the optimization of modular neural networks in pattern recognition. *Expert Systems with Applications* 42 (14): 5839–5847.

Hatamlou, A., S. Abdullah, and Z. Othman. 2011. Gravitational search algorithm with heuristic search for clustering problems. In conference data minimum optimaztion no. June, pp. 190–193.
Hernández, E., O. Castillo, J. Soria, 2020. Optimization of fuzzy controllers for autonomous mobile robots using the grey wolf optimizer. *Hybrid Intelligent Systems in Control, Pattern Recognition and Medicine*, 289–299.
Karambakhsh, A., B. Sheng, P. Li, P. Yang, Y. Jung, D.D. Feng. 2020. Hybrid convolutional neural network for active 3D object recognition. *IEEE Access 8*: 70969–70980.
Kh-Madhloom, J., S.A. Diwan, and A.A. Zainab. 2020. Smile detection using convolutional neural network and fuzzy logic. *Journal of Information Science and Engineering* 36 (2): 269–278.
Lagunes, M.L., O. Castillo, F. Valdez, J. Soria. 2020. Comparison of fuzzy controller optimization with dynamic parameter adjustment based on of Type-1 and Type-2 Fuzzy logic. *Hybrid Intelligent Systems in Control, Pattern Recognition and Medicine*, 47–56.
Le Cun, Y., B. Boser, J.S. Denker, D. Henderson, R. Howard, W. Hubbard, and L. Jackel. 1990. Handwritten digit recognition with a backprop-agation neural network. In *Advances in neural information processing systems* 2, D. S. Touretzky, Ed. San Mateo, CA: Morgan Kaufmann, pp. 396–404.
Le Cun Jackel, L.D., B. Boser, J.S. Denker, D. Henderson, R.E. Howard, W. Hubbard, B. Le Cun, J. Denker, and D. Henderson. 1990. Handwritten digit recognition with a back-propagation network. *Advances in Neural Information Processing Systems*, pp. 396–404.
LeCun, Y., and Y. Bengio. 1998. Convolution Networks for Images, Speech, and Time-Series. *Igarss* 2014 (1): 1–5.
Mirjalili, S., S.Z. Mohd Hashim, and H. Moradian Sardroudi. 2012. Training feedforward neural networks using hybrid particle swarm optimization and gravitational search algorithm. *Application Mathematical Computer*, 218(22): 11125–11137.
Nair, V. and G.E. Hinton. 2010. Rectified linear units improve restricted boltzmann machines. In Proceedings 27th international conference mathematical learning, no. 3, pp. 807–814.
Olivas, F., F. Valdez, P. Melin, A. Sombra, and O. Castillo. 2019. Interval type-2 fuzzy logic for dynamic parameter adaptation in a modified gravitational search algorithm. *Information Sciences* 476: 159–175.
Poma, Y., P. Melin C. González, G. Martinez. 2020. Optimal recognition model based on convolutional neural networks and fuzzy gravitational search algorithm method. https://doi.org/10.1007/978-3-030-34135-0_6.
Rashedi, E., H. Nezamabadi-pour, and S. Saryazdi. 2009. GSA: A gravitational search algorithm. *Information Science (Ny)* 179 (13): 2232–2248.
Rosenblatt, F. 1958. The perceptron: A probabilistic model for information storage and organization in the brain. *Psychological Review* 65: 386–408. https://doi.org/10.1037/h0042519.
Sánchez, D., P. Melin, J. Carpio, and H. Puga. 2017. Comparison of optimization techniques for modular neural networks applied to human recognition. In Nature-Inspired Design of Hybrid Intelligent Systems (pp. 225–241). Springer, Cham.
Sánchez, D., and P. Melin. 2014. Optimization of modular granular neural networks using hierarchical genetic algorithms for human recognition using the ear biometric measure. *Engineering Applications of Artificial Intelligence* 27: 41–56.
Sanchez, M.A., O. Castillo, J.R. Castro, and P. Melin. 2014. Fuzzy granular gravitational clustering algorithm for multivariate data. *Information Sciences* 279: 498–511.
Sánchez, D., P. Melin, and O. Castillo. 2017b. Optimization of modular granular neural networks using a firefly algorithm for human recognition. *Engineering Applications of Artificial Intelligence* 64: 172–186.
Sombra, A., F. Valdez, P. Melin, and O. Castillo. 2013. A new gravitational search algorithm using fuzzy logic to parameter adaptation. In *2013 IEEE Congress on Evolutionary Computation no.* 3, pp. 1068–1074.
Venkatesan, R. and B. Li. 2017. *Convolutional neural networks in visual computing: A concise guide*. CRC Press.

Verma, O.P. and R. Sharma. 2012. Newtonian gravitational edge detection using gravitational search algorithm. In International conference communication system networks technology, pp. 184–188.
Yang, J., K. Yu, Y. Gong, and T. H. Beckman. 2009. Linear spatial pyramid matching using sparse coding for image classification. *IEEE computer society conference* on *computer* vision and *pattern recognition*, pp. 1794–1801.
Zangeneh, E., M. Rahmati, Y. Mohsenzadeh. 2020. Low resolution face recognition using a two-branch deep convolutional neural network architecture. *Expert Systems Application,* 139.

Optimization of Membership Function Parameters for Fuzzy Controllers in Cruise Control Problem Using the Multi-verse Optimizer

Lucio Amézquita, Oscar Castillo, José Soria, and Prometeo Cortes-Antonio

Abstract In this paper we propose the application of metaheuristics to optimize control systems that use fuzzy logic, involving the multi-verse optimizer in various situations that involve control. The systems that we study are benchmark fuzzy controllers, like the case of cruise control, which focuses on controlling the velocity that a vehicle has to achieve and maintain in an optimal environment for its implementation, without constraints of the outside world like air friction or the use of inclination. This is by using a simple one input-output fuzzy system that is being optimized in its membership functions parameters; also another system used to prove the functionality of the algorithm is a case of approximation for the tipper system, were the objective is to approximate the system using a two input-one output fuzzy system. The main goal is to introduce the multi-verse optimizer in control problems as a great choice for these systems.

Keywords Multi-verse optimizer · Metaheuristics · Control system · Fuzzy logic · Study · Optimization · Cruise control · Mamdani · Sugeno

1 Introduction

The metaheuristics are part of the computational techniques that are nowadays more commonly used to solve problems related to make better the solutions to a problem and this is what we call optimization. More commonly we can say that these are

L. Amézquita · O. Castillo (✉) · J. Soria · P. Cortes-Antonio
Tijuana Institute of Technology, Tijuana, Mexico
e-mail: ocastillo@tectijuana.mx

L. Amézquita
e-mail: lucio.amezquita19@tectijuana.edu.mx

P. Cortes-Antonio
e-mail: prometeo.cortes@tectijuana.edu.mx

O. Castillo and P. Melin (eds.), *Fuzzy Logic Hybrid Extensions of Neural and Optimization Algorithms: Theory and Applications*, Studies in Computational Intelligence 940,
https://doi.org/10.1007/978-3-030-68776-2_2

stochastic methods that implement optimization to find better solutions for specific problems (Engelbrecht 2007), like the case of finding the best parameter configuration for a fuzzy system.

We can find many areas were optimization is applied in computer science, like artificial neural networks (Melin 2005), fuzzy systems (Jang 1997), or related areas like machine learning (Khomh 2018) or artificial vision (Carranco 2020); if we go to the fuzzy logic area, there are applications like fuzzy controllers, were optimization can find the best configurations for the membership functions or the number of membership functions to solve the problem.

There is a wide variety of metaheuristic algorithms, which have different inspirations from behaviors of nature beings or artificial behaviors like Flower Pollination Algorithm (FPA) (Yang 2014), Particle Swarm Optimization (PSO) (Engelbrecht 2007) or Grey Wolf Optimizer (GWO) (Mirjalili et al. 2014), just to mention examples.

Fuzzy logic has some history, it was proposed by Zadeh in conjunction to fuzzy set theory (Zadeh 1968, 1973), in fuzzy logic we can see one of his main applications in fuzzy inference systems or FIS, which use the fuzzy set theory, fuzzy if-then rules and fuzzy reasoning (Jang 1997). Basically, fuzzy systems are used in various situations where human reasoning is needed, this because has tolerance for imprecision (Mahmoodabadi and Jahanshahi 2016) thanks to the fuzzy sets that assign a numeric value instead of only true-false decisions that come from traditional logic.

Now, some of the areas where fuzzy logic has been implemented very often is control, this on the need to make systems on multiple areas, like vehicles, to promote better decisions based on the controller design, from PID to more complex fuzzy controllers that can also be fuzzy-PID, this to handle better the inputs of the system so it can have a more complete output for the system that is being controlled (Izadbakhsh 2017).

We can notice on this paper that the use of an optimization algorithm such as multi-verse optimizer (MVO) can be a great candidate for fuzzy control systems such as the problem of cruise control on vehicles, so it can approximate the best parameters for the membership functions so it can outperform a PID based controller without moving too much the parameters of the system, so we can have proof that this algorithm can handle these type of problems easily an competitively.

This paper is organized as follows: Sect. 2 describes Mamdani and Sugeno fuzzy systems, Sect. 3 mentions control systems and some applications, Sect. 4 describes some metaheuristics and the MVO algorithm, Sect. 5 has the results of the cases of study and Sect. 6 outlines the conclusions to this work.

2 Fuzzy Systems

In traditional set theory, an element needs to belong or not of a set, like in binary logic, were we have the 1 or 0 that represent this reasoning. If we compare this to the human reasoning, many times we can have some level of decision, were we can say that we are not completely sure about a choice, and here come a level of uncertainty (Engelbrecht 2007).

When Zadeh proposed fuzzy logic and fuzzy sets (Zadeh 1968, 1973), he referred as an approximate reasoning; because with a fuzzy set, an element can have a certain degree for a choice, or better called a degree of membership. With fuzzy logic it can allow to reason with a level of uncertainty to have choices more approximated to human reasoning, because it allows to model common sense.

Fuzzy systems can model human reasoning because of its elements that conform him: fuzzy rules, membership functions and a reasoning mechanism. A basic representation of fuzzy systems can be appreciated in Fig. 1.

There are three types of fuzzy inference systems that have been used in many situations: Mamdani, Sugeno and Tsukamoto. From these three, in this paper we will explain only two, that are used on the experiments made for the cases of study.

2.1 Mamdani Model

One of the most used models for fuzzy inference systems is the one called Mamdani, which was proposed as an attempt to control a steam engine and boiler combination. In this problem they used a set of linguistic control rules that were obtained by human

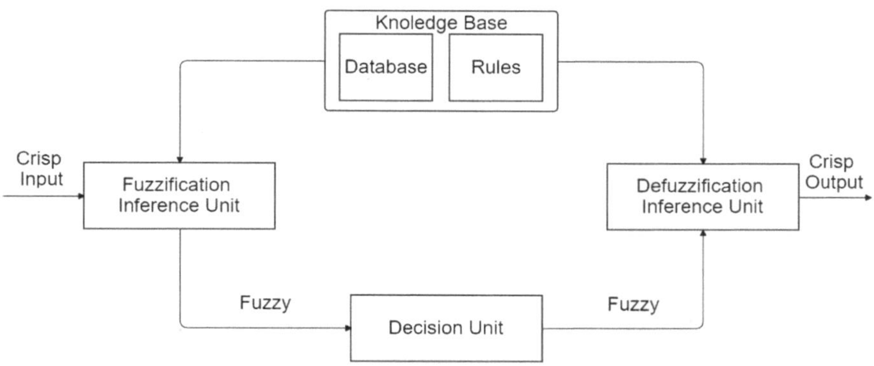

Fig. 1 Basic representation of a fuzzy inference system

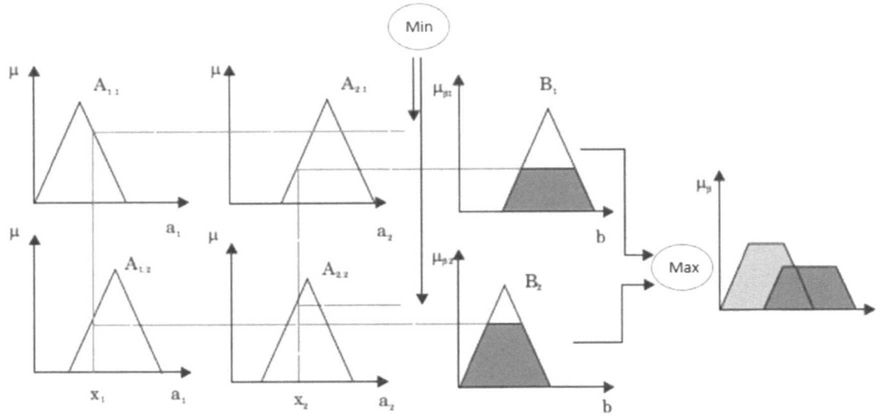

Fig. 2 Example of Mamdani FIS using min and max as operators

operators of the engine (Jang 1997), this model was proposed in 1975 by Ebrahim Mamdani. We can observe in Fig. 2 how a two-rule system derives the output z.

There are some steps to be followed to obtain the output from this model: Have a set of fuzzy rules, then use input membership function to make the input fuzzy, now establish rule strength by combining the inputs according to fuzzy rules, then we determine the consequent of rule combining the strength of the rule and the output membership function, from here we combine all the consequents and we obtain a defuzzified output.

The defuzzification is the way a crisp value is extracted from a fuzzy set as a representative value; there are five methods to do this: Centroid of area, Bisector of area, Mean of maximum, Smallest of maximum and Largest of maximum. In Fig. 3 we can find an example for centroid of area.

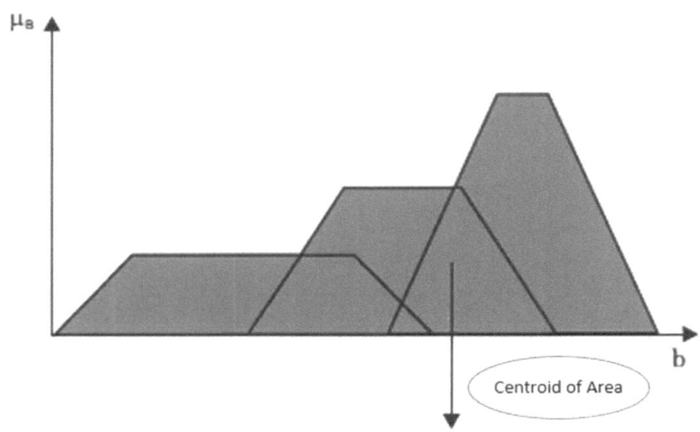

Fig. 3 Defuzzification using centroid of area

2.2 *Sugeno Model*

The Sugeno fuzzy model or Takagi-Sugeno fuzzy model was proposed by Takagi, Sugeno and Kang in 1985, this was developed a systematic approach to generate fuzzy rules from a respective data set. In essence, the way it changes from Mamdani model is that, the fuzzy rules have an output that is a polynomial, meaning that it does not need a defuzzification method, its output can be directly used after the polynomial evaluation.

In the case of Sugeno, we have two types: first-order and zero-order. The fist-order Sugeno fuzzy model uses a first-order polynomial in the consequent; the zero-order Sugeno fuzzy model uses a constant as the consequent, and is the most used in practice between the two types.

The format of the rules is given as:

$$\text{IF } x \text{ is } A \text{ and } y \text{ is } B \text{ THEN } z = f(x, y)$$

where A and B are the fuzzy sets in antecedent and $z = f(x, y)$ is a crisp function for the consequent.

3 Control Systems

Control systems have been used for almost two thousand years, although, modern control systems do not have that much time. Since the use of electricity, more ways to achieve the objective of controlling a system have been accomplished, like the use of PID controllers (Proportional-Integral-Derivative controller).

A control system can manage the behavior of other devices or systems in control loops, and can be from a simple heater system to more complex applications like a space shuttle launch. For a constant control of a system, feedback controllers are used to make this automatically, where the control system compares the obtained value of a system with a desired value, this difference or error then is used to adjust the system to its optimal output. In Fig. 4 we can observe a generic control system.

There are two types of control that are the most used: the open-loop and the closed-loop. The open-loop control system just sets some level of drive to a plan without seeing the need to inspect the systems behavior to that input, that is why is called open-loop. The most used applications of this type of systems are and electric bulb, automatic washing machine, electric hand drier; just to mention some cases. In Fig. 5 we can find a basic block model of this system.

A closed-loop control system is a control system that implements a feedback loop to the system to adjust the output of the plant. The feedback means that some part of the output is used in the input, so it can maintain the stability of the control system; by adding the error signal to the input, the generated output of the system will be corrected, that is why closed-loop systems are affected less by external disturbances. In Fig. 6 we can observe a basic block model for this system.

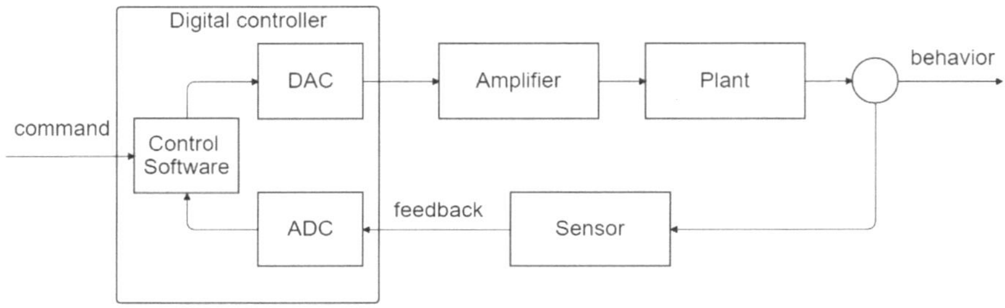

Fig. 4 Generic control system

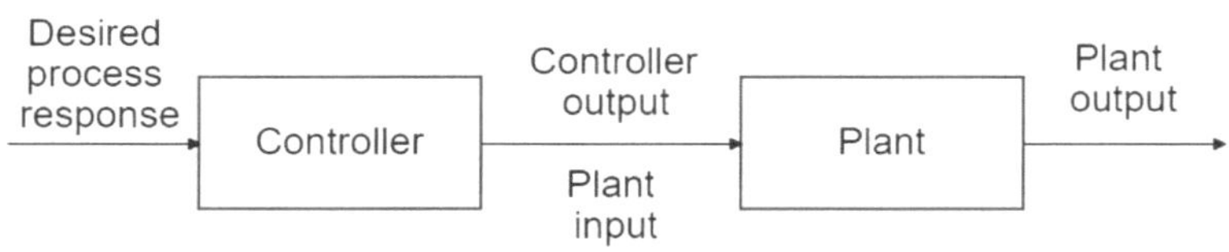

Fig. 5 Basic block model of open-loop control system

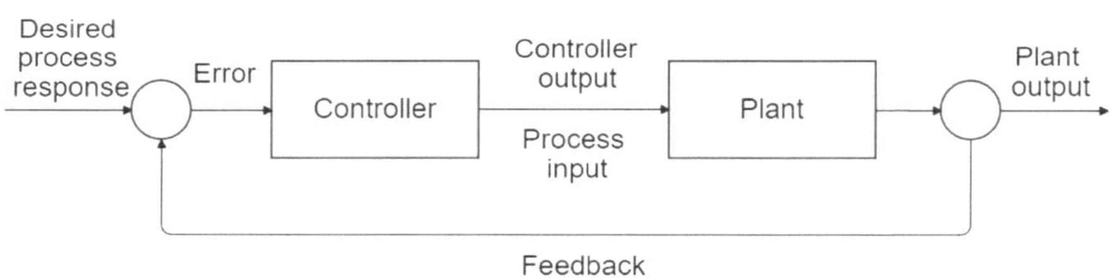

Fig. 6 Basic block model of closed-loop control system

For most control systems to be simulated by computer, there is the need to use a plant. A plant is the combination of process and actuator, that is referred with a transfer function that indicates the relation between an input and output of a system that has no feedback (Wescott 2006).

4 Metaheuristics and Multi-verse Optimizer

In the area of Computer Science there is a wide variety of methods used to solve distinct problems, such as the metaheuristics. Metaheuristics are stochastic methods used in optimization problems; they have distinct inspirations that can be from living organisms to artificial behaviors. These methods can be observed in two main forms: single-based and population based.

Single-based solution algorithms have the characteristic that, the algorithm has an initial randomized solution and this changes over the time or execution of the algorithm, looking for a solution, which can be exposed for a mayor problem in this form: can stay still in a local optimum and have no further changes over execution of the algorithm.

Population-base solution algorithms have a different approach, they start with multiple random solutions instead of only one, these change over the execution of the algorithm, bringing more candidates to solution to the main problem; these algorithms are more often inspired by groups of nature, like the colonies of bees.

Between the optimization algorithms we can mention one of the most used over multiple problems in control: Particle Swarm Optimization (PSO) (Engelbrecht 2007). PSO algorithm has inspiration over Swarm Intelligence, which comes from the study of colonies and the social behavior of the individuals in the swarm. This algorithm is a stochastic method of optimization which has its inspiration over the birds, where the particles are every bird on the flocks, representing each one a solution to the problem to solve. Here every particle or bird "flies" in a limited search space toward the best solution, and every particle can be influenced by other particles which have a better fitness or solution. One application over neural networks and fuzzy systems is the case of Gaxiola et al. (2016) that use the algorithm to optimize the fuzzy system parameters to adjust the weights of the neural network.

In swarm intelligence we have many bio-inspired algorithms such as Bee Colony Optimization (BCO) (Teodorović 2009); one application of this algorithm is the case of Amador-Angulo and Castillo (2018) that use a modified version of the algorithm to find the optimal distribution of membership functions in fuzzy controllers for nonlinear plants.

In the areas of organized behaviors of pack of animals we can observe the Grey Wolf Optimizer (GWO) (Mirjalili et al. 2014), which has inspiration over the packs of grey wolfs. One application we can mention is the case of Hernández (2020) where the algorithm is used in a control application over an autonomous mobile robot to optimize a fuzzy inference system controller to maintain the best route with the minimal error.

Another optimization algorithm that its nature inspired is the case of the Firefly Algorithm(FA), which uses the behavior of fireflies to work (Pérez 2017); to mention one application we can observe the results of Lagunes (2020) where it uses the algorithm to optimize fuzzy controllers with type-1 and type-2 fuzzy logic using the water tank problem and temperature shower control problem.

In the role of nature inspirations over plants we have the Flower Pollination Algorithm (FPA) proposed by Yang (2014), which inspiration comes from the pollination of flowers. One application is the case of Carvajal et al. (n.d.), where the algorithm optimized a fuzzy controller for an autonomous mobile robot with two inputs and two outputs to control the robot trajectory.

Other algorithms that are inspired on social behavior and physics are the Imperialist Competitive Algorithm (ICA) (Atashpaz-Gargari and Lucas 2007) and Harmony Search Algorithm (HAS) (Geem et al. 2001). One example of the application of ICA is the case of Bernal (2020), where it used the algorithm in conjunction with a type-2 fuzzy system to adjust the decades of the ICA; other case for the HAS is the work of Peraza (2020) where is used to adjust parameters on some control problems like the movement of an autonomous robot.

4.1 Multi-verse Optimizer

The Multi-verse Optimizer (MVO) algorithm was proposed by Mirjalili et al. (2016) in 2015, has inspirations over various concepts of cosmology and the big bang theory, these concepts are the white hole, black hole and wormhole; these objects interact over multiple universes which represent the solutions of the problem, and with the help of these types of holes, it can develop multiple solutions to the problem. In Fig. 7 we can see a representation of the concepts.

The way it works is that the algorithm moves objects or parts of the solutions in the universes, the most common way to interact between universes is the white and black holes, but the wormhole works as a type of mutation in the algorithm, where it makes some randomized movements between the less fitted universes and the best solution. In the case of the holes that interact in MVO there are certain conditions to be met, there has to be a high inflation rate or fitness so the probability to have white holes is greater but it has lower probability to have black holes, the universes that have higher inflation rate send objects through white holes so in universes with less inflation rate can receive through black holes objects; in addition to this, the wormhole acts to send random movements to the best universe (Steinhardt and Turok 2005).

The universes can be represented by a matrix like in (1), where U is the set of universes or solutions, d represents the variables or parameters and n is the number of universes or solutions that are candidates. Each parameter can also be represented like in (2).

Fig. 7 Reference image for white, black and worm holes

$$U = \begin{bmatrix} x_1^1 & x_1^2 & \cdots & x_1^d \\ x_2^1 & x_2^2 & \cdots & x_2^d \\ \vdots & \vdots & \vdots & \vdots \\ x_n^1 & x_n^2 & \cdots & x_n^d \end{bmatrix} \tag{1}$$

$$x_i^j = \begin{cases} x_k^j & r1 < NI(U_i) \\ x_i^j & r1 \geq NI(U_i) \end{cases} \tag{2}$$

where each parameter represent as follows: x_i^j indicates the jth parameter of the ith universe, U_i represents the ith universe, $NI(U_i)$ is the normalized inflation rate of the ith universe, $r1$ is a random number between [0, 1], and x_k^j indicates the jth parameter of the kth universe selected by a roulette wheel.

In the black hole and white hole, we have a way to represent it in Pseudocode, where $-NI$ is used for minimization problems, if it is for maximization NI is used. In Fig. 8 we have a representation of MVO in a chart of blocks, where the white dots represent the wormholes moving object to the best universe so far, and the use of black and white holes are between the universes of same dot color, also in In Fig. 9 we can appreciate this Pseudocode.

In the mechanism of the wormhole, which can be used to achieve exploitation of the solutions, each universe must present wormholes to move objects between universes and the best universe at random, the way to represent these wormholes is as mentioned in (3).

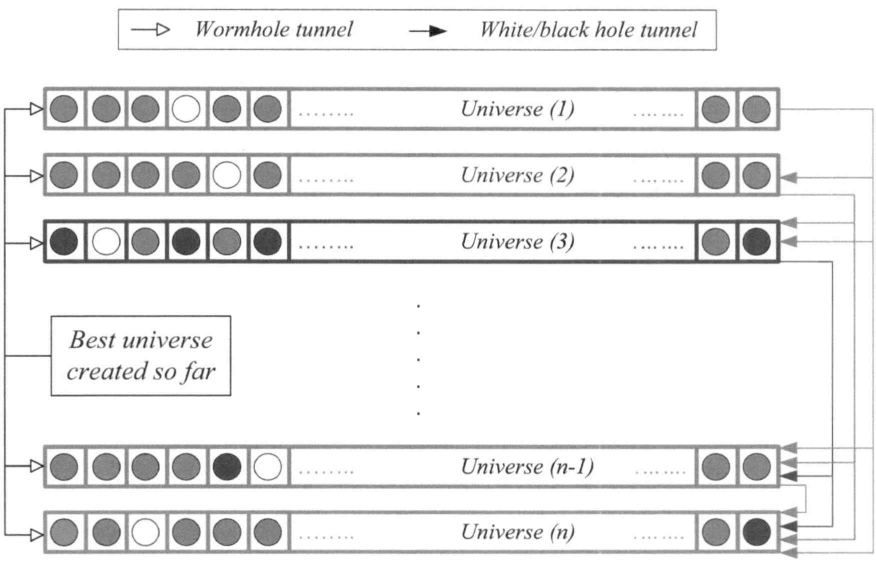

Fig. 8 Representation of the MVO algorithm (Mirjalili et al. 2016)

```
SU=Sorted universes
NI=Normalize inflation rate (fitness) of the universes
for each universe indexed by i
    Black_hole_index=i;
    for each object indexed by j
        r1=random([0,1]);
        if r1<NI(Ui)
            White_hole_index= RouletteWheelSelection(-NI);
            U(Black_hole_index,j)= SU(White_hole_index,j);
        end if
    end for
end for
```

Fig. 9 Pseudocode for white hole and black hole mechanism

$$x_i^j = \begin{cases} \begin{cases} X_j + TDR \times \left(\left(ub_j - lb_j\right) \times r4 + lb_j\right) & r3 < 0.5 \\ X_j - TDR \times \left(\left(ub_j - lb_j\right) \times r4 + lb_j\right) & r3 \geq 0.5 \end{cases} & r2 < WEP \\ x_i^j & r2 \geq WEP \end{cases} \tag{3}$$

In this equation X_j indicates the jth parameter of the best universe so far, *TDR* and *WEP* are coefficients named Travel Distance Rate and Wormhole Existence Probability respectively, lb_j shows the lower limit of the jth variable, ub_j is the upper limit of the jth variable, x_i^j indicates the jth parameter of the ith universe, and $r2, r3, r4$ are random numbers between [0, 1]. From this equation we can obtain also a Pseudocode for its operation shown of Fig. 10.

We have two more concepts to define in equations, these are *TDR* and *WEP*, they change with the pass of iterations to work from exploration to exploitation; as iterations pass, *TDR* increases to perform exploitation of solution. In (4) and (5) we

```
for each universe indexed by i
    for each object indexed by j
        r2=random([0,1]);
        if r2<Wormhole_existance_probability
            r3= random([0,1]);
            r4= random([0,1]);
            if r3<0.5
                U(i,j)=Best_universe(j) + Travelling_distance_rate
            * (( ub(j) - lb(j)) * r4 + lb(j));
            else
                U(i,j)=Best_universe(j) - Travelling_distance_rate
            * (( ub(j) - lb(j)) * r4 + lb(j));
            end if
        end if
    end for
end for
```

Fig. 10 Pseudocode for wormhole mechanism

can observe how both parameters are represented in the algorithm.

$$WEP = min + l \times \left(\frac{max - min}{L} \right) \tag{4}$$

$$TDR = 1 - \frac{l^{1/p}}{L^{1/p}} \tag{5}$$

Here *min* is the minimum, *max* is the maximum, l indicates the current iteration, and L is the maximum of iterations. In the case of *TDR*, p is the precision of the exploitation in the iterations; the higher the p value, the faster and more precise the local search or exploitation. Finally, we can see the complete algorithm in a flowchart that is presented in Fig. 11.

4.2 *Applications of MVO*

Since the appearance of the algorithm in 2015 many scientists have done tests on different areas of Computer Science. One of the most notable is the case of the creator Mirjalili et al. (2016), where he did tests with benchmark functions and ones taken from CEC 2005 special session, the tests were done 30 times for each function and it applied the Wilcoxon's nonparametric statistical test. Also, for the practical applications, it used five constrained real engineering problems.

With time, other scientists did another incorporations to the algorithm, like the case of Hu et al. (2016), where he used the Levi flights in MVO calling it LFMVO, which was tested with 23 benchmark functions and 40 executions per function. In the area of neural networks we have the case of Faris et al. (2016) used the MVO algorithm to train a Multi-layer Perceptron(MLP).

If we go to cases where hybridization is done, we have Jangir et al. (2017) were it presents a new technique hybrid Particle Swarm Optimization (PSO)-Multi-verse Optimizer (MVO). Another variant was presented by Al-Madi et al. (2019), showing a binary version of the algorithm to solve problems with discrete variables called BMVO. More recent studies are applied over this method like Cao et al. (2020) where they propose a K-means MVO to construct DNA code storage in a better way than MVO.

5 Test and Results

For tests over the MVO algorithm we first started with the benchmark functions that the original author (Mirjalili et al. 2016) used, from the 19 functions we used the first 13 that correspond to unimodal and multi-modal functions, meaning that the

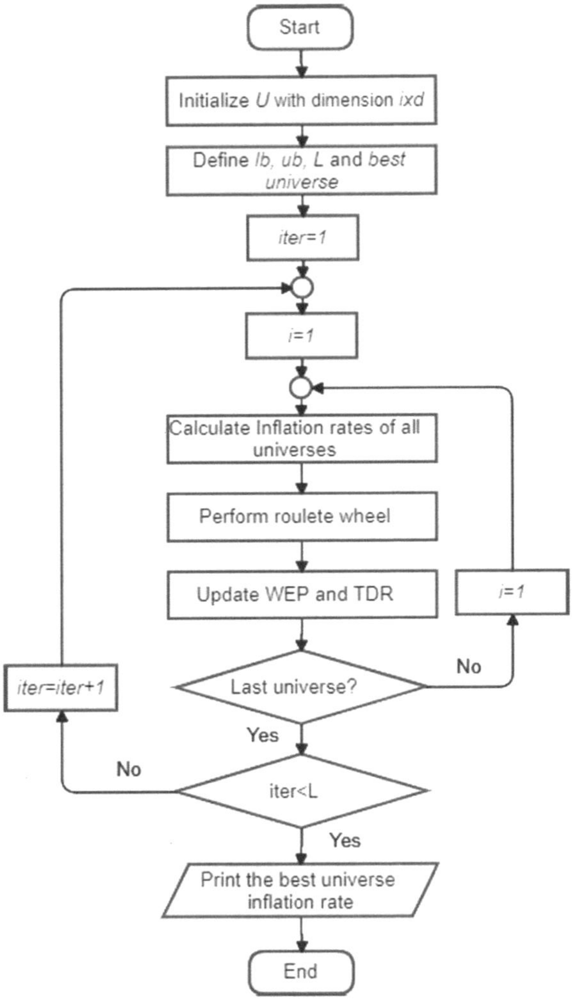

Fig. 11 General flowchart of MVO

functions 1–7 have one global optimum, and the functions 8–13 have global optimum and multiple local optimum so we can test with multiple dimensions.

These tests were done with the algorithm running 30 times like Mirjalili used in his tests, so we can verify his results. The statistical tests were done to compare with the results presented in Mirjalili et al. (2016) with PSO, GA, GSA and GWO. The parameters for the algorithm go from 30 universes, maximum iterations of 500, WEP values go from 0.2 to 1 and TDR from 0.6 to 0 using Eqs. (4) and (5) respectively.

Table 1 Benchmark functions for MVO tests with min = 50 and dim = 50

Function	Range
$F_1(x) = \sum_{i=1}^{n} x_i^2$	[−100, 100]
$F_2(x) = \sum_{i=1}^{n} \lvert x_i \rvert + \prod_{i=1}^{n} \lvert x_i \rvert$	[−10, 10]
$F_3(x) = \sum_{i=1}^{n} \left(\sum_{j-1}^{i} x_j \right)^2$	[−100, 100]
$F_4(x) = \max_{\mathrm{i}} \{ \lvert x_i \rvert, 1 \le i \le n \}$	[−100, 100]
$F_5(x) = \sum_{i=1}^{n-1} \left[100\left(x_{i+1} - x_i^2\right)^2 + (x_i - 1)^2 \right]$	[−30, 30]
$F_6(x) = \sum_{i=1}^{n} ([x_i + 0.5])^2$	[−100, 100]
$F_7(x) = \sum_{i=1}^{n} \left(i x_i^4 + random[0, 1] \right)$	[−1.28, 1.28]
$F_8(x) = \sum_{i=1}^{n} -x_i \sin\left(\sqrt{\lvert x_i \rvert}\right)$	[−500, 500]
$F_9(x) = \sum_{i=1}^{n} \left[x_i^2 - 10\cos(2\pi x_i) + 10 \right]$	[−5.12, 5.12]
$F_{10}(x) = -20exp\left(-0.2\sqrt{\frac{1}{n} \sum_{i=1}^{n} x_i^2} \right) - \exp\left(\frac{1}{n} \sum_{i=1}^{n} \cos(2\pi x_i) \right) + 20 + e$	[−32, 32]
$F_{11}(x) = \frac{1}{4000} \sum_{i=1}^{n} x_i^2 - \prod_{i=1}^{n} \cos\left(\frac{x_i}{\sqrt{i}} \right) + 1$	[−600, 600]
$F_{12}(x) = \frac{\pi}{n} \left\{ 10sin(\pi y_1) + \sum_{i=1}^{n-1} (y_i - 1)^2 \left[1 + 10\sin^2(\pi y_{i+1}) \right] + (y_n - 1)^2 \right\} +$ $\sum_{i=1}^{n} u(x_i, 10, 100, 4) y_i = 1 + \frac{x_i + 1}{4} u(x_i, a, k, m) = \begin{cases} k(x_i - a)^m & x_i > a \\ 0 & -a < x_i < a \\ k(-x_i - a)^m & x_i < -a \end{cases}$	[−50, 50]
$F_{13}(x) =$ $0.1 \left\{ \sin^2(3\pi x_1) + \sum_{i=1}^{n} (x_i - 1)^2 \left[1 + \sin^2(3\pi x_i + 1) \right] + (x_n - 1)^2 \left[1 + \sin^2(2\pi x_n) \right] \right\} +$ $\sum_{i=1}^{n} u(x_i, 5, 100, 4)$	[−50, 50]

Table 2 Average test results in benchmark functions for MVO

Function/Ave	MVO	GWO	GSA	PSO	GA
F1	**1.32E + 00**	2.32E + 03	2.98E + 03	3.55E + 00	2.72E + 04
F2	1.29E + 02	1.44E + 01	1.10E + 01	**8.72E + 00**	6.87E + 01
F3	**2.31E + 03**	7.28E + 03	1.14E + 05	2.38E + 03	4.85E + 04
F4	1.92E + 01	**1.31E + 01**	3.23E + 01	2.15E + 01	6.30E + 01
F5	**3.94E + 02**	3.43E + 06	7.58E + 03	1.13E + 03	6.54E + 07
F6	**1.40E + 00**	5.01E + 03	7.46E + 04	8.66E + 01	4.96E + 04
F7	**6.91E−02**	4.08E−01	2.12E + 01	5.77E−01	1.87E + 01
F8	**−1.23E + 04**	−1.07E + 04	−4.64E + 03	−6.73E + 03	−1.07E + 04
F9	2.84E + 02	**8.91E + 01**	1.28E + 02	9.98E + 01	2.73E + 02
F10	6.18E + 00	9.45E + 00	**1.65E + 00**	4.30E + 00	1.86E + 01
F11	**8.04E−01**	2.25E + 01	1.02E + 03	6.24E + 02	3.53E + 02
F12	**4.19E + 00**	3.20E + 06	7.42E + 05	1.34E + 01	2.21E + 08
F13	**4.24E−01**	7.82E + 06	6.67E + 06	2.11E + 01	4.49E + 08

5.1 Benchmark Function Test and Results

The test functions for the 13 benchmark functions are listed in Table 1, where the fist 7 are unimodal and the last 6 are multimodal. Using the parameters mentioned previously, we managed to obtain similar results as the author, proving the competitivity of the algorithm over exploitation in the unimodal functions and exploration on the multi-modal functions; in Table 2 we can observe the results of the algorithm comparing with other algorithms.

From the results of tests, we can verify the functionality of the algorithm over the initial benchmark functions where it was tested by the original author, concluding In this subsection that the algorithm can work normally over these constraints applied, so the model can be applied to other problems in diverse scenarios like the ones in the next subsection.

5.2 Applications Test and Results

In this subsection we are presenting two problems were the algorithm was used, so we can have more proof over real-life applications, mainly over control problems. The following problems are applications over fuzzy logic, were the main objective is to optimize membership functions for the fuzzy inference system, so it can work over a reference to approximate the fuzzy system.

5.2.1 Tipper problem

This problem is used in the first examples of MATLAB for fuzzy systems design in his toolbox, is mainly used for reference design of fuzzy models. The objective over the application of this system is to approximate another fuzzy model that is made from two inputs and one output, like the original model, but with changes over the number and shape of membership functions.

In the reference model, which is a Mamdani Fuzzy inference system, we can observe the membership functions and surface in Fig. 12; it has 3 gaussian membership functions for input 1, for input 2 has 2 trapezoidal membership functions, and for output has 3 triangular membership functions, using only 3 fuzzy rules. For the model that we used in the algorithm to approximate the surface of the fuzzy system we implemented two model: one for Mamdani and another for Sugeno Fuzzy inference system (Table 3).

For the Mamdani model we used a different arrangement of membership functions, for input 1 we used 3 triangular membership functions, for input 2 also we used 3 triangular membership functions, and for output 1 we used 5 triangular membership

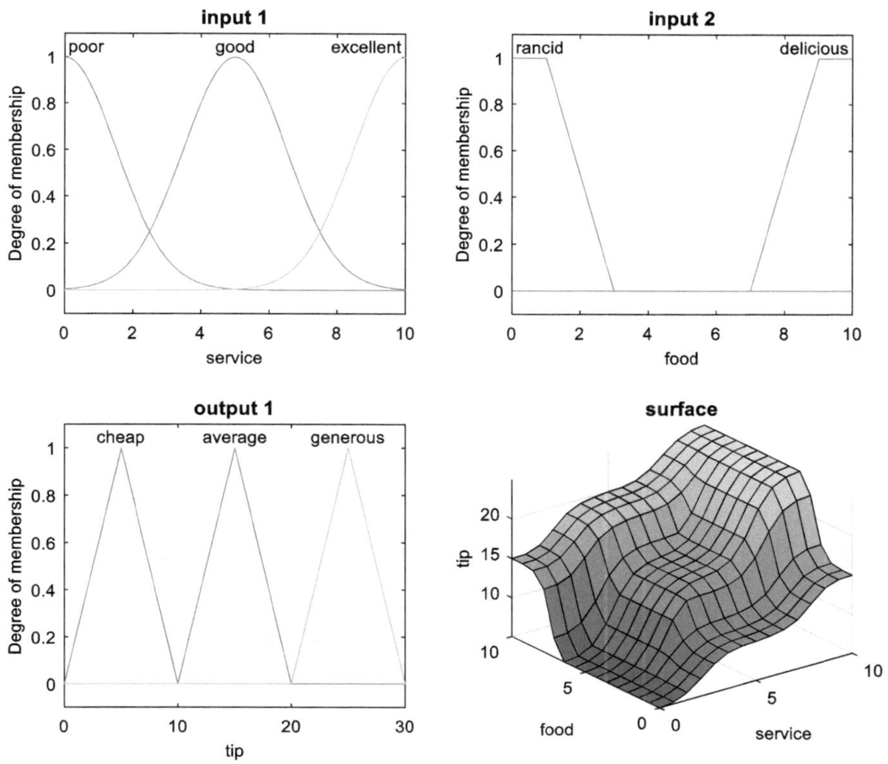

Fig. 12 Reference tipper problem

Table 3 Fuzzy rules set for Mamdani and Sugeno model

No.	Rule
1	If (service is poor) and (food is rancid) then (tip is poor) (1)
2	If (service is poor) and (food is normal) then (tip is poor) (0.5)
3	If (service is poor) and (food is delicious) then (tip is average) (0.5)
4	If (service is good) and (food is rancid) then (tip is mid poor) (0.5)
5	If (service is good) and (food is normal) then (tip is average) (1)
6	If (service is good) and (food is delicious) then (tip is mid generous) (0.5)
7	If (service is excellent) and (food is rancid) then (tip is average) (0.5)
8	If (service is excellent) and (food is normal) then (tip is generous) (0.5)
9	If (service is excellent) and (food is delicious) then (tip is generous) (1)

functions; also, we used 9 fuzzy rules. In Fig. 13 we can observe the base model to optimize in MVO so it can approximate to original model, in Table 4 we have the set of rules.

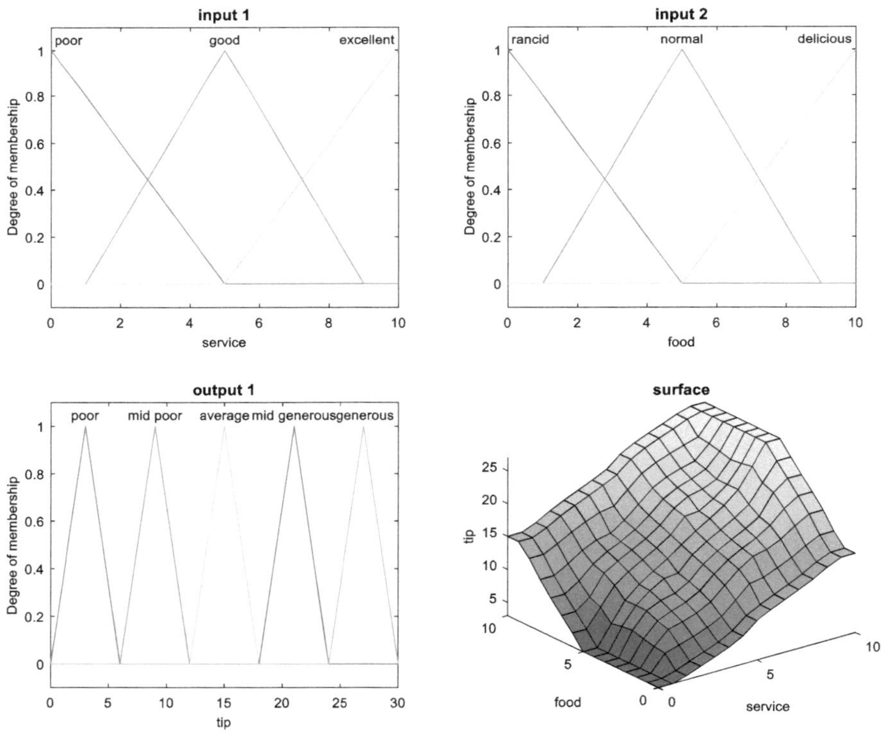

Fig. 13 Base Mamdani model for optimization

Table 4 Parameters for Algorithm Tipper case

Parameter	Value
Universes	50
Dimensions	14
Max iterations	200
Lower bound	0
Upper bound	10
WEP max	1
WEP min	0.2
Tests done	30

In Sugeno model we used the same arrangement of membership functions for inputs except for output, were we used 5 five constants, making this model a zero order Sugeno fuzzy inference system; also, we used the same 9 fuzzy rules but with the new output. In Fig. 14 we can observe the base model to optimize in MVO s o i t can approximate to original model.

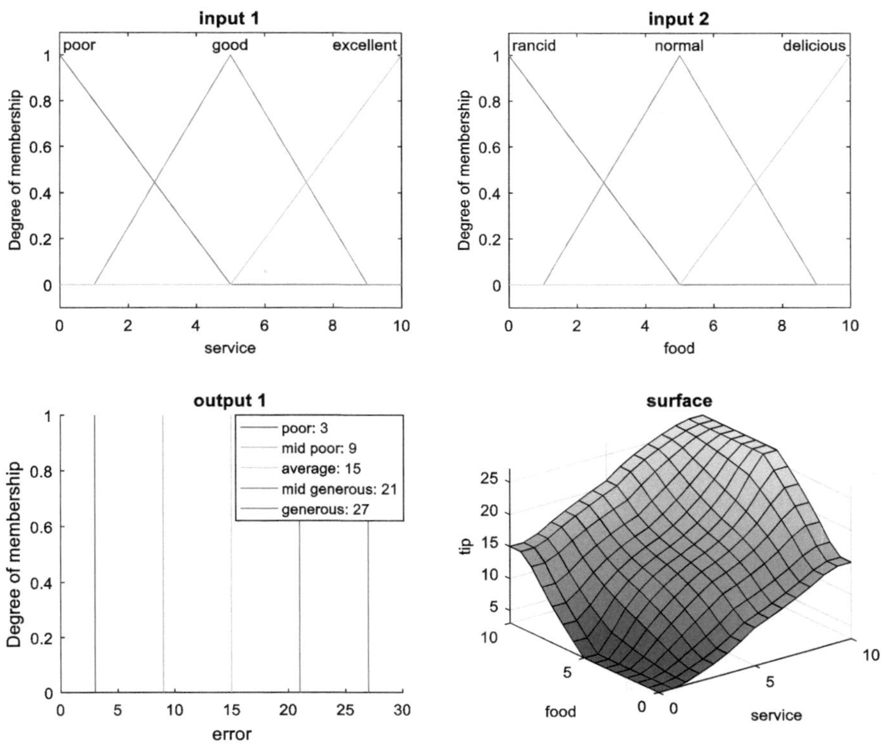

Fig. 14 Base Sugeno model for optimization

Table 5 Sorting of parameters for inputs MVO

Universe	Input 1						
	P1	P2	P3	P4	P5	P6	P7
mf Param	mf1-p2	mf2-p1	mf1-p3	mf2-p2	mf3-p1	mf2-p3	mf3-p2
Input 2							
Universe	P8	P9	P10	P11	P12	P13	P14
mf Param	mf1-p2	mf2-p1	mf1-p3	mf2-p2	mf3-p1	mf2-p3	mf3-p2

Table 6 RMSE results for Tipper problem

System/Ave	Base system	MVO	GWO	GSA	PSO	GA
Mamdani	1.93130	**1.33170**	1.35331	1.41512	1.36504	1.46503
Sugeno	1.79500	**1.36979**	1.39171	1.46279	1.41147	1.48007

For the optimization of each model we used 14 dimensions in the algorithm, so we can optimize only the inputs of each model, giving 7 values for each one of the inputs, these values can go from [0, 10], were 2 values are for the first membership function, 3 for the second membership function and 2 for the third membership function; we used the same shapes for input 2; before using the parameters on the membership functions, the parameters are sorted. As for the rest of the parameters for the MVO algorithm they go as shown on Table 4, and the sorting of parameters for the membership functions are shown on Table 5.

Using the mentioned parameters, we managed to approximate both models so they can be very similar in terms of the surface of the fuzzy inference system, in the case of Mamdani the results were more sharp, but with good results; in Sugeno the results were more smooth in terms of the surface, but not better than Mamdani approximation. The way we compared was with RMSE in both cases, used in the cost function minimizing this value, in Table 6 we have the average of these 30 tests comparing with other 5 algorithms, also in Fig. 15 we have the results of both models with its convergence curve.

5.2.2 Cruise Control Problem

This is a common problem in control, were a vehicle speed is controlled to achieve a desired speed. There are not many constraints because the model is designed for an ideal space where is not considered air friction or any inclination, so there are not perturbances to this problem. The initial model comes with a PID controller that considers proportional gain, integral gain and derivative gain. The values set for implementation of PID controller in this case are 800 for Kp, 40 for Ki and 10 for Kd, where Kp, Ki and Kd are the values for proportional gain, integral gain and derivative gain. The system equation and the output can be appreciated bellow.

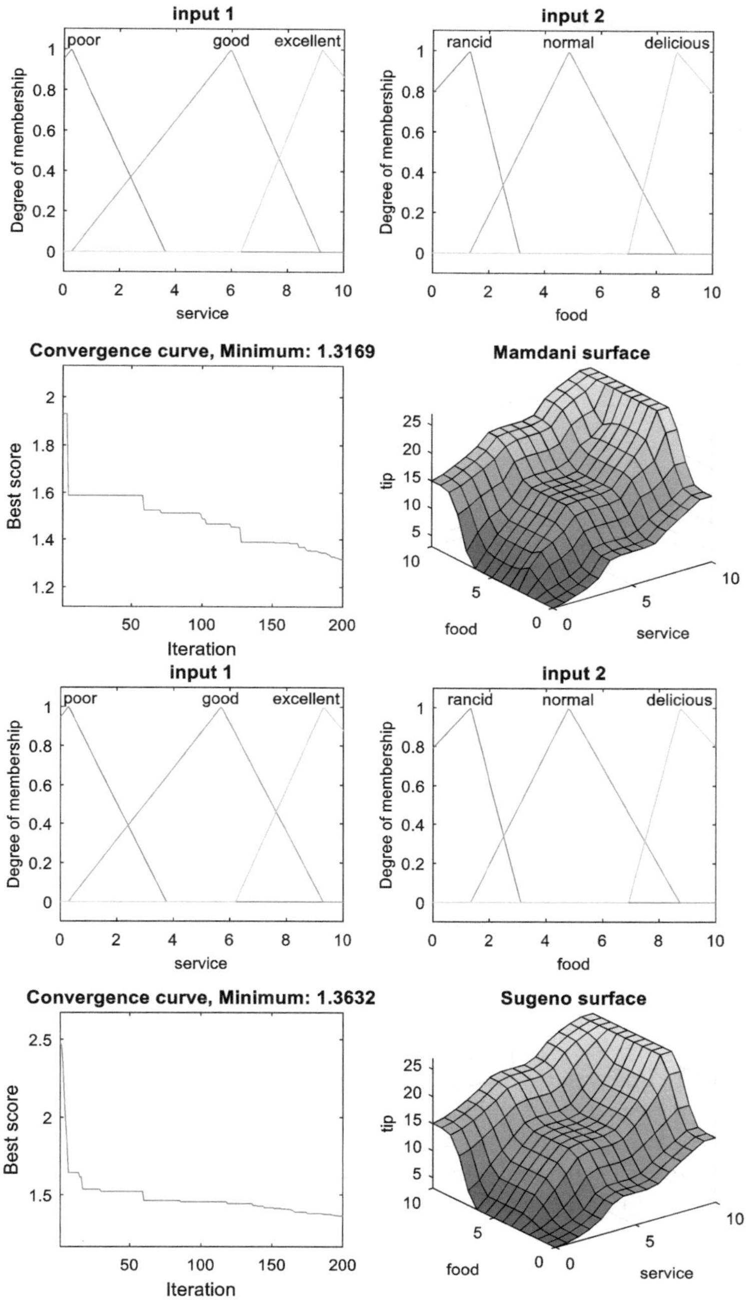

Fig. 15 Output and test results for tipper problem

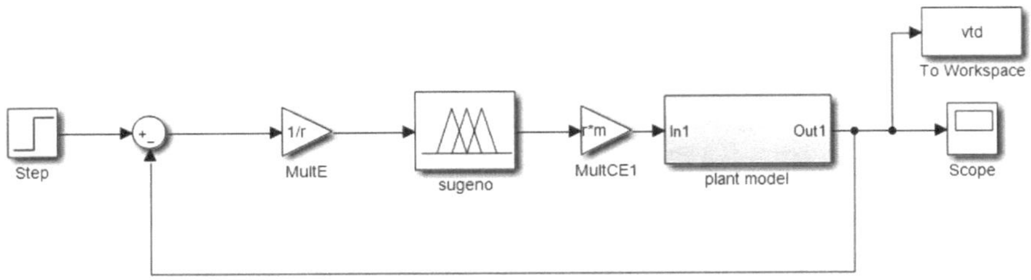

Fig. 16 Simulink model for cruise control

$$m\dot{v} + bv = u \tag{6}$$

$$y = v \tag{7}$$

The main objective of this experiment is to achieve a desired speed of 10 m/s in a vehicle with a mass of 1000 kg, this by using a fuzzy inference system which controls the behavior of the vehicle. For implementation of this model we used a Simulink model and a fuzzy inference system, a representation of the system is shown on Fig. 16. For the fuzzy model we used a Mamdani and a Sugeno model to analyze behavior of the algorithm. In both cases we use the output speed to compare in a reference speed and obtain the error to use un the fuzzy controller and adjust the output of the system from this feedback.

For the base Mamdani model we used one input and one output, were the input used 5 membership functions, the first and last are trapezoidal and the rest are triangular shaped; mostly the same shapes were used for the output, using 5 membership functions with triangular and trapezoidal shape. As for the Sugeno model, we used the same input but, in the output, we used 5 constants, making a zero order Sugeno model. The membership functions for both models are shown on Fig. 17 as for the fuzzy rules are in Table 7.

For the optimization of this case, both the input and the output membership functions were optimized, so we used for Mamdani 26 dimensions and for Sugeno we used 18 dimensions, this because the output for Sugeno model uses only 5 parameters of the algorithm instead of 13 like in Mamdani. Each input and output range from $[-1, 1]$, using 13 parameters each, the first and las membership functions use 2 values each and the three membership functions on the center use 3 values each, same for the output in Mamdani, except in Sugeno. The parameters used in MVO algorithm can be observed in Table 8, and the sorting of parameters for the membership functions can be observed in Table 9.

While using the previous parameters, the cruise control system managed to work well, because it could approximate the reference speed of the model in a great time. For the tests and cost function we used RMSE evaluating the parameters obtained from the algorithm and making the simulation implementing the parameters on the

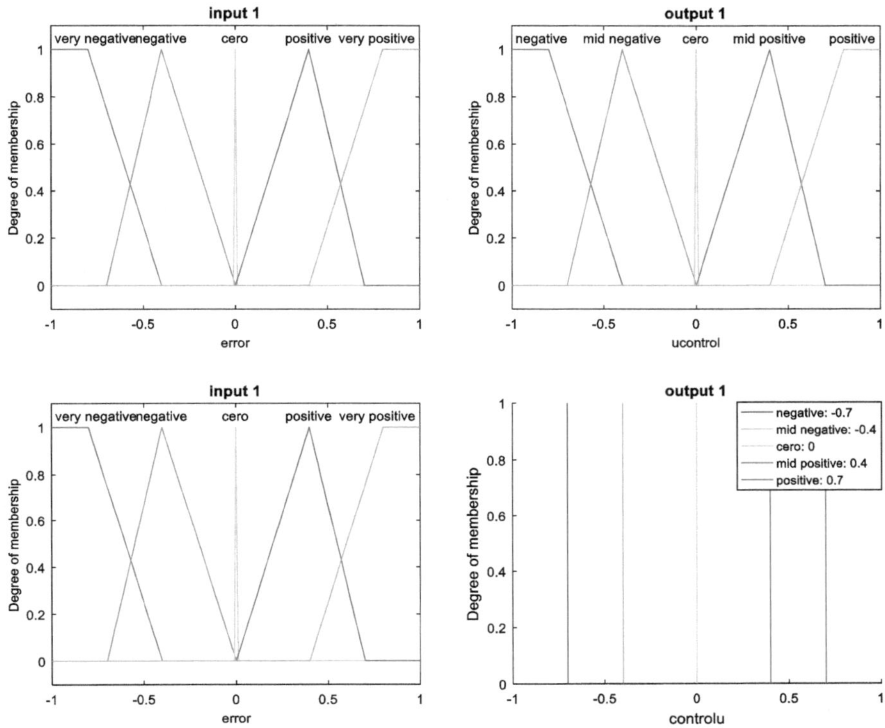

Fig. 17 Membership functions for Mamdani and Sugeno model

Table 7 Fuzzy rules set for Mamdani and Sugeno model

No.	Rule
1	If (error is very negative) then (ucontrol is negative) (1)
2	If (error is negative) then (ucontrol is mid negative) (1)
3	If (error is cero) then (ucontrol is cero) (1)
4	If (error is positive) then (ucontrol is mid positive) (1)
5	If (error is very positive) then (ucontrol is positive) (1)

Table 8 Parameters for algorithm cruise control case

Parameter	Value
Universes	10
Dimensions	26–18
Max iterations	100
Lower bound	−1
Upper bound	1
WEP max	1
WEP min	0.2
Tests done	30

Table 9 Sorting of parameters for inputs MVO cruise control case

Universe	Input 1 Mamdani/Sugeno						
	P1	P2	P3	P4	P5	P6	P7
mf Param	mf1-p3	mf2-p1	mf1-p4	mf2-p2	mf3-p1	mf2-p3	mf3-p2
Universe	P8	P9	P10	P11	P12	P13	
mf Param	mf4-p1	mf3-p3	mf4-p2	mf5-p1	mf4-p3	mf5-p2	
Output 1 Mamdani							
Universe	P14	P15	P16	P17	P18	P19	P20
mf Param	mf1-p3	mf2-p1	mf1-p4	mf2-p2	mf3-p1	mf2-p3	mf3-p2
Universe	P21	P22	P23	P24	P25	P26	
mf Param	mf4-p1	mf3-p3	mf4-p2	mf5-p1	mf4-p3	mf5-p2	
Output 1 Sugeno							
Universe	P14	P15	P16	P17	P18		
mf Param	mf1-p1	mf2-p2	mf3-p3	mf4-p4	mf5-p5		

Table 10 RMSE results for cruise control case

System/Ave	Base system	MVO	GWO	GSA	PSO	GA
PID	**79.7439**	–	–	–	–	–
Mamdani	72.64880	**58.53123**	58.53209	59.74792	58.53230	58.74731
Sugeno	74.58200	**58.49715**	58.49763	58.73710	58.49933	58.63267

fuzzy inference system for both cases. In Table 10 we have the results comparing against the PID model, Mamdani and Sugeno model obtained in the tests and the other algorithms, in Fig. 18 we have the outputs of the models also.

6 Conclusions

In this paper we introduced the MVO algorithm in the control area so we can analyze its functionality over this type of problems. One of the activities done in this work was the study of the behavior of MVO over benchmark functions as the author did; as for the applications mentioned, we started the tests with fuzzy systems in some simple cases as the approximation of a fuzzy inference system model but with different configuration of membership functions and a set of rules. One of the tasks was to design the cost function for the algorithm, so we would distribute the parameters obtained from the dimensions of the algorithm to the parameters of the membership functions but, being very careful with the distribution so the membership functions would overlap by ordering the parameters before applying them to the fuzzy system.

From obtaining the first good results, we proceeded to a simple control case, which was the cruise control problem, were we designed the fuzzy model and did

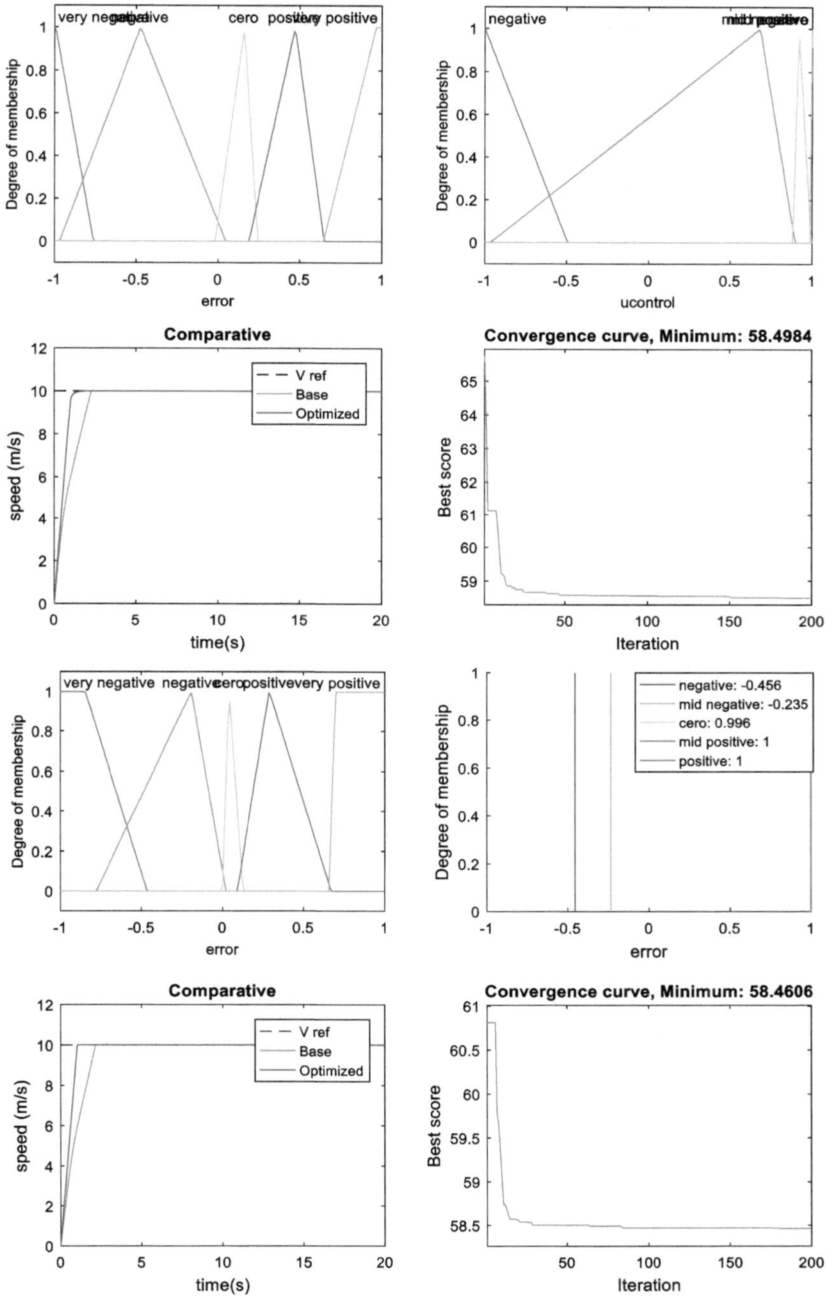

Fig. 18 Output and test results for cruise control case

the same distribution of the parameters using the same idea as the tipper problem, but implementing it on the output also. From the results, we managed to outperform the PID controller, and there was no need to use more iterations on the algorithm, so the convergence was fast.

With the results obtained from the tests, we conclude that the Multi-verse Optimizer is a great choice for control cases and we would further perform more tests on more complex models so we can confirm its applications over this area, managing to outperform more common algorithms like PSO. In future work we could consider other applications for the proposed method, like the ones in (Castillo 1998; Castillo and Melin 2003; Sanchez et al. 2014).

References

Al-Madi, N., H. Faris, and S. Mirjalili. 2019. Binary multi-verse optimization algorithm for global optimization and discrete problems. *The International Journal of Machine Learning and Cybernetics* 10: 3445–3465. https://doi.org/10.1007/s13042-019-00931-8.

Amador-Angulo, L., and O. Castillo. 2018. A new fuzzy bee colony optimization with dynamic adaptation of parameters using interval type-2 fuzzy logic for tuning fuzzy controllers. *Soft Computing* 22: 571–594. https://doi.org/10.1007/s00500-016-2354-0.

Atashpaz-Gargari, E., and C. Lucas. Imperialist competitive algorithm: An algorithm for optimization inspired by imperialistic competition. In *2007 IEEE congress on evolutionary computation*, CEC 2007. pp. 4661–4667. https://doi.org/10.1109/CEC.2007.4425083.

Bernal, E., O. Castillo, J. Soria, and F. Valdez. 2020. Parameter adaptation in the imperialist competitive algorithm using generalized Type-2 Fuzzy logic. In *Intuitionistic and Type-2 Fuzzy logic enhancements in neural and optimization algorithms: Theory and applications* ed by O. Castillo, P. Melin, and J. Kacprzyk (eds.), pp. 3–10. Springer International Publishing, Cham. https://doi.org/10.1007/978-3-030-35445-9_1.

Cao, B., S. Zhao, X. Li, and B. Wang. 2020. K-means multi-verse optimizer (KMVO) algorithm to construct DNA storage codes. *IEEE Access* 8: 29547–29556. https://doi.org/10.1109/ACCESS.2020.2970838.

Carranco, C., P. Encalada, J. Gavilanes, G. Delgado, and M.V. Garcia. 2020. An approach of a control system for autonomous driving based on artificial vision techniques and NAO Robot. In *Advances in intelligent systems and computing*, pp. 196–206. Springer. https://doi.org/10.1007/978-3-030-32022-5_19.

Carvajal, O.R., O. Castillo, and J. Soria. Optimization of membership function parameters for fuzzy controllers of an autonomous mobile robot using the flower pollination algorithm. The *Journal* of *Intelligent* and *Robotic Systems*. https://doi.org/10.14313/JAMRIS_1-2018/6.

Castillo, O., and P. Melin. 1998. A new fuzzy-fractal-genetic method for automated mathematical modelling and simulation of robotic dynamic systems. In *1998 IEEE international conference on fuzzy systems* (FUZZ-IEEE 1998) Proceedings. Volume 2, 1182–1187.

Castillo, O., and P. Melin. 2003. Intelligent adaptive model-based control of robotic dynamic systems with a hybrid fuzzy-neural approach. *Applied Soft Computing* 3 (4): 363–378.

Engelbrecht, A.P. 2007. *Computational intelligence: An introduction.* Wile.

Faris, H., I. Aljarah, and S. Mirjalili. 2016. Training feedforward neural networks using multi-verse optimizer for binary classification problems. *Applied Intelligence* 45: 322–332. https://doi.org/10.1007/s10489-016-0767-1.

Gaxiola, F., P. Melin, F. Valdez, J.R. Castro, and O. Castillo. 2016. Optimization of type-2 fuzzy weights in backpropagation learning for neural networks using GAs and PSO. *Applied Soft Computing Journal* 38: 860–871. https://doi.org/10.1016/j.asoc.2015.10.027.

Geem, Z.W., J.H. Kim, and G.V. Loganathan. 2001. A New heuristic optimization algorithm: Harmony search. *Simulation* 76: 60–68. https://doi.org/10.1177/003754970107600201.

Hernández, E., O. Castillo, and J. Soria. 2020. Optimization of fuzzy controllers for autonomous mobile robots using the grey wolf optimizer. In *Hybrid intelligent systems in control, pattern recognition and medicine* ed by O. Castillo and P. Melin, pp. 289–299. Springer International Publishing, Cham (2020). https://doi.org/10.1007/978-3-030-34135-0_20.

Hu, C., Z. Li, T. Zhou, A. Zhu, and C. Xu. 2016. A multi-verse optimizer with levy flights for numerical optimization and its application in test scheduling for network-on-chip. *PLoS ONE* 11: 1–22. https://doi.org/10.1371/journal.pone.0167341.

Izadbakhsh, A. 2017. FAT-based robust adaptive control of electrically driven robots without velocity measurements. *Nonlinear Dynamics* 89: 289–304. https://doi.org/10.1007/s11071-017-3454-9.

Jang, J.S.R., C.T. Sun, and E. Mizutani. 1997. *Neuro-fuzzy and soft computing: a computational approach to learning and machine intelligence.* Prentice Hall.

Jangir, P., S.A. Parmar, I.N. Trivedi, and R.H. Bhesdadiya. 2017. A novel hybrid particle swarm optimizer with multi verse optimizer for global numerical optimization and optimal reactive power dispatch problem. *Engineering Science and Technology, an International Journal* 20: 570–586. https://doi.org/10.1016/J.JESTCH.2016.10.007.

Khomh, F., B. Adams, J. Cheng, M. Fokaefs, and G. Antoniol. 2018. *Software engineering for machine-learning applications: The road ahead.* https://doi.org/10.1109/MS.2018.3571224.

Lagunes, M.L., O. Castillo, F. Valdez, and J. Soria. 2020. Comparison of fuzzy controller optimization with dynamic parameter adjustment based on of Type-1 and Type-2 Fuzzy logic. In Hybrid intelligent systems in control, pattern recognition and medicine, ed by O. Castillo, P. Melin, pp. 47–56. Springer International Publishing, Cham. https://doi.org/10.1007/978-3-030-34135-0_4.

Mahmoodabadi, M.J., and H. Jahanshahi. 2016. Multi-objective optimized fuzzy-PID controllers for fourth order nonlinear systems. *Engineering Science and Technology, an International Journal* 19: 1084–1098. https://doi.org/10.1016/j.jestch.2016.01.010.

Melin, P., and O. Castillo. 2005. *Hybrid intelligent systems for pattern recognition using soft computing.* Springer, Berlin, Heidelberg. https://doi.org/10.1007/b97585.

Mirjalili, S., S.M. Mirjalili, and A. Lewis. 2014. Grey wolf optimizer. *Advanced Engineering Software* 69: 46–61. https://doi.org/10.1016/j.advengsoft.2013.12.007.

Mirjalili, S., S.M. Mirjalili, and A. Hatamlou. 2016. Multi-Verse Optimizer: a nature-inspired algorithm for global optimization. *Neural Computing and Applications* 27: 495–513. https://doi.org/10.1007/s00521-015-1870-7.

Peraza, C., F. Valdez, and O. Castillo. 2020. Harmony search with dynamic adaptation of parameters for the optimization of a benchmark controller. In *Intuitionistic and Type-2 Fuzzy logic enhancements in neural and optimization algorithms: Theory and applications* ed by O. Castillo, P. Melin, and J. Kacprzyk, pp. 157–168. Springer International Publishing, Cham. https://doi.org/10.1007/978-3-030-35445-9_14.

Pérez, J., F. Valdez, and O. Castillo. 2017. Modification of the bat algorithm using type-2 fuzzy logic for dynamical parameter adaptation. In Studies in computational intelligence. pp. 343–355. Springer. https://doi.org/10.1007/978-3-319-47054-2_23.

Sanchez, M.A., O. Castillo, J.R. Castro, and P. Melin. 2014. Fuzzy granular gravitational clustering algorithm for multivariate data. *Information Sciences* 279: 498–511.

Steinhardt, P.J., and N. Turok. 2005. *The Cyclic model Simplified.* https://doi.org/10.1016/j.newar.2005.01.003.

Teodorović, D. 2009. Bee colony optimization (BCO). *Studies in Computational Intelligence* 248: 39–60. https://doi.org/10.1007/978-3-642-04225-6_3.

Wescott, T. 2006. *Applied control theory for embedded systems.* Newnes.

Yang, X.S. 2014. *Nature-inspired optimization algorithms.* Elsevier Inc. https://doi.org/10.1016/C2013-0-01368-0.

Zadeh, L.A. 1968. Fuzzy algorithms. *Inform. Control* 12: 94–102. https://doi.org/10.1016/S0019-9958(68)90211-8.

Zadeh, L.A. 1973. Outline of a new approach to the analysis of complex systems and decision processes. *IEEE Transactions* on *Systems*, *Man*, and *Cybernetics*. SMC-3, 28–44. https://doi.org/10.1109/TSMC.1973.5408575.

Performance Analysis of a Distributed Steady-State Genetic Algorithm Using Low-Power Computers

Anabel Martínez-Vargas, M. A. Cosío-León, Andrés J. García-Pérez, and Oscar Montiel

Abstract In this chapter, we describe the implementation and evaluation of a distributed steady-state genetic algorithm on low-power computers. Specifically, we integrate the NVIDIA Jetson card and the ESP32 cards to form a master-slave model. NVIDIA Jetson card is the master, whereas the ESP32 cards are the slaves. We evaluated the distributed steady-state genetic algorithm on two challenging combinatorial problems: the n-Queens problem and the travelling salesman problem. To compare the performance of the distributed steady-state genetic algorithm on those well-known problems, we implemented a sequential steady-state genetic algorithm. The simulation results indicate that the distributed steady-state genetic algorithm can escape local minima in the travelling salesman problem; hence, the solutions have a better quality of fitness than the sequential genetic algorithm ones. In contrast, for the n-Queens problem, both genetic algorithms' performance is remarkably similar.

Keywords NVIDIA jetson · ESP32 · Master-slave model · Distributed steady-state genetic algorithm · Travelling salesman problem · n-Queens problem

A. Martínez-Vargas · M. A. Cosío-León · A. J. García-Pérez
Universidad Politécnica de Pachuca, Zempoala, Hidalgo, Mexico
e-mail: anabel.martinez@upp.edu.mx

M. A. Cosío-León
e-mail: ma.cosio.leon@upp.edu.mx

A. J. García-Pérez
e-mail: andresjgarciaperez@micorreo.upp.edu.mx

O. Montiel (✉)
Instituto Politécnico Nacional, Centro de Investigación y Desarrollo de Tecnología Digital (CITEDI-IPN), Tijuana, BC, Mexico
e-mail: oross@ipn.mx

O. Castillo and P. Melin (eds.), *Fuzzy Logic Hybrid Extensions of Neural and Optimization Algorithms: Theory and Applications*, Studies in Computational Intelligence 940,
https://doi.org/10.1007/978-3-030-68776-2_3

1 Introduction

According to the complexity of a problem, it may be solved by an exact method or a metaheuristic method. An exact method obtains optimal solutions, whereas a metaheuristic method generates high-quality solutions in a reasonable time for practical use. However, a metaheuristic method does not guarantee to find an optimal solution. Those last main features of metaheuristics enable them to tackle large-size problem instances (Talbi 2009).

The metaheuristic that emulates intelligent bio-inspired processes in the form of a computational algorithm is known as a bio-inspired algorithm. In this context, special attention has been paid in bio-inspired population-based metaheuristic approaches. The first contribution in this area was the genetic algorithm (GA) inspired by Darwinian evolutionary principles (Del-Ser et al. 2019). It has now found widespread applications in almost all branches of science and engineering (Valenzuela et al. 2019; Sepúlveda et al. 2018; Slowik and Kwasnicka 2020).

Work in Slowik and Kwasnicka (2020) analyzed the number of publications in the Web of Science (WoS) database during the years 2000–2018. It found that the number of papers published related to GA was equal to 98596. It also studied its popularity in the selected scientific databases such as Google Scholar, Springer, IEEE Xplore, ACM, Scientific, Science Direct, Sage, Taylor, and WoS. The total number of papers was equal to 1,304,205 for GA. Finally, it mentions the number of practical applications of GA patented by corporations like Caterpillar Inc., Yamaha Motor Co. Ltd., Fujitsu Limited, International Business Machines Corporation, Lsi Logic Corporation, Honda Research Institute Europe Gmbh, Prometheus Laboratories Inc., Siemens. The total number of patents registered in the Google Patents database (in years 2000–2018) for GAs was equal to 43,284. Even though GA is one of the oldest and most known metaheuristics, the amount of publications and patents suggests it is still active in the research community.

GA has two different models of population management: the generational model and the steady-state model. The generational model is the most used in the literature. There, after each generation, the whole population is replaced by the offspring. On the other hand, in the steady-state model, the whole population is not changed at once, but rather a part of it Eiben and Smith (2015). It produces one or two offspring in each cycle. It is an overlapping system due to parents and offspring compete for survival (Lozano et al. 2008).

There is evidence that steady-state GAs perform better than generational GAs. This is due to steady-state GAs have higher selection pressure and changes in the exploration/exploitation balance (Lozano et al. 2008). Despite those advantageous features, few works are reported in the literature related to analyzing steady-state GAs and their applications. For example, work in Martínez-Vargas et al. (2019) applies a steady-state GA to find optimal locations for unmanned aerial vehicles. This is done to provide the largest number of communication services in disaster zones. Another application is reported in work (Dyer et al. 2012). There, authors use a steady-state GA to design a missile system. Regarding the analysis of the behavior of the steady-state GA performance, work in Oliveto et al. (2020) finds out that larger population sizes improve its performance and states its optimal mutation rate for the standard uni-modal OneMax function. In this context, work in Corus and Oliveto (2019) concludes that larger populations increase the probability that crossover finds improved solutions quickly, reducing the expected running time.

The distributed evolutionary computation has recently received considerable attention due to the increasing complexity of real-world optimization problems. It has two types of models Gong et al. (2015): population-distributed model and dimension distributed model. The first model distributes individuals of the population to multiple processors or computing nodes. The second model distributes the partitions of the problem dimensions.

In this chapter, we propose a distributed steady-state GA on low-power computers. We implemented a population-distributed model, particularly the master-slave model. The physical platform is based on the NVIDIA Jetson card as the master and a set of ESP32 cards as the slaves (see Fig. 1) building a local area network of low power devices (LAN_{lp}). NVIDIA Jetson card not only performs crossover, mutation, selection operations but also sends individuals to slave-devices (ESP32 cards) for fitness evaluations. This last task forms most of the computing load. The NVIDIA Jetson card communicates to ESP32 cards when it sends the individuals, and they return the corresponding fitness values to it in each iteration. Some efforts have been reported in distributed generational GAs in single-objective optimization (Jianli et al. 2020) and multi-objective optimization (Moreno et al. 2017). This paper contributes to state-of-the-art with a distributed steady-state GA and its performance analysis on a low power architecture. Based on an extensive bibliographic review, this is the first work that explores a distributed steady-state GA. To test its performance, we use small instances of two very well-known combinatorial problems: the n-Queens problem and the travelling salesman problem (TSP).

Fig. 1 The master-slave model hardware and LAN_{lp}

The chapter is organized as follows: Sect. 2 describes the distributed steady-state GA and its application on the travelling salesman problem and the n-Queens problem. Section 3 is devoted to the master-slave low power architecture. Section 4 shows computational experiments obtained from sequential and distributed GAs when they are applied to the travelling salesman problem and the n-Queens problem. Finally, Sect. 5 concludes the chapter and gives direction for future research.

2 Distributed Steady-State Genetic Algorithm

Our proposed distributed steady-state GA is a population-distributed model, specifically a master-slave model (see Fig. 2), so-called Global parallel GA (Cantu-Paz 2000). The master performs crossover, mutation, and selection operations. It also sends individuals to slaves for fitness evaluation. The evaluations of individuals are independent. Therefore, there is no communication among slaves. The communication only occurs when the master sends individuals to slaves, and consequently, they return the fitness values to the master in each cycle (Gong et al. 2015).

An advantage of this model is computation distribution on multiple devices, avoiding massive loss of information. This is due to multiple slave-nodes perform the same task with different individuals. So, slaves-nodes can be replaced if they fail, and new ones would perform missing individual computations. On the other hand, a disadvantage of the master-slave model is related to network communication costs. Hence, an efficient communication protocol selection is relevant. In this work, we choose the UDP protocol since it has a low cost on packet transmission. The main issue in this model is when the master crashes, it can stop the whole execution.

Algorithm 1 shows our distributed steady-state GA proposed. In the beginning, the master generates the initial population randomly. It performs 'tournament selection' (Eiben and Smith 2015) to create two parents. In that process, fitness evaluation is required, so the master sends individuals to the slave-nodes. After that, the master performs 'cut-and-cross fill' crossover (see Algorithm 2) (Eiben and Smith 2015) if a

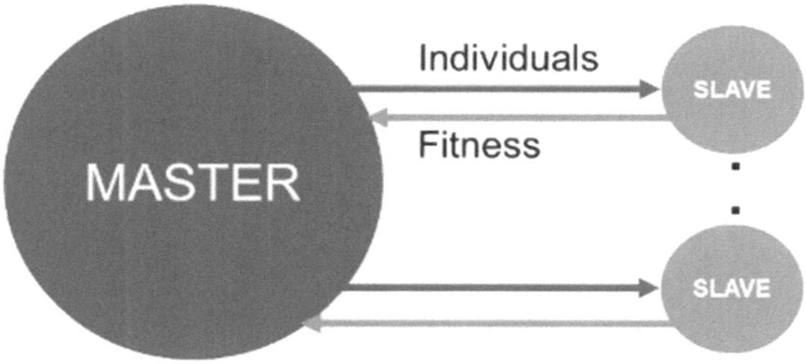

Fig. 2 Master-slave model

random number from a uniform distribution over the range [0,1] is less than or equal a crossover probability. Then, the master makes 'swap mutation' (Eiben and Smith 2015) to the two-resulting offspring; if a random number from a uniform distribution over the range [0,1] is less than or equal a mutation probability. 'Swap mutation' selects two positions (genes) at random in the individual, and their allele values are swapped. Finally, the master executes a replacement strategy based on family competition replacement methods known as 'elitist recombination' (see Algorithm 3) (Lozano et al. 2008). The replacement strategy involves fitness evaluation; then, the master sends individuals to the slaves-nodes. About the ending criteria, the execution stops when there are 2500 cycles without improvement in fitness.

Algorithm 1 Distributed steady-state GA (DSSGA)

procedure DSSGA. Inputs are population size, crossover probability, and mutation probability

Master-node randomly generates an initial population

loop *Do until some stopping condition is met*:

Master-node sends individuals to slave-nodes to execute 'tournament selection'
Slave-nodes perform fitness evaluation
Slave-nodes send fitness evaluation to master-node
Master-node executes 'cut-and-cross fill' crossover
Master-node executes 'swap' mutation
Master-node sends individuals to slave-nodes to execute 'elitist recombination' replacement
Slave-nodes perform fitness evaluation
Slave-nodes send fitness evaluation to master-node

End loop
End procedure

Algorithm 2 Cut-and-cross fill crossover

procedure crossover

Select a random position, the crossover point, $i \in \{1, \ldots, n\}$
Cut both parents into two segments at this position
Copy the first segment of parent 1 into child 1 and the first segment of parent 2 into child 2
Scan parent 2 from left to right and fill the second segment of child 1 with values from parent 2, skipping those that it already contains Do the same for parent 1 and child 2

End procedure

Algorithm 3 Elitist recombination

procedure replacement

Evaluate parents and offspring
Select the two best individuals out of the two parents and two offspring.
Call those best individuals, *best1* and *best2*.
Replace parents with *best1* and *best2* respectively

End procedure

2.1 Application of Distributed Steady-State Genetic Algorithm in the n-Queens Problem

The goal of the n-Queens problem is to place n queens on a chessboard of size n x n such that no queen is attacked by another one. One queen attacks another queen if they share the same row, column, or diagonal, as Fig. 3 shows. The n-Queens problem is a generalization of the original proposal of putting eight non-attacking queens on a chessboard, posed in 1848 by M. Bezzel, a German chess player (Bell and Stevens 2009). It has several applications, such as parallel memory storage schemes, very-large-scale integration (VLSI) testing, traffic control, and deadlock prevention (Bell and Stevens 2009).

An individual has the permutation representation given in Fig. 3. There, an individual is encoded as an n-tuple $(q_1, q_2,\ldots, q_n)$, it is a permutation of the set $(1, 2,\ldots, n)$. Each q_i (index position) is the row occupied by the queen positioned in the ith column (value). This representation avoids horizontal and vertical collisions; hence,

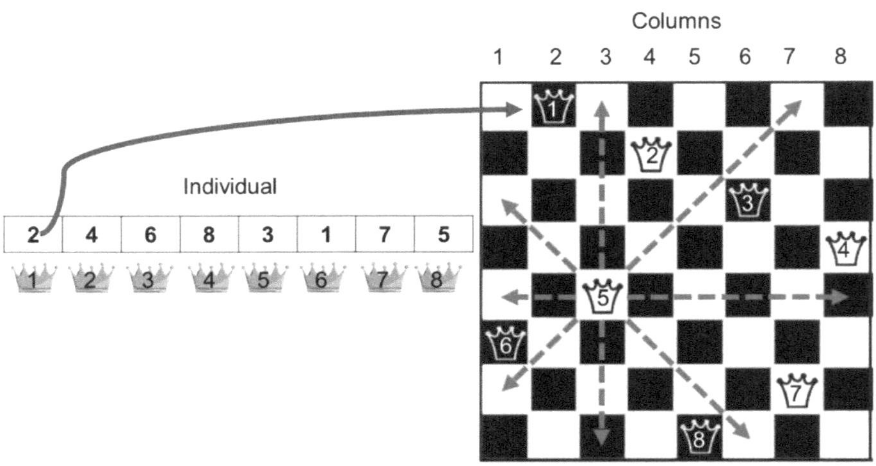

Fig. 3 An individual for the n-Queens problem

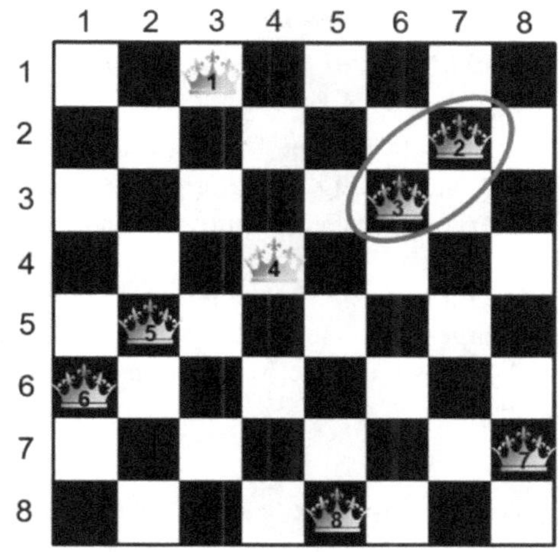

Fig. 4 A diagonal collision for the *n*-Queens problem

there are no restrictions. Therefore, the objective function is defined as the number of diagonal collisions along the chessboard (Osaba et al. 2014). Two queens, q_i and q_j, collide diagonally if (Osaba et al. 2014):

$$\left|q_i - q_j\right| = |i - j| \qquad \forall i, j : 1, 2, \ldots, n; i \neq j \tag{1}$$

Consider the individual (3, 7, 6, 4, 2, 1, 8, 5). Now, take the second and third queens, i.e., $i = 2$ and $j = 3$. Then, $q_2 = 7$ and $q_3 = 6$. By applying Eq. (1), we have |7–6| = |2–3|, i.e.,|1| = |–1|. Then, there is a collision between q_2 and q_3 as Fig. 4 depicts.

We aim to minimize the number of diagonal collisions along the chessboard as the objective function in (2) shows:

$$Minimize\ f(i, j) = \sum_{i=1}^{n}\sum_{j=1}^{n} \frac{\left(\left|q_i - q_j\right| = |i - j|\right)}{2} \tag{2}$$

The *n*-Queens problem has three variants (Jianli et al. 2020):

- Variant I: obtain one valid solution.
- Variant II: obtain a family of valid solutions.
- Variant III: obtain all valid solutions.

In this work, we deal with variant II. The *n*-Queens problem is well suited for the distributed steady-state GA, as pointed out in work (Jianli et al. 2020). The authors mentioned that accelerating fitness function is the key to improve the performance algorithm and evaluation.

2.2 Application of Distributed Steady-State Genetic Algorithm in the Travelling Salesman Problem

The goal is to find the shortest route for a salesperson to take in visiting N cities, starting and ending in the origin city. Each city is visited once (Haupt and Haupt 2004). This problem requires a round trip, i.e., a closed-loop (Fogel 2000). The TSP has some practical applications in engineering, like the optimal layout of a gas pipeline, design of an antenna feed system, and configuration of transistors on very-large-scale integration (VLSI) circuit (Haupt and Haupt 2004).

We apply the objective function proposed in work (Haupt and Haupt 2004) which states that:

$$Minimize\ f(x, y) = \sum_{n=1}^{N} \sqrt{(x_n - x_{n+1})^2 + (y_n - y_{n+1})^2} \tag{3}$$

where (x_n, y_n) are the coordinates of the nth city visited. This version of the TSP is an undirected weighted graph, and we consider only positive values of distances in the objective function.

We propose to represent the individual as a permutation, as shown in Fig. 5. There, an individual is an ordered list in which the cities are visited exactly once. Specifically, Fig. 5 represents a tour of nine cities. The genetic operations are applied to the individuals enclosed in the blue dashed rectangle shown in Fig. 5. We use zero value in the individual in the first position to indicate the origin city and a zero value in the last place of the individual to indicate the return to the origin city; in this way, we have a round trip, i.e., the first element in the list is also the last one (Fogel 2000).

The TSP is a challenging problem well suited for testing our distributed steady-state GA. The distributed steady-state GA must comply with the operating credo

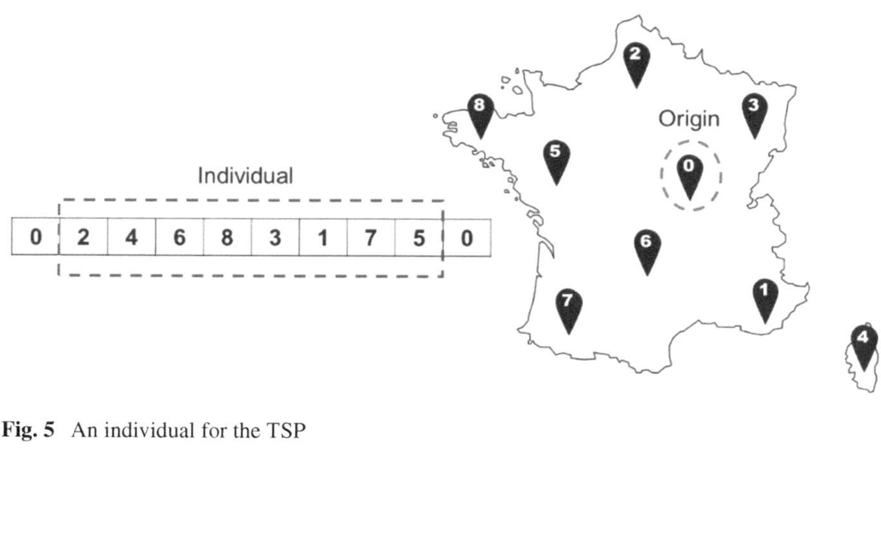

Fig. 5 An individual for the TSP

of evolutionary practitioners: "Solutions should be good enough and generated fast enough to be useful" (Fogel 2000).

3 Master-Slave Low Power Architecture

We start defining a set of terms, supporting the relevance of the analysis here proposed for real-world applications.

The first set of elements on a sensing network structure is called the fog layer. Devices in this layer acquire, process, and send data to the cloud. On top of this layer is the edge computing layer, having more robust devices. Both layers move the data processing tasks from the centralized cloud server to the devices at the edge of the network, to reduce latency. Considering the above, the performance analysis of devices doing data processing tasks in edge layers is paramount today.

To reproduce a real-world arrangement on the layers mentioned above, we implement network interactions among fog and edge devices based on previously described master-slave hardware architecture.

3.1 Rationale on Master-Slave Architecture Starting Procedure

The general sequence diagram is explained using Figs. 6, 7, and 8. They show the interactions among the total elements in this architecture: the users, the Telegram bots, the Telegram API, the master-device, and the slave-devices. The first step is the starting command for the Telegram bot, which requires a command sequence to do

Fig. 6 Master-slave architecture, initialization sequence diagram

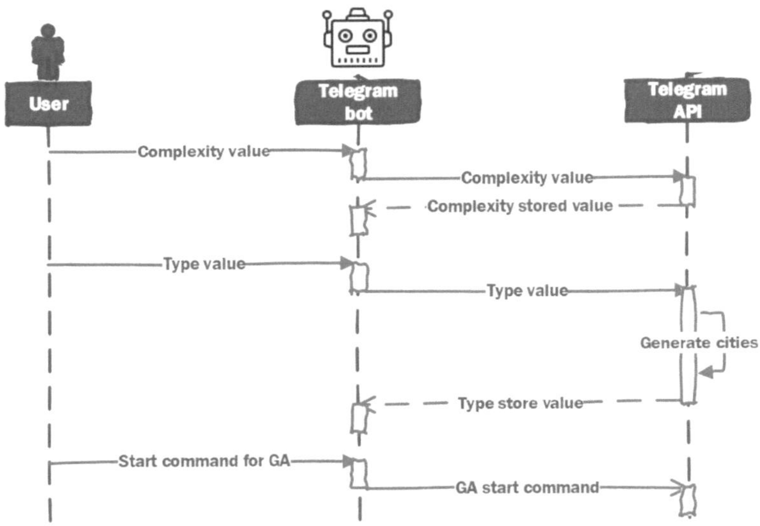

Fig. 7 Master-slave architecture, configuration sequence diagram

tasks. Telegram API builds the answer key, which has the user request, along with this, a demon starts a UDP server in the Master-device (see Fig. 6).

A parallel asynchronous task is to turn on ESP32 devices. Once the ESP32 device is running, the second processor starts and launches a UDP client. This is done to establish a connection with the UDP server onto the master-device (see Fig. 6).

Then the slave-device waits until a fitness function evaluation is required, (see Fig. 6).

Three relevant parameters to configure the architecture are (see Fig. 7):

- The complexity parameter. It is the size of the problem (number of cities for TSP, or number of queens for n-Queens problem). In the case of the TSP, a specific instance name is required (for example, the words: Qatar, Uruguay, and Western) or a number to define the problem size. They are the possible tests that can be carried out. The complexity is saved into the master-device. In TSP cases, it will create a daemon to initialize a matrix with the city's coordinates.
- The GA parameter so-called type, to be executed. The available options are one or two (one performs the GA to solve the n-Queens problem, and two performs the GA to solve the TSP problem).
- The Ready parameter that executes the selected GA. Once the GA stars, its first step is to initialize the population; after, the GA processes are executed in a loop until the algorithm's termination criterion is reached. For the function evaluation, a daemon in the master-devices is created; it manages this process (see Algorithm 1).

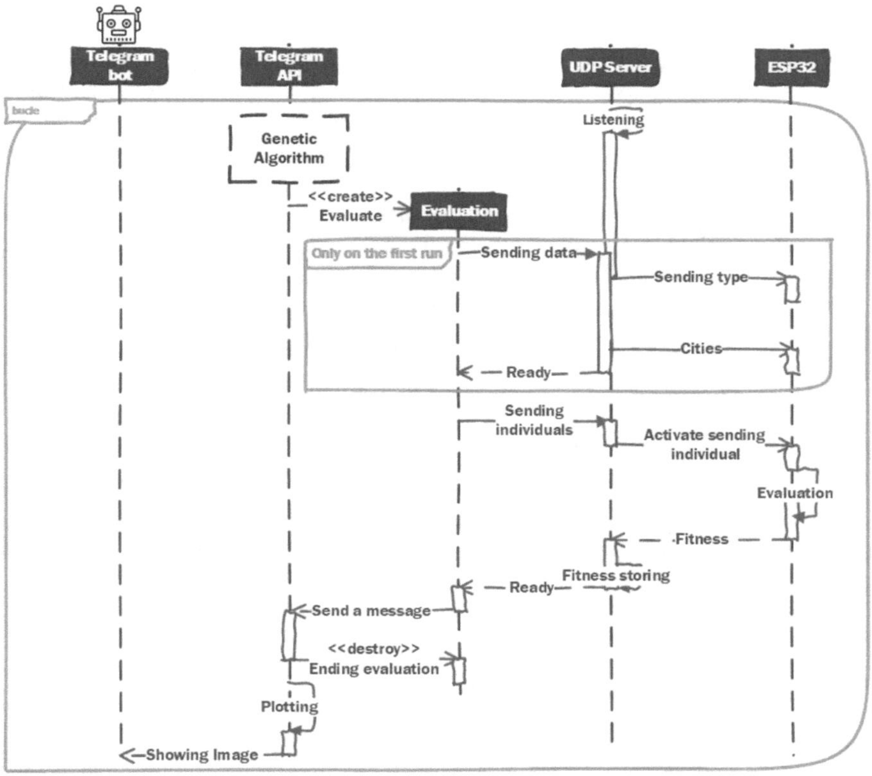

Fig. 8 Master-slave architecture, execution GA sequence diagram

3.2 Function Evaluation Task on Slave-Devices

The Master-device starts the function evaluation process, sending data packets to slave-devices through the UDP server. These packages contain type and complexity parameters. If type is equal to two, the UDP server also broadcasts all city points. The UDP server transmits a ready signal to the evaluation daemon and starts to send the individuals from the registry matrix to each slave-device.

Once the slave-device receives all data packets, it changes its status to ready-to-evaluate and the second processor starts the evaluation process. The evaluation process closes to receiving a ready signal. Then, the first processor sends the fitness value obtained by the evaluation process to the master-device, where the UDP server obtains data and store it according to arrival time. Once all the packages from the LAN_{lp} are received, the UDP server sets state flag ready, and the evaluation daemon in slave-device receives a n auto-destroy command (see Fig. 8).

3.3 Fail-Safe Algorithm on Master-Device

If any fitness-value is missed because the timeout is exceeded, the fail-safe algorithm is executed. Two scenarios can occur: one or more slave-devices do not answer, or none of them do. For the first scenario, missing individual evaluations are re-transmitted to available slave-devices, and a restart signal is sent to slave-device that does not work. For the second scenario, a restart signal is sent on broadcast, and the UDP server on the master-device waits for answers from the LAN_{lp}. If the LAN_{lp} starts-up, then the evaluation task on ESP32 devices begins.

At the end of each iteration, the best individual is displayed using a graph on the Telegram application.

4 Computational Results

The next subsections are devoted to results. We tested some instances for TSP on the sequential and distributed GAs to evaluate their performance. This is also done for the *n*-Queens problem. Finally, we applied statistical tests to analyze if there is a significant difference between GAs.

4.1 Experimental Setup

The distributed steady-state GA was coded in Python 3.7 on the master-device and C++ over an Arduino framework for the slave-devices. A sequential GA was developed to compare the performance. As described in Sect. 1, the distributed steady-state GA runs on the physical low-power architecture. A sequential GA was executed on a computer laptop, which characteristics are: AMD Ryzen 3 3300U with Radeon graphics card, Vega 6 Processor 2.1 GHz, 6 MB cache and 4 cores, 12 GB SDRAM DDR4-2400 (1 × 4 GB, 1 × 8 GB), SATA 1 TB and 5400 rpm, and SSD M.2 of 128 GB.

Three different TSP instances were taken from World TSP (http://www.math.uwaterloo.ca/tsp/world/countries.html): Uruguay (734 cities), Qatar (194 cities), and Western Sahara (29 cities). The Euclidean distance specifies the cost of travel between cities rounded to the nearest whole number (the TSPLIB EUC 2D-norm). The distributed steady-state GA and the sequential GA were executed 30 times with separate random seeds for each instance.

On the other hand, for the *n*-Queens problem, its size was varied incrementally with step size = 10 queens, beginning in 10-Queens, and ending in 100-Queens.

This applied to the sequential GA and distributed steady-state GA. The algorithm parameter values were set as it is shown in Table 1.

Table 1 Parameters used for distributed steady-state GA and sequential GA

Parameter	Value
Population size	500
Crossover probability	0.85
Mutation probability TSP	$1/N$
Mutation probability n-queens	$1/n$
Tournament size k	2

The stop condition for both algorithms occurs when there have elapsed 2500 cycles without improvement in fitness.

The performance of GAs solving the TSP and n-Queens problem was measured using the average distance, in the sense of a goal function's value of the best solutions on sequential GA (relative percentage deviation, PRD) and the best solutions of the distributed GA.

Besides, we use the average relative percentage deviation (APRD), which is like PRD but using the average solutions of the distributed GA (Bozejko and Makuchowski 2009), (see Eq. 5), instead of the best value (see Eq. 4). Negative values imply a better performance of distributed GA, and zero means equal performance.

$$PRD = \frac{Best_{Alg} - BKS}{BKS} \times 100 \tag{4}$$

$$APRD = \frac{Avg_{Alg} - BKS}{BKS} \times 100 \tag{5}$$

where Avg_{Alg} is the average value obtained by the distributed GA. $Best_{Alg}$ is the best value obtained by the distributed GA. BKS is the best-known value obtained by the reference algorithm, the sequential GA.

Another metric considered is the relative standard deviation (RSD). The meaning of RSD results is whether the standard deviation is a small or large value when compared to the mean of the analyzing results. That is how precise the average of the results is.

$$RSD = \frac{std\ dev}{\bar{x}} \tag{6}$$

where *std dev* is the standard deviation. $\bar{x}$ is the sample mean.

4.2 n-Queens Problem Experimental Arrangement Results

Table 2 reports fitness for each experimental arrangement of the n-Queens problem. The first column is the number of Queens. The second column labeled as fitness has

Table 2 Fitness and time values of both GA resolving *n*-Queens problem

Queens	Sequential		Distributed		Queens	Sequential		Distributed	
	Fitness	Time (s)	Fitness	Time (s)		Fitness	Time (s)	Fitness	Time (s)
10	1	70.191	**0**	**99.881**	60	20	1642.117	**17**	**2192.370**
10	**0**	**81.643**	**0**	**111.015**	60	19	1644.697	20	2190.669
10	1	82.501	**0**	**107.344**	60	18	1868.771	20	2338.547
10	1	78.028	**0**	**114.090**	60	18	1867.567	18	2337.546
10	**0**	**80.058**	1	109.549	60	20	1930.001	20	2082.001
10	**0**	**74.880**	1	95.191	60	21	1931.076	19	2020.771
10	**0**	**86.015**	**0**	**106.643**	60	**17**	**2040.37**	18	1796.697
10	**0**	**82.344**	1	107.501	60	20	2038.669	18	2019.567
10	**0**	**89.090**	1	103.028	60	20	2186.547	20	1794.117
10	1	84.549	**0**	**105.058**	60	18	2185.546	21	2083.076
10	*0*	*82.3383*	*0*	*107.339*	*60*	*17*	*2040.37*	*17*	*2192.37*
20	3	323.139	**2**	**358.798**	70	26	2719.251	26	2899.251
20	4	332.591	3	343.923	70	26	2717.957	26	2897.957
20	**2**	**318.085**	3	335.217	70	24	2545.579	24	2725.579
20	3	313.075	3	297.946	70	26	2545.072	26	2725.072
20	**2**	**308.798**	3	335.982	70	**23**	**2317.746**	**23**	**2317.746**
20	3	293.923	3	415.941	70	**23**	**2145.657**	**23**	**3520.510**
20	3	285.217	3	373.139	70	25	3342.420	25	2773.473
20	3	247.946	4	382.591	70	24	3340.510	24	2325.657
20	3	285.982	**2**	**368.085**	70	26	2592.944	26	2772.944

(continued)

Table 2 (continued)

Queens	Sequential		Distributed		Queens	Sequential		Distributed	
	Fitness	Time (s)	Fitness	Time (s)		Fitness	Time (s)	Fitness	Time (s)
20	3	365.941	3	363.075	70	24	2593.473	24	3522.420
20	*2*	*313.4415*	*2*	*363.4415*	*70*	*23*	*2141.702*	*23*	*2919.128*
30	7	1661.867	6	487.364	80	29	3413.222	29	3611.222
30	6	398.819	8	691.467	80	**28**	**3403.676**	**28**	**3601.676**
30	**5**	**410.612**	7	778.188	80	31	3161.996	31	3359.996
30	7	343.823	8	598.523	80	29	3245.624	29	3443.624
30	8	390.983	6	561.856	80	29	2879.375	29	3077.375
30	6	424.864	7	486.319	80	**28**	**2962.809**	**28**	**3160.809**
30	8	628.967	6	461.319	80	31	2976.510	31	3260.581
30	7	715.688	**5**	**473.112**	80	29	3062.581	29	3174.510
30	8	536.023	7	406.323	80	30	2856.343	30	3054.343
30	6	499.356	8	453.483	80	32	2939.087	32	3137.087
30	*5*	*410.612*	*5*	*473.112*	*80*	*28*	*3183.243*	*28*	*3381.2425*
40	**10**	**1309.374**	**10**	**854.037**	90	32	3793.652	31	4418.922
40	**10**	**1232.929**	**12**	807.523	90	34	3700.106	34	4574.272
40	12	1135.88	**10**	**890.414**	90	32	3739.070	34	4714.246
40	11	1134.049	**10**	**923.275**	90	**29**	**3640.493**	33	4471.357
40	**10**	**755.072**	11	922.859	90	33	4058.595	35	4685.556
40	**10**	**754.037**	**10**	**1461.374**	90	31	4184.922	32	4027.652

(continued)

Table 2 (continued)

Queens	Sequential		Distributed		Queens	Sequential		Distributed	
	Fitness	Time (s)	Fitness	Time (s)		Fitness	Time (s)	Fitness	Time (s)
40	12	707.523	**10**	**1332.929**	90	34	4340.272	34	3934.106
40	**10**	**790.414**	12	1235.808	90	34	4480.246	32	3973.070
40	**10**	**823.275**	11	855.072	90	33	4237.357	**29**	**3874.493**
40	11	822.859	**10**	**855.072**	90	35	4451.556	33	4292.595
40	*10*	*944.1835*	*10*	*1052.8502*	*90*	*29*	*3640.493*	*29*	*3874.493*
50	14	1593.634	**8**	**2347.814**	100	41	5182.562	41	5443.562
50	14	1447.052	**17**	9295.814	100	**38**	**5432.711**	**38**	**5693.711**
50	16	1444.959	16	5855.829	100	39	5012.325	39	5273.325
50	14	1509.364	15	1831.487	100	40	5216.460	40	5477.460
50	16	1509.133	17	1957.895	100	39	5037.642	39	5298.642
50	15	1528.355	14	2103.735	100	39	5072.564	39	5333.564
50	15	1529.946	16	1884.257	100	39	5056.597	39	5317.597
50	**13**	**1881.195**	15	1884.257	100	40	5333.192	40	5594.192
50	17	1880.984	16	1938.275	100	39	5067.620	39	5328.620
50	14	1860.456	17	2050.930	100	40	5435.980	40	5696.980
50	*13*	*1881.195*	*8*	*2347.814*	*100*	*38*	*5432.711*	*38*	*5693.711*

the best value of executed arrangement, for this study, the lowest collision values. It also has the execution time that GAs spend to reach the best fitness value.

The row in italics shows the best solution for the *n*-Queens experiment. When there is more than one best solution, the average time is reported.

The trend of the collision results is shown in Fig. 9, as *n* size increases, collisions also do, following a linear pattern. The number of collisions diagonally in both GAs has a similar occurrence; hence the performance is similar in most of the experiments. Same behavior shows the execution time in Fig. 10, except for one, where the result changed, but the pattern is similar in both GAs.

Due to sequential and distributed GAs found optimal values on six executions out of ten executions for the 10-Queens problem, the BKS of collisions is zero; then, the use of PRD and APRD was not possible.

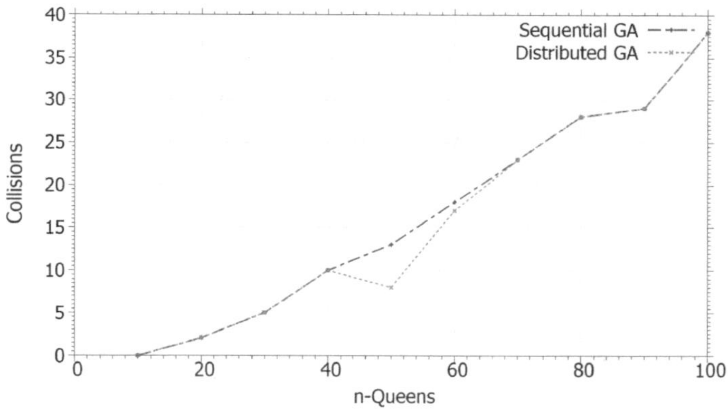

Fig. 9 *n*-Queens versus collisions on both GAs

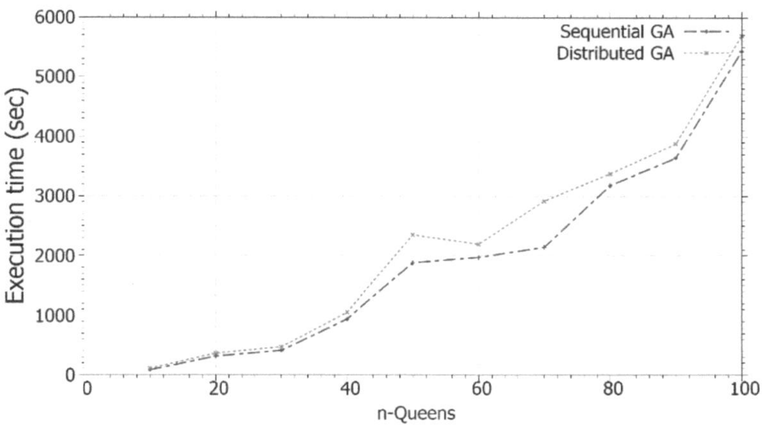

Fig. 10 *n*-Queens versus execution time on both GAs

By analyzing the execution time, the standard deviation is 33.38% of the mean of 105 (see Table 3, on 10-Queens arrangement PRD and APRD rows), which is a large value when compared to the mean and its positive sign, similar behavior has APRD, 41.46%, so the sequential GA is better than the distributed GA considering execution time.

PRD and APRD in Table 3 show that as the size of the n-Queens problem increases, the performance of distributed GA also increases, in comparison to the sequential GA results (see Table 3). An example is the 100-Queens problem resolved by distributed GA, which has a PRD-fitness of 0 and PRD-time of 4.80; meanwhile, APRD-fitness is 3.68 and APRD-time is 0.24. RSD here shows the entire algorithm performance on resolving the n-Queens problem arrangement, RSD-fitness for both GAs was 0.70 while RSD-time was slightly different 0.83 and 0.82.

Mostly PRD values are zero; meanwhile, APRD and RSD values are near zero; these results allow us to conclude that both algorithms' performance is very similar even when a physical architecture of the laptop computer is less robust, and the full potential o f the graphic card is not being used.

We used the Mann-Whitney-Wilcoxon test to gather statistical evidence focused on the median analysis of both GAs and their median difference using the fitness column from Table 3. The null hypothesis is $H_0 : \eta_1 - \eta_2 = 0$, and the alternative hypothesis $H_1 : \eta_1 - \eta_2 \neq 0$; so we mainly determined if the performance of both GAs is equivalent based on their median values. η_1 is the fitness values median of the sequential GAs, and η_2 is the fitness values median of the distributed GAs. By considering 95% of confidence level and $\alpha = 0.05$. The test produced a $p-$ value of 1. The difference between the population medians is between -3 and 3, yielding a 95.01% of confidence. So, we accept the H_0 due to the difference between the medians of 0 is not statistically significant.

A second test was performed with the same aim and configuration. η_1 is the execution time values median of the sequential GAs, and η_2 is the execution–time values median of the distributed GAs in Table 3. By considering 95% of confidence level and $\alpha = 0.05$. The test produced a p–value of 0.44. The difference between the population medians of -89.6305 is between -472.766, and 275.727, providing 95.01% of confidence. So, we accept the H_0 due to the difference between the medians is not statistically significant.

4.3 Travelling Salesman Problem Results

Table 4 shows the experimental results of fitness from 30 executions on both GA, including three selected instances described in Sect. 4.1. The last row of the table represents the average of executed experiments of each GA solving chosen cases (see Table 4).

Solving Western Sahara, a small size instance, the distributed GA outperforms sequential GA; the former algorithm gathers a fitness of 32378.949 meanwhile the second one 28377.41, both on execution 21. In the same execution (row 21), the

Table 3 Metric results of both GAs resolving *n*-Queens problem

Metric	10				Metric	60			
	Sequential		Distributed			Sequential		Distributed	
	Fitness	Time (s)	Fitness	Time (s)		Fitness	Time (s)	Fitness	Time (s)
Average	0.4	80.9299	0.4	105.93	Average	19.1	1933.5361	19.1	2085.5361
Std Dev	0.5163	5.4936	0.5163	5.4935	Std Dev	1.2866	190.5862	1.2866	190.5862
PRD	–	–	**–**	**33.38**	PRD	–	–	0	7.44
APRD	–	–	**–**	**41.46**	APRD	–	–	12.35	2.21
20					70				
Average	2.9	307.4697	2.9	357.4697	Average	24.7	2668.0609	24.7	2848.0609
Std Dev	0.5676	31.84137	0.5676	31.8413	Std Dev	1.2516	409.2719	1.2516	409.2719
PRD	–	–	0	16.19	PRD	–	–	0	8.42
APRD	–	–	45	15.76	APRD	–	–	7.39	33.22
30					80				
Average	6.8000	601.1002	6.8000	539.7954	Average	29.6000	3090.1223	29.6000	3288.1223
Std Dev	1.0328	390.4874	1.0328	117.9687	Std Dev	1.3499	206.5079	1.3499	206.5079
PRD	–	–	36.00	15.22	PRD	–	–	0.00	6.68
APRD	–	–	36.00	31.46	APRD	–	–	5.71	10.97
40					90				
Average	10.6	946.5412	10.6	1013.844	Average	32.7	4062.6269	32.7	4296.6269
Std Dev	0.8432	228.6165	0.8432	236.0880	Std Dev	1.7669	322.3035	1.7669	322.30357
PRD	–	–	0	13.26	PRD	–	–	0	6.42
APRD	–	–	6	34.45	APRD	–	–	12.75	18.02

(continued)

Table 3 (continued)

Metric	10				Metric	60			
	Sequential		Distributed			Sequential		Distributed	
	Fitness	Time (s)	Fitness	Time (s)		Fitness	Time (s)	Fitness	Time (s)
50					100				
Average	14.80	1618.50	15.1	3115.029	Average	39.4	5184.7653	39.4	5445.7653
Std Dev	2.6853	2493.8380	2.6853	2491.2850	Std Dev	0.8432	163.8922	0.8432	163.8922
PRD	–	–	−38.46	24.80	PRD	–	–	**0**	**4.80**
APRD	–	–	16.15	65.59	APRD	–	–	**3.68**	**0.24**

Table 4 Fitness values of sequential and distributed GAs

Execution	Western sahara		Uruguay		Qatar	
	Sequential	Distributed	Sequential	Distributed	Sequential	Distributed
1	38489.55	34019.97	1341622	1316575	59331.24	55227.22
2	39190.429	33452.46	1339957	1319032	52412.41	**43977.05**
3	35237.18	35557.1	1322769	1291832	53367.87	52925.51
4	42569.48	33695.28	1356695	1372347	61379.63	50772.81
5	35544.71	31494.17	1308616	1309720	53325.1	54483.76
6	42954.62	37882.35	1320305	1335403	54488.35	53379.11
7	39117.31	39330.32	1357118	1345952	57026.37	53100.18
8	42558.41	36199.03	1344545	1388324	49290.57	53739.16
9	35557.1	39167.27	1314178	1342117	61379.63	52180.22
10	34845.43	39677.2	1331619	1327486	55240.43	59560.77
11	36229.09	34404.65	1323182	1307483	52480.03	53266.2
12	37899.369	33105.009	1341712	1333591	49435.09	57856.85
13	39403.509	32985.79	1343345	1328135	55225.25	53780.33
14	35200.13	33439.7	1337120	1357468	52857.04	53543.55
15	39303.55	33545.56	1318841	**1291231**	46681.51	46856.35
16	34674.59	31757.72	1360831	1313338	48930.12	57269.14
17	35916.7	35515.65	1314880	1347224	49055.69	50824.58
18	33452.46	34212.79	1316245	1331738	54963.31	60505.76
19	32617.099	38789.28	1308018	1352929	49893.72	52626.87
20	36624.61	32392.36	1328006	1349707	59063.86	54937.96
21	**32378.949**	**28377.41**	**1292978**	1328091	**45424.99**	52727.34
22	35952.2	37517.24	1301113	1328617	55791.61	51426.92
23	34930.64	36221.83	1345168	1392911	50763.54	48406.81
24	42879.1	33807.19	1347899	1353659	48202.68	48003.58
25	36893.55	33880.39	1351260	1295737	60204.59	54083.31
26	42026.799	32607.67	1338294	1333144	52989.62	53327.4
27	39826.74	35074	1327652	1327650	53833.6	54613.49
28	36054.45	35159.26	1322036	1332811	51357.01	58632.44
29	40767.73	33363.32	1341396	1319018	58745.46	56122.49
30	34019.97	37820.509	1358082	1318455	49606.95	50674.35
Average	**36426.85**	**34116.38**	**1334369**	**1330177**	**53157.36**	**53353.25**

Table 5 Performance metrics (fitness)

Metrics	Western sahara		Uruguay		Qatar	
	Sequential	Distributed	Sequential	Distributed	Sequential	Distributed
A	36426.85	34116.38	1334369	1330177	53157.36	53353.25
B	32378.949	28377.41	1292978	1291231	45424.99	43977.05
C	3150.6414	2627.6638	17911.59	24751.44	4339.794	3620.979
D	–	−12.35	–	−0.13	–	−3.18
E	–	5.36	–	2.87	–	17.45

runtime was 20.8020 and 72.0548 (see Table 6), respectively. Due to the stop condition on the sequential algorithm, we hypothesized that it was stagnant in a locally optimal, reaching its stop condition without improving its result. On the other hand, distributed GA was able to explore a larger space without stagnant, finding a better solution compared with the sequential GA.

PRD on Table 5, shows a better performance of the distributed GA than the sequential GA, having on the three instances negative results: −12.35, −0.13, −3.18. In contrast, APRD shows positive results, 5.36, 2.87, 17.4. We are concluding that the distributed algorithm has the capacity to explore different regions than the sequential GA can explore and sometimes find better solutions (see Table 5). There, the average is represented by A, the best value is represented by B, the standard deviation is represented by C, the PRD best value is represented by D, and the APRD best value is represented by E.

The quality of a solution can be visualized using quantitative analysis; a second option is analyzing the solution's patterns of the resulting graph. The pattern is defined by the number of crossing lines; there is an inverse correlation between the number of crossing lines and the solution quality. Figure 11 shows Western Sahara set of points, whose optimal solution is a 2D-graph, without crossing lines (see http://www.math.uwaterloo.ca/tsp/world/countries.html). In our experiments, both algorithms' results show two 2D-graphs with a similar structure and number of crossing lines (see Figs. 12 and 13). We expected that using other stop condition in both GAs could find the optimal solution. In Figs. 11, 12, 13, 14, 15, 16, 17 and 18, the axis names "coordinate X" and "coordinate Y" indicate the location of each city in the plane.

The set of figures from 14, 15 up to 16, describe the biggest analyzed instanced named Uruguay, and results of both GAs solving it, also share characteristics, as similar graph structure, an extensive amount of crossing lines. The best solution is a 2D-graph, without crossing lines (see http://www.math.uwaterloo.ca/tsp/world/countries.html).

Figure 17 shows the medium size instance in our computational experiment, it is Qatar, which size is 194 cities. It also shares the same pattern on results as before described graphs offered by GAs. Comparing Figs. 18 and 19, there is a subtle difference, having a cleaner graph on the later figure than in the former one.

Table 6 explores the amount of time in finding a solution. We select the best values found by each algorithm over the three selected instances. Execution 15 from the sequential GA solving Western Sahara spends 6.8510 s., reaching a fitness of

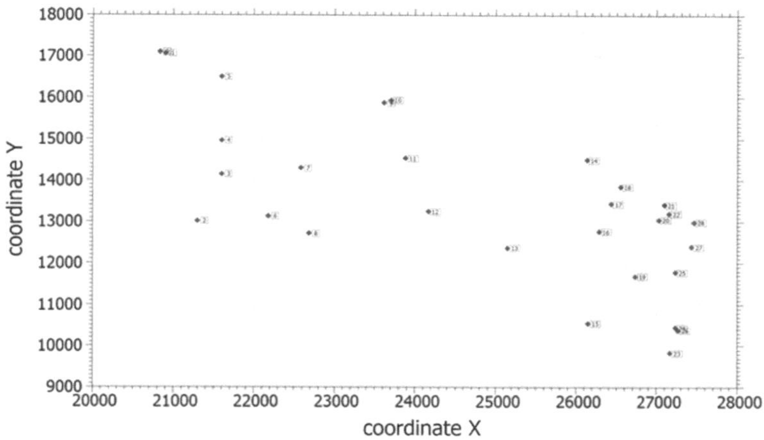

Fig. 11 Western Sahara instance size is 29 cities

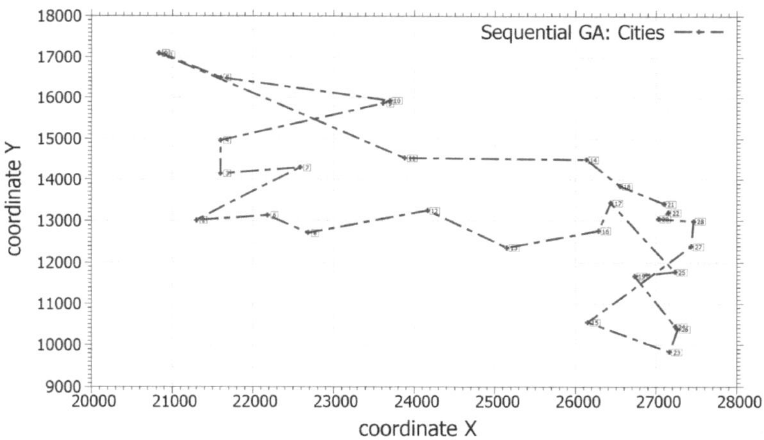

Fig. 12 Western Sahara—the best solution of the sequential GA

39303.55: meanwhile distributed algorithm spends 30.2018 s., reaching a fitness of 33545.56. When GAs solve Uruguay instance in both GAs the runtime has the same pattern. However, comparing GAs runtimes from the best found, 131.5400 and 686.8409 s., the runtime increases in the distributed GA. Here we proposed a question: Why the sequential algorithm is prone to stagnant in local optimal than the distributed GA?

On the other hand, PRD and APRD values show a better performance of the sequential algorithm because both metrics do not include the effect of the relation time-fitness as Table 7 shows. There, the best value is represented by A, the average

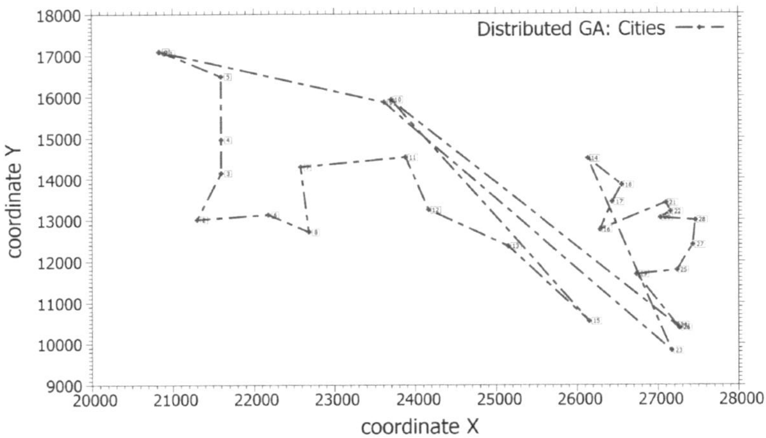

Fig. 13 Western Sahara—the best solution of the distributed GA

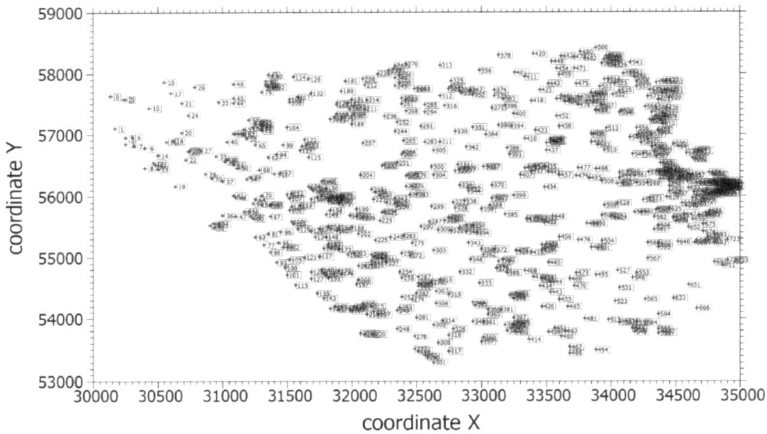

Fig. 14 Uruguay instance size is 734 cities

is represented by B, the standard deviation is represented by C, the PRD best value is represented by D, and the APRD best value is represented by E.

In this set of experiments, we also use the Mann-Whitney-Wilcoxon test to gather statistical evidence focused on the median analysis of both GAs and their median difference using the fitness column of entire instances in Table 6. The null hypothesis is $H_0 : \eta_1 - \eta_2 = 0$, and the alternative hypothesis $H_1 : \eta_1 - \eta_2 \neq 0$ to determine if both GAs performance based on median values is equal or different η_1 is the fitness values median of the sequential GAs, and η_2 is the fitness values median of the distributed GA.

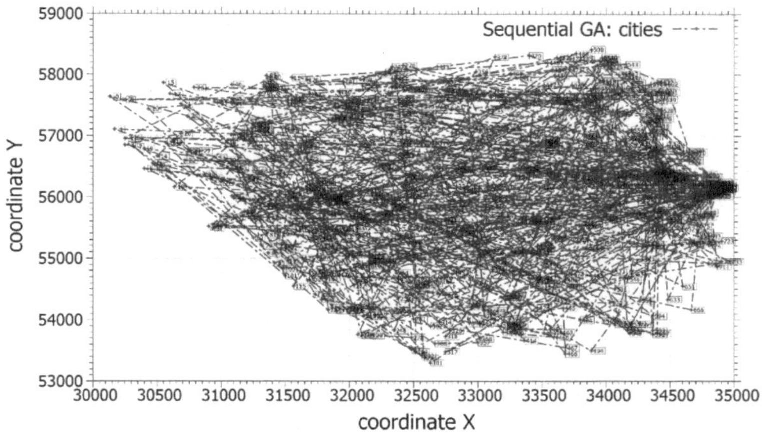

Fig. 15 Uruguay–the best solution of the sequential GA

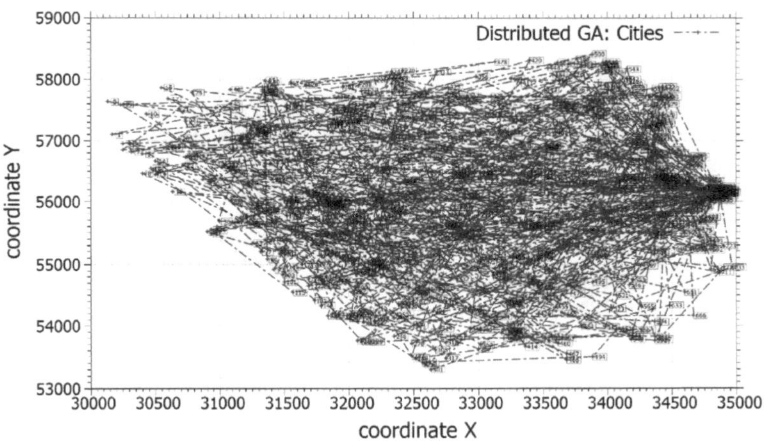

Fig. 16 Uruguay–the best solution of the distributed GA

We considered a 95% of confidence level and $\alpha = 0.05$. The test provided a *p*–value of 0.540. The difference between the population medians of 1381.49 is between −3944.02, and 6556.5, with a 95.01% confidence. So, we accept the H_0 due to the difference between the medians is not statistically significant.

A second test was performed with the same aim and configuration. η_1 is the execution time values median of the sequential GAs, and η_2 is the execution time value median of the distributed GAs in Table 6.

Similarly, we considered a 95% of confidence level and $\alpha = 0.05$. The test provided a *p*–value of 0. The difference between the population medians of −170.215 is between −306.899, and −81.9589, with a 95.01% confidence. So, we reject the H_0 because the difference between the medians is statistically significant. So, we have

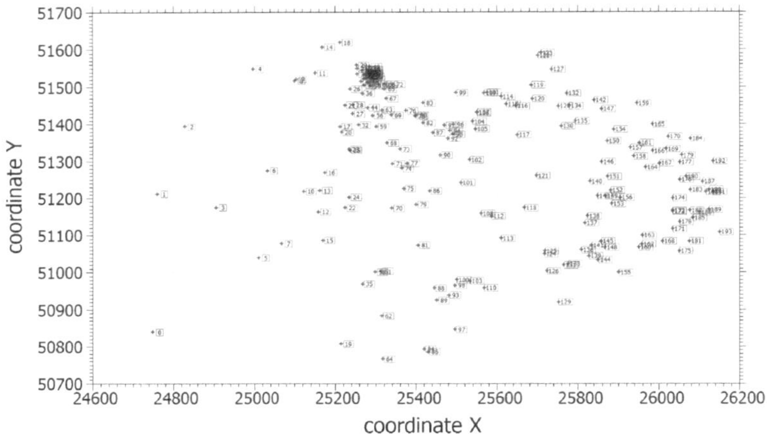

Fig. 17 Qatar instance size is 194 cities

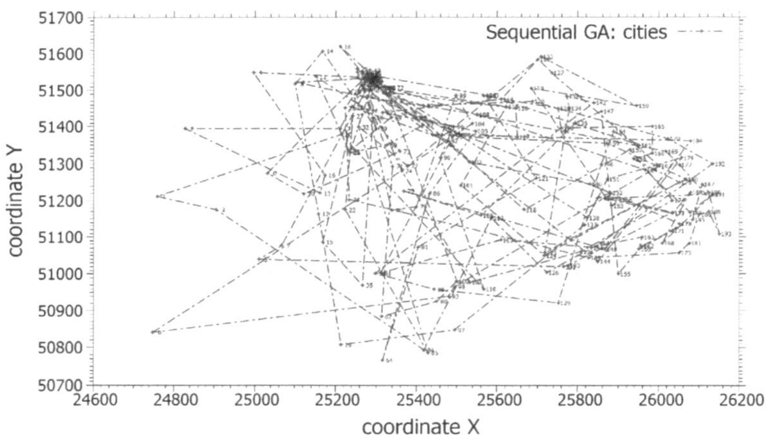

Fig. 18 Qatar—the best solution of the sequential GA

enough statistical evidence on results to conclude that distributed and sequential GA performed differently; the same conclusions gather on PRD and APRD metrics (see Table 7).

5 Conclusions and Future Work

In regards to the GAs solving the n-Queens problem, the time execution and the collision results on the small size arrangement of the n-Queens problem show that

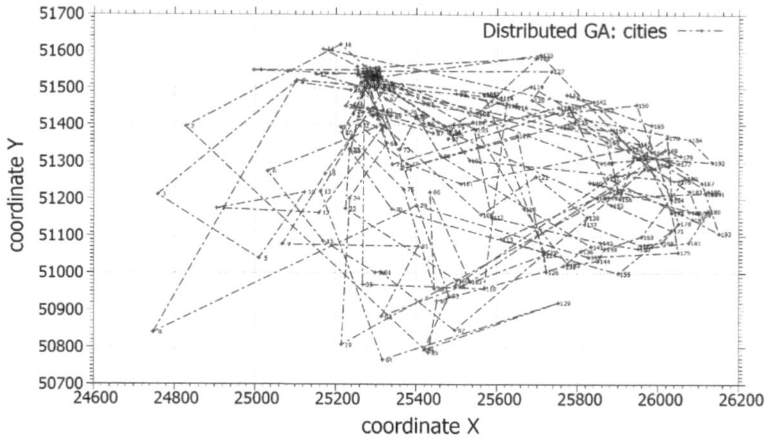

Fig. 19 Qatar—the best solution of the distributed GA

the sequential GA execution time is marginally better than in the distributed GA. Meanwhile, collisions do not offer enough statistical evidence to select one of the GAs as the best; both algorithms provided reasonable solutions since the difference between the medians is not statistically significant. By analyzing the rest of the results of both GAs solving the *n*-Queens instances, both algorithms' performance is remarkably similar. Both GAs performed quite similar even when the laptop physical architecture is less robust, and the full potential of the graphic card was not used in the distributed architecture.

The second analysis comprises the results of GAs solving the TSP problem. Here, our conclusions differ from the first analysis. The Mann-Whitney-Wilcoxon test based on the execution time resulting values, provide us enough statistical evidence to conclude that distributed and sequential GAs are different, which is consistent with PRD and APRD metrics (see Table 7). Meanwhile, the test based on the fitness resulting values, medians (Table 4), PRD, and APRD (Table 5) metrics analysis do not provide enough evidence to decide that the performance of both GAs is different as the execution time resulting values and medians suggest. These two results are opposed, so they must be analyzed together to understand their effects when trying to decide based on GAs performance. However, Derrac et al. in (2011) evidence that: "In a pairwise analysis, if we try to extract a conclusion involving more than one pairwise comparison, we are losing the control on the Family-Wise Error Rate (FWER), defined as the probability of making one or more false discoveries among all the hypotheses when performing multiple pairwise tests." So, considering the above described, as future work, we propose conducting a more exhaustive statistical analysis to explain both algorithms' performance. Besides this, we will answer the following questions:

Using the full potential of the graphic card on the distributed architecture and increasing the LAN_{lp} size, what will be the effect on distributed GA performance?

Table 6 Execution time values of sequential and distributed GA

Exec	Western sahara		Uruguay		Qatar	
	Sequential	Distributed	Sequential	Distributed	Sequential	Distributed
1	57.4380	181.9628	146.2810	760.5459	42.6720	143.5879
2	44.2890	142.5158	170.0770	879.5259	123.7810	386.9149
3	29.4390	97.9658	187.7540	967.9109	77.4980	248.0659
4	14.0560	51.8168	278.5960	1422.1209	62.2090	202.1989
5	17.9160	63.3968	192.5710	991.9959	105.4310	331.8649
6	8.9730	36.5678	320.7960	1633.1209	51.8410	171.0949
7	9.6060	38.4668	154.4450	801.3659	38.8870	132.2329
8	13.4520	50.0048	272.8280	1393.2809	117.4020	367.7779
9	16.8580	60.2228	560.8030	2833.1559	62.2090	202.1989
10	19.4610	68.0318	196.4150	1011.2159	60.0690	195.7789
11	12.3420	46.6748	150.7120	782.7009	80.0160	255.6199
12	16.5320	59.2448	139.4630	726.4559	113.4840	356.0239
13	8.9150	36.3938	406.0870	2059.5759	51.2700	169.3819
14	11.9070	45.3698	195.4500	1006.3909	70.6240	227.4439
15	**6.8510**	**30.2018**	166.1910	860.0959	221.8740	681.1939
16	8.2660	34.4468	**131.5400**	**686.8409**	105.8500	333.1219
17	10.2200	40.3088	162.7390	842.8359	110.2300	346.2619
18	18.3650	64.7438	602.5310	3041.7959	110.0140	345.6139
19	18.9380	66.4628	285.8190	1458.2359	114.0520	357.7279
20	8.0680	33.8528	214.0750	1099.5159	40.1660	136.0699
21	20.8020	72.0548	371.5060	1886.6709	283.1470	865.0129
22	11.3020	43.5548	208.9150	1073.7159	60.6400	197.4919
23	12.4030	46.8578	398.9590	2023.9359	154.6730	479.5909
24	11.1730	43.1678	292.1000	1489.6409	185.0150	570.6169
25	11.8630	45.2378	172.6560	892.4209	**37.2260**	**127.2499**
26	11.9130	45.3878	369.2520	1875.4009	85.8080	272.9959
27	10.5970	41.4398	768.1740	3870.0109	73.7190	236.7289
28	11.6320	44.5448	320.3950	1631.1159	119.0780	372.8059
29	7.7320	32.8448	227.9890	1169.0859	51.0530	168.7309
30	13.1890	49.2158	227.7840	1168.0609	193.1900	595.1419
Average	**12.1275**	**46.0313**	**220.9295**	**1133.7884**	**82.9120**	**264.3079**

Table 7 Performance metrics (time (s))

Metrics	Western sahara		Uruguay		Qatar	
	Sequential	Distributed	Sequential	Distributed	Sequential	Distributed
A	6.8510	30.2018	131.5400	686.8409	37.2260	127.2499
B	12.1275	46.0313	220.9295	1133.7884	82.9120	264.3079
C	10.8092	32.4275	150.6658	753.3289	58.5661	175.6982
D	–	340.83	–	422.15	–	241.83
E	–	571.89	–	761.93	–	610.01

What will be the effect of selecting other stop conditions on both algorithms' performance?

According to the results, the distributed GA requires a tuning methodology, so, which could be the best-suited tuning methodology for asymmetric scenarios as proposed in this work?

Why is the sequential algorithm prone to stagnant in local optimal than the distributed GA?

Acknowledgements Anabel Martínez-Vargas, M.A. Cosío-León, and Andrés J. García-Pérez thank CITEDI-IPN for sharing the know-how and the NVIDIA Jetson card provided through the SIP project 2020053. Anabel Martínez-Vargas thanks PRODEP for providing ESP32 cards by the grant number 149557.

References

Bell, J., and B. Stevens. 2009. A survey of known results and research areas for n-queens. *Discrete Mathematics* 309: 1–31. https://doi.org/10.1016/j.disc.2007.12.043.

Bozejko, W., and M. Makuchowski. 2009. A fast hybrid tabu search algorithm for the no-wait job shop problem. *Computer and Industrial Engineering* 56: 1502–1509.

Cantu-Paz, E. 2009. Efficient and accurate parallel genetic algorithms. Springer Science & Business Media.

Corus, D., and P.S. Oliveto. 2019. On the benefits of populations for the exploitation speed of standard steady-state genetic algorithms. In *Proceedings of the Genetic and Evolutionary Computation Conference*, 1452–1460. New York, NY, USA: Association for Computing Machinery. https://doi.org/10.1145/3321707.3321783.

Del-Ser, J., E. Osaba, D. Molina, X.-S. Yang, S. Salcedo-Sanz, D. Camacho, S. Das, P.N. Suganthan, C.A.C. and F. Herrera. 2019. Bio-inspired computation: Where we stand and what's next. *Swarm Evolutionary Computation* 48: 220–250. https://doi.org/10.1016/j.swevo.2019.04.008.

Derrac, J., S. García, D. Molina, and F. Herrera. 2011. A practical tutorial on the use of nonparametric statistical tests as a methodology for comparing evolutionary and swarm intelligence algorithms. *Swarm and Evolutionary Computation* 1: 3–18. https://doi.org/10.1016/j.swevo.2011.02.002.

Dyer, J.D., R.J. Hartfield, G.V. Dozier, and J.E. Burkhalter. 2012. Aerospace design optimization using a steady state real-coded genetic algorithm. *Applied Mathematics and Computation* 218: 4710–4730. https://doi.org/10.1016/j.amc.2011.07.038.

Eiben, A.E., and J.E. Smith. 2015. *Introduction to evolutionary computing*. Incorporated: Springer Publishing Company.

Fogel, D.B. 2000. What is evolutionary computation? *IEEE Spectrum* 37: 26–32.

Gong, Y.-J., W.-N. Chen, Z.-H. Zhan, J. Zhang, Y. Li, Q. Zhang, and J.-J. Li. 2015. Distributed evolutionary algorithms and their models: A survey of the state-of-the-art. *Applied Soft Computing* 34: 286–300. https://doi.org/10.1016/j.asoc.2015.04.061.

Haupt, R.L., and S.E Haupt. 2004. Advanced applications. In Practical Genetic Algorithms, 151–186. John Wiley & Sons, Ltd. https://doi.org/10.1002/0471671746.ch6.

Jianli, C., C. Zhikui, W. Yuxin, and G. He. 2020. Parallel genetic algorithm for N-Queens problem based on message passing interface-compute unified device architecture. Computational Intelligence. https://doi.org/10.1111/coin.12300.

Lozano, M., F. Herrera, J.R. Cano. 2008. Replacement strategies to preserve useful diversity in steady-state genetic algorithms. *Information Sciences (Ny.* 178: 4421–4433. https://doi.org/10.1016/j.ins.2008.07.031.

Martínez-Vargas, A., G.L. Rodríguez-Cortés, and O. Montiel-Ross. 2019. Comparative representations of a genetic algorithm to locate unmanned aerial vehicles in disaster zones. *Engineering Letters* 27: 374–384.

Moreno, J.J., G. Ortega, E. Filatovas, J.A. Martínez, and E.M. Garzón. 2017. Using low-power platforms for evolutionary multi-objective optimization algorithms. *Journal Supercomputer* 302–315.

Oliveto, P.S., D. Sudholt, and C. Witt. 2020. A tight lower bound on the expected runtime of standard steady state genetic algorithms. In *Proceedings of the 2020 Genetic and Evolutionary Computation Conference*, 1323–1331. New York, NY, USA: Association for Computing Machinery. https://doi.org/10.1145/3377930.3390212.

Osaba, E., F. Diaz, R. Carballedo, E. Onieva, and P. Lopez. 2014. A study on the impact of heuristic initialization functions in a genetic algorithm solving the n-queens problem. In *Proceedings of the Companion Publication of the 2014 Annual Conference on Genetic and Evolutionary Computation*, 1473–1474. New York, NY, USA: Association for Computing Machinery. https://doi.org/10.1145/2598394.2602269.

Sepúlveda, C., O. Montiel, J.M. Cornejo, and R. Sepúlveda. 2018. Estimation of Population Pharmacokinetic Model Parameters Using a Genetic Algorithm. In *Fuzzy logic in intelligent system design*, ed. P. Melin, O. Castillo, J. Kacprzyk, M. Reformat, and W. Melek, 214–221. Cham: Springer International Publishing.

Slowik, A., and H. Kwasnicka. 2020. Evolutionary algorithms and their applications to engineering problems. *Neural Computer Applied* 1–17.

Talbi, E.-G. 2009. Common concepts for metaheuristics. In Metaheuristics, 1–86. John Wiley & Sons, Ltd. https://doi.org/10.1002/9780470496916.ch1.

Valenzuela, V.M., C.A. Brizuela, M.A. Cosío-León, and A.D. Romero-Ocaño. 2019. A combination of two simple decoding strategies for the no-wait job shop scheduling problem. In Proceedings of the Genetic and Evolutionary Computation Conference, 864–871. New York, NY, USA: Association for Computing Machinery. https://doi.org/10.1145/3321707.3321870.

Ensemble Recurrent Neural Networks for Complex Time Series Prediction with Integration Methods

Martha Pulido and Patricia Melin

Abstract In this paper the development of the general architecture of a method to design recurrent ensemble neural networks for time series prediction is presented. Therefore, experiments are shown for the ensemble recurrent neural network, as well as the integration of the responses of the ensemble recurrent neural network, with integration by average, weighted average integration, type-1, type-2 and Generalized Type-2 fuzzy systems. The time series used to test the proposed architecture is that of petroleum. The simulation results show the effectiveness of the proposed method.

Keywords Recurrent neural networks · Fuzzy logic · Time series prediction

1 Introduction

Time series is a sequence of observations of a variable taken at various instants of time (Brockwell and Davis 2002; Davey et al. 1999; Cowpertwait and Metcalfe 2009). Various techniques have been used to predict future data (Aci and Avcı 2016; Hoffmann et al. 2020; Turabee et al. 2020; Zhang et al. 2020). Among these techniques we find neural networks (Sharkey 1996, 1999; Sollich and Krogh 1996; Pulido et al. 2009; Soto et al. 2014). There are different types of neural networks that allow us to handle different amounts of information such as modular networks, recurrent neural networks (Sudipto and Raghava 2006; Walid and Alamsyah 2016; Yao et al. 2017; Zhang and Man 1998; Zhang and Xio 2000; Min et al. 2004; Mikolov et al. 2011; Petnehazi 2019; Rohitash and Mengjie 2012; Jerome et al. 1994).

In this paper we combine techniques such as recurrent neural networks, and fuzzy logic (Melin et al. 2007, 2011; Mendel 2001; Castillo and Melin 2002, 2007. This

M. Pulido · P. Melin (✉)
Tijuana Institute of Technology, Tijuana, Mexico
e-mail: pmelin@tectijuana.mx

M. Pulido
e-mail: marthapulido_84@hotmail.com

O. Castillo and P. Melin (eds.), *Fuzzy Logic Hybrid Extensions of Neural and Optimization Algorithms: Theory and Applications*, Studies in Computational Intelligence 940,
https://doi.org/10.1007/978-3-030-68776-2_4

work describes a approach for designing monolithic and ensemble time series prediction. The integration of the recurrent neural network responses is performed with type-1, type-2 and generalized type-2 fuzzy systems (Castillo and Melin 2001a, b; Castillo and Melin 2007, 2008; Castro et al. 2011; Castillo 2012; Ontiveros-Robles et al. 2018; Castillo et al. 2020a, b; Ramirez et al. 2020; Gonzalez et al. 2017a, b; Petroleum Database 2020).

The main contribution of the paper is the proposed architecture of the ensemble recurrent neural network and after obtaining this architecture the results produced by the modules are integrated integration by average, weighted average integration, type-1, type-2 and Generalized Type-2 fuzzy systems. The inputs to the fuzzy system are the responses according to the number of network modules that correspond to the number of inputs of the fuzzy system. The maximum number of inputs taken into account are 3 inputs and one output with two Gaussian membership functions and these are granulated into two linguistic variables that are low and high forecast, and thereby obtains the forecast error for the corresponding series. The proposed hybrid ensemble of recurrent neural networks with fuzzy response integration shows good results.

This paper is organized as follows: In Sect. 2 the problem statement and the proposed method are presented, Sect. 3 shows the simulation results of the proposed method and Sect. 4 offers the conclusions.

2 Problem Statement and Proposed Method

This paper is concerned with the study of a fuzzy integration method that can be applied to ensemble recurrent neural networks with applications to complex time series, in addition to developing alternative methods for the integration of ensemble recurrent neural network as are the average and weighted average. Figure 1 presents

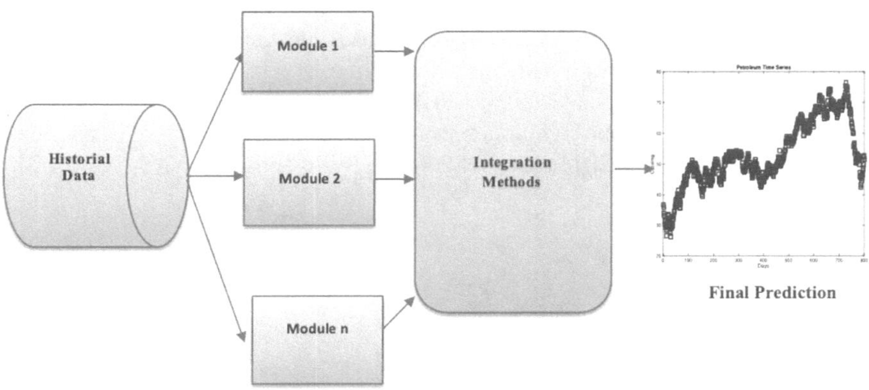

Fig. 1 General architecture of the proposed method

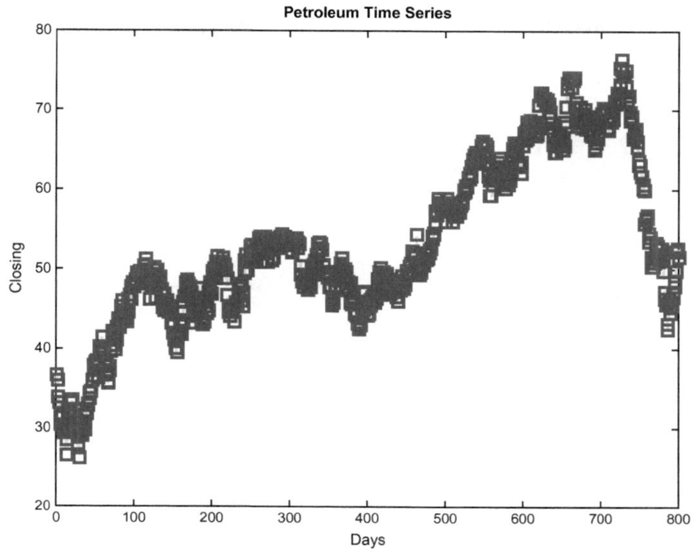

Fig. 2 Series data of petroleum time series

the general architecture used in this work. This Section describes the problem statement and the proposed method for time series prediction based on ensemble recurrent neural networks with fuzzy response integration.

Data of the Petroleum time series: The data consist of 800 points that correspond to a period from 04/01/2016to 11/01/2019 (Gonzalez et al. 2017a) (as shown in Fig. 2).

The proposed method consists in the following steps.

2.1 *Analyze the Time Series*

Plot and analyze the time series to observe the behavior, trend, seasonality and irregular changes (Petroleum time series).

2.2 *Creation of the Recurrent Neural Network*

In Fig. 3 we show the structure of the recurrent neural network.

The Recurrent Neural Network consists of multiple fixed activation function units, one for each time step. Each unit has an internal state which is called the hidden state of the unit. This hidden state signifies the past knowledge that that the network currently holds at a given time step. This hidden state is updated at every time step to

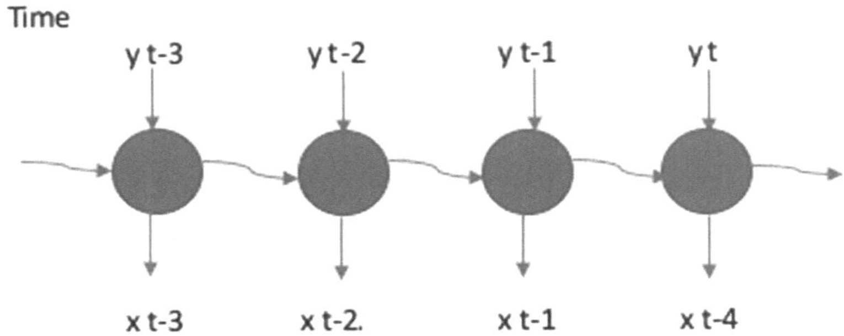

Fig. 3 Recurrent neural network

signify the change in the knowledge of the network about the past. The hidden state is updated using the following the recurrence relation:

$$y_t = fW(x_t, h_{t-1}) \tag{1}$$

where h_t represent the new hidden state, h_{t-1} the old hidden state, x_t, the concurrent inpunt and fW the fixed function wit trainable weights. At each time step, the new hidden state is calculated using the recurrence relation as given above. This new generated hidden state is used to generate indeed a new hidden state and so on.

2.3 *Integration by Average*

Integration by average: this method is used in the ensembles of networks. This integration method is the simplest and most straightforward, consists in the sum of the results generated by each module is divided by the sum of the number of modules, and the disadvantage is that there are cases in which the prognosis is not good.

2.4 *Integration by Weighted Average*

Integration of Weighted Average: this method is an extension of the integration by average, with the main difference that the weighted average assigns importance weights to each of the modules. These weights are assigned to a particular module based on several factors; the most important is the knowledge product of experience. This integration method belongs to the well now aggregation operators.

2.5 *Integration by Gating Network*

Each module constitutes an expert in a specific task. A special module formed by a network of doors ("Gating Network") is in charge of coordinating the operation of the modules during learning, taking responsibility for determining which of the k modules is the one that must learn to generate the desired response, as well as generating it during the operation of the complete system. Each of the k outputs of said module is a scalar that will do the gate work to adjust the influence or weight that each of the k modules will have in the formation of the final output of the system. In addition, an adder module appears that produces the final output.

2.6 *Type-1 and Type-2 Fuzzy System Integration*

The aggregation of the responses of the recurrent ensemble neural network is performed with type-1 and type-2 fuzzy systems. In this work the fuzzy system consisting of 3 inputs depending on the number of modules of the neural network ensemble and one output is used. Each input and output linguistic variable of the fuzzy system uses 2 Gaussian membership functions. The performance of the type-2 fuzzy aggregators is analyzed under different levels of uncertainty to find out the best design of the membership functions for the rules of the fuzzy system. Previous tests have been performed only with the input fuzzy system and the fuzzy system changes according to the responses of the neural network to give us better prediction error. In the type-2 fuzzy system we also change the levels of uncertainty to obtain the best prediction error (as shown in Fig. 4).

Figure 5 represents the 8 possible rules of the fuzzy system; we have 5 inputs in the fuzzy system with 2 membership functions, and the outputs with 2 membership functions. These fuzzy rules are used for both the type-1 and type-2 fuzzy systems.

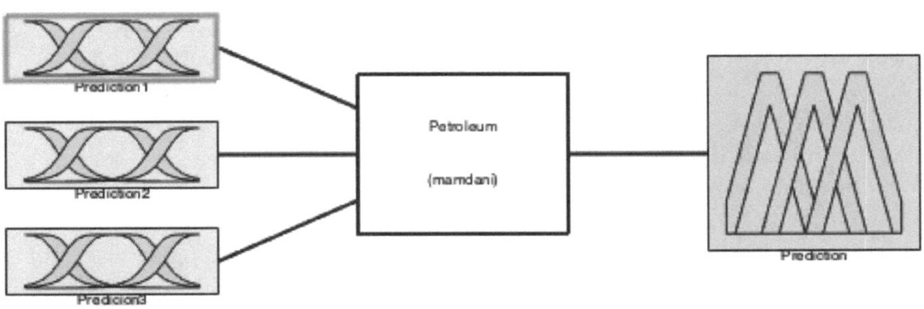

Fig. 4 Fuzzy inference system for integration of the ensemble neural network

1. If (Prediction1 is Pron1Low) and (Prediction2 is Pron2Low) and (Prediction3 is Pron3Low) then (Prediction is Low)
2. If (Prediction1 is Pron1High) and (Prediction2 is Pron2High) and (Prediction3 is Pron3High) then (Prediction is High)
3. If (Prediction1 is Pron1Low) and (Prediction2 is Pron2Low) and (Prediction3 is Pron3High)) then (Prediction is Low)
4. If (Prediction1 is Pron1High) and (Prediction2 is Pron2High) and (Prediction3 is Pron3Low) then (Prediction is High)
5. If (Prediction1 is Pron1Low) and (Prediction2 is Pron2High) and (Prediction3 is Pron3High) then (Prediction is High)
6. If (Prediction1 is Pron1High) and (Prediction2 is Pron2Low) and (Prediction3 is Pron3Low) then (Prediction is Low)
7. If (Prediction1 is Pron1Low) and (Prediction2 is Pron2High) and (Prediction3 is Pron3Low) then (Prediction is Low)
8. If (Prediction1 is Pron1High) and (Prediction2 is Pron2Low) and (Prediction3 is Pron3High) then (Prediction is High)

Fig. 5 Rules of the type-2 fuzzy system

2.7 Generalized Type-2 Fuzzy System

Figure 6 shows a fuzzy system consisting of 3 inputs which are number of modules for ensemble recurrent neural network ensemble and one output. Each input and output linguistic variable of the Generalized Type-2 fuzzy system uses 2 Gaussian membership functions, one output and 8 rules and shown Fig. 6.

3 Simulation Results

This section shows results of the recurrent neural network ensemble, as well as results for the integration by average, integration by weighted average, type-1, type-2 and Generalized Type-2 Integration fuzzy systems. The main goal is finding the best ensemble recurrent network architecture for each of the time series: In this case the method is illustrated with the Petroleum Time Series.

The best architecture of ensemble recurrent neural network was for module1, in the first layer we used 18 neurons and 19 neurons in second layer. In module 2 we used 20 neurons in the first layer and 20 neurons in the second, and in module 3 we used 18 neurons in the first layer and 20 neurons in the second. The training method used was the Bayesian Regulation (BR); we also applied 3 delays to the recurrent neural network. (as shown in Table 1 and Fig. 7)

In this table, the best error for average integration was in the experiment number 9 with an 0.0297, weighted average integration was in the experiment number 6 with an 0.0298, Gating network integration was in the experiment number 5 with an 0.0263 and type-1 fuzzy integration was in the experiment number 4 with 7 0.1748, which is shown in Table 2.

In the Table 3 shows to integrate the results, a type-2 fuzzy system was implemented using different levels of the uncertainty footprint where the best results that were obtained are as follows: a degree uncertainty of 0.2 yielded a forecast error of 0.3248, with a degree of uncertainty of 0.3 the error was of 0.3303, with a degree of uncertainty of 0.4 the error is of 0.3312 and with a degree of uncertainty of 0.5 the error was of 0.3303.

Table 4 shows the results of the generalized type-2 fuzzy system, where the best result obtained was following: the experiment number 5 with an error of 0.1600.

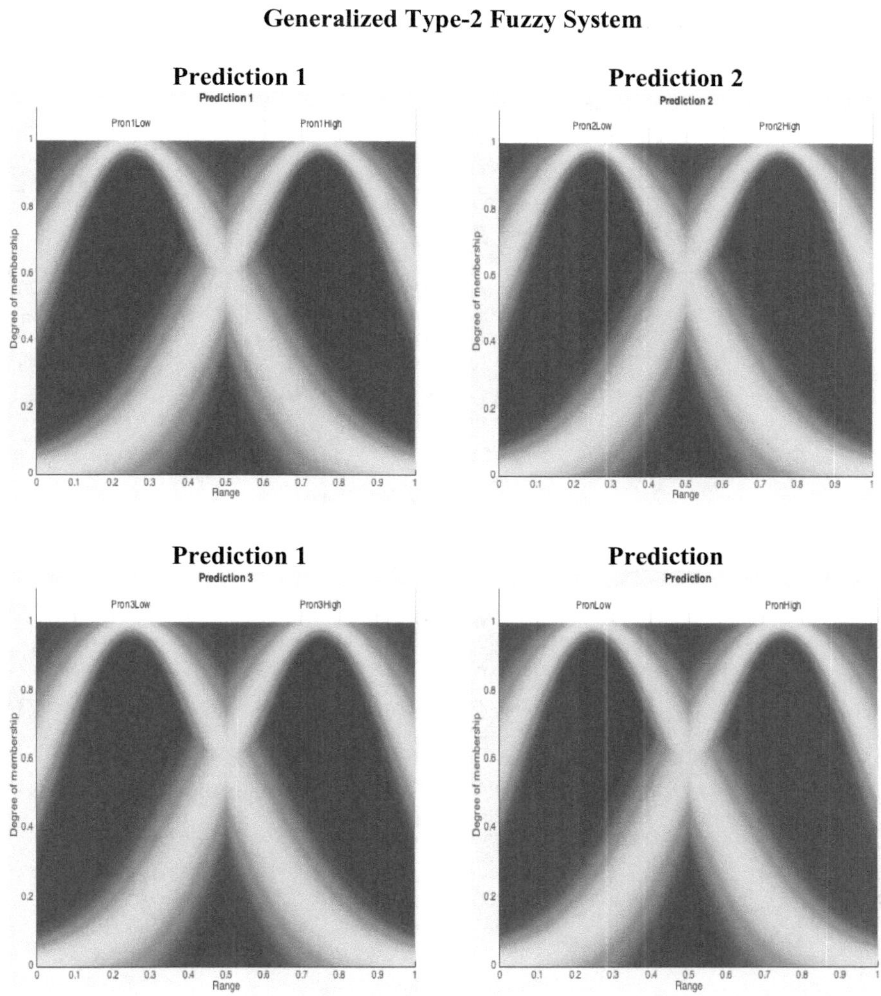

Fig. 6 Generalized type-2 fuzzy inference system for integration of the ensemble RNN

4 Conclusions

The objective of this paper was to find the best ensemble recurrent neural network architecture, the petroleum time series was used and the results obtained were as follows: The best architecture of ensemble recurrent neural network with 2 layers using 3 delays (as shown Table 1), Different integration methods were also implemented and the results obtained are as follows:, the best error for average integration was 0.0297, weighted average was 0.0298, Gating network integration was 0.0298,

Table 1 Results of training ensemble recurrent neural network

	Number of layers	Number neurons	Target error	Max Epoch	Epoch	Duration	Error module	Error total
Training 1	3	20,32,20 19,30,20 22,38,23	0.000000001 0.000000001 0.000000001	500 1500 200	1 1 1	00:0014	0.0297 0.0683 0.0302	0.1080
Training 2	3	30,42,42 21,46,40 26,40,43	0.0000001 0.00000001 0.000001	200 500 500	1 2 1	00:0012	0.115 0.0295 0.0461	0.1564
Training 3	3	30,17,32 21,22,36 26,26.35	0.0000001 0.00000001 0.000001	200 500 600	1 1 1	00:0014	0.1085 0.0300 0.1166	0.1773
Training 4	3	24,26,29 26,24,25 28,29,27	0.0000001 0.00000001 0.000001	200 500 600	1 1 1	00:0016	0.0437 01133 0.0299	0.1133
Training 5	3	28,36,22 30,36 20,28,22	0.0000001 0.00000001 0.000001	200 500 600	1 1 1	00:0014	0.0515 0.0300 0.0298	0.0915
Training 6	3	23,23 26,26 24,24	0.0000001 0.00000001 0.000001	200 500 600	1 1 1	00:0015	0.0298 0.0296 0.0301	0.0694
Training 7	3	25,25 27,28 27,27	0.0000001 0.00000001 0.000001	200 500 600	1 1 1	00:00:09	0.0299 0.0515 0.0547	0.0996
Training 8	3	28,29 30,30 33,31	0.0000001 0.00000001 0.000001	200 500 600	1 1 1	00:00:13	0.0301 0.0301 0.515	0.0773
Training 9	**3**	**18,19** **20,20** **18,20**	**0.0000001** **0.00000001** **0.000001**	**200** **500** **600**	**1** **1** **1**	**00:00:16**	**0.0298** **0.0298** **0.0296**	**0.0694**
Training 10	3	22,22 20,20 24,21	0.0000001 0.00000001 0.000001	200 500 600	1 1 1	00:00:17	0.0298 0.0299 0.0502	0.0764

type-1 fuzzy integration was 0.1748, type-2 fuzzy system using different levels footprint the results were as follows: a degree uncertainty of 0.2 yielded a forecast error of 0.348, with a degree of uncertainty of 0.3 the error was of 0.3303, with a degree of uncertainty of 0.4 the error is of 0.3312 and with a degree of uncertainty of 0.5 the error was of 0.3303 and Generalized type-2 fuzzy system, where the best result obtained was following: the experiment number 5 with an 0.1600. Analyzing the results, we can say that the architecture obtained from the ensemble recurrent neural network generates excellent results for this type of time series problems. In this case, we show the petroleum time series, since the recurrent neural networks analyze the

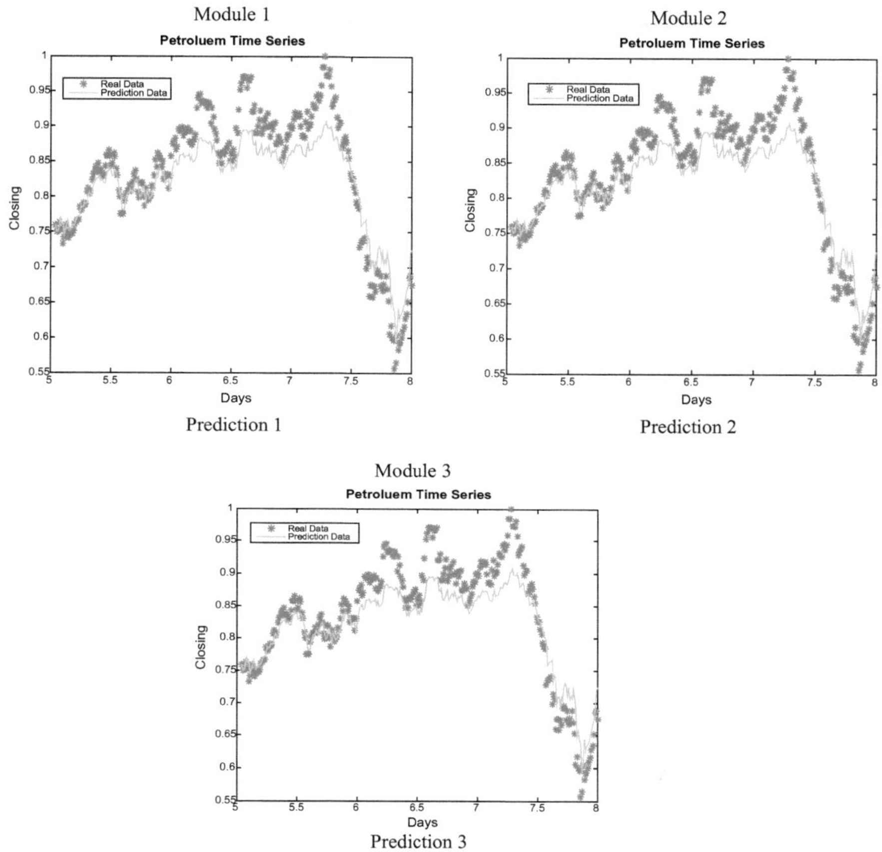

Fig. 7 Simulation of the training of ensemble recurrent neural network

data to be able to take actions. preventive and corrective in terms of time series. Regarding the integration fuzzy system, we can say that the Type-2 Generalized provided the best result because the system handles more uncertainty. As future work we would like to consider other types of applications, like the ones described in (Sanchez et al. 2014; Sánchez and Melin 2014; Sánchez et al. 2017; Castillo and Melin 1998, 2003).

Table 2 Results of the integration methods

Experiments	Average integration	Weighted average integration	Gating network	Type-1 fuzzy integration
Experiment 1	0.0419	0.00489	0.0285	0.1750
Experiment 2	0.0582	0.0734	0.0266	0.1768
Experiment 3	0.0840	0.0718	0.0284	0.1817
Experiment 4	0.0556	0.0702	0.0268	0.1751
Experiment 5	0.0346	0.0380	**0.0263**	0.1783
Experiment 6	0.0298	**0.0298**	0.0295	0.1750
Experiment 7	0.0422	0.0490	0.0261	**0.1748**
Experiment 8	0.0347	0.0380	0.0266	0.1749
Experiment 9	**0.0297**	0.0298	0.0296	0.1750
Experiment 10	0.0343	0.0377	0.0248	0.1756

Table 3 Results of type-2 fuzzy integration

Experiments	Prediction error with footprint uncertainty 0.2	Prediction error with footprint uncertainty 0.3	Prediction error with footprint uncertainty 0.4	Prediction error with footprint uncertainty 0.5
Experiment 1	0.3258	0.3299	0.3315	0.3325
Experiment 2	0.3265	**0.3303**	0.3317	**0.3303**
Experiment 3	0.3232	0.3285	0.3308	0.3321
Experiment 4	0.3263	0.3281	0.3322	0.3329
Experiment 5	0.3257	0.3299	0.3315	0.3325
Experiment 6	0.3257	0.3299	0.3315	0.3325
Experiment 7	0.3278	0.3309	0.3320	0.3328
Experiment 8	0.3267	0.3304	0.3318	0.3326
Experiment 9	0.3257	0.3299	0.3315	0.3325
Experiment 10	**0.3248**	0.3294	**0.3312**	0.3323

Table 4 Results of generalized type-2 fuzzy integration

Experiments	Generalized type-2 prediction error
Experiment 1	0.1602
Experiment 2	0.1605
Experiment 3	0.1605
Experiment 4	0.1604
Experiment5	0.1600
Experiment 6	0.1601
Experiment 7	0.1613
Experiment 8	0.1600
Experiment 9	**0.1600**
Experiment 10	0.1607

Acknowledgment We would like to express our gratitude to the CONACYT, Tijuana Institute of Technology for the facilities and resources granted for the development of this research.

References

Aci, M., and M. Avcı. 2016. Artificial neural network approach for atomic coordinate prediction of carbon nanotubes. *Applied Physics A*. 1–14.

Brockwell, P.D., and R.A. Davis. 2002. *Introduction to time series and forecasting*, 1–219. New York: Springer-Verlag.

Castillo, O., and P. Melin. 1998. A new fuzzy-fractal-genetic method for automated mathematical modelling and simulation of robotic dynamic systems, In 1998 IEEE International Conference on Fuzzy Systems (FUZZ-IEEE 1998) Proceedings, vol. 2, 1182–1187.

Castillo, O., and P. Melin. 2001a. Simulation and forecasting complex economic time series using neural networks and fuzzy logic, Proceedings of the International Neural Networks Conference vol. 3, 1805–1810.

Castillo, O., and P. Melin. 2001b. Simulation and forecasting complex financial time series using neural networks and fuzzy logic, Proceedings the IEEE the International Conference on Systems, Man and Cybernetics, vol. 4, 2664–2669.

Castillo, O., and P. Melin. 2002. Hybrid intelligent systems for time series prediction using neural networks, fuzzy logic, and fractal theory neural networks. *IEEE Transactions on Neural Networks* 13(6): 1395–1408.

Castillo, O., and P. Melin. 2003. Intelligent adaptive model-based control of robotic dynamic systems with a hybrid fuzzy-neural approach. *Applied Soft Computing* 3(4): 363–378.

Castillo, O., and P. Melin. 2007. Comparison of hybrid intelligent systems, neural networks and interval type-2 fuzzy logic for time series prediction, Proceedings IJCNN, 3086–3091.

Castillo, O., and P. Melin. 2008. Type-2 fuzzy systems in, type-2 fuzzy logic theory and application, 30–43. Springer.

Castillo, O. 2012. Type-2 fuzzy logic in intelligent control applications. Berlin, Germany: Springer.

Castillo, O., P. Melin, and J. Kacprzyk. 2020a. Intuitionistic and type-2 fuzzy logic enhancements in neural and optimization algorithms: Theory and applications. Studies in Computational Intelligence, vol. 862. Springer. ISBN 978-3-030-35444-2.

Castillo, O., P. Muhuri, P. Melin, and P. Pulkkinen. 2020b. Emerging issues and applications of type-2 fuzzy sets and systems. Engineering Applications of Artificial Intelligence.

Castro, J., O. Castillo, P. Melin, O. Mendoza, and A. Rodríguez. 2011. An interval type-2 fuzzy neural network for chaotic time series prediction with cross-validation and akaike test. *Soft Computing for Intelligent Control and Robotics* 269–285.

Cowpertwait, P., and A. Metcalfe. 2009. *Time series, introductory time series with R*, 2–5. Dordrecht Heidelberg London New York: Springer.

Davey, N., S. Hunt, and R. Frank. 1999. *Time series prediction and neural networks*. Hatfield: University of Hertfordshire.

Gonzalez, C., P. Melin, J. Castro, O. Mendoza, and O. Castillo. 2017a. General type-2 fuzzy edge detection in the preprocessing of a face recognition system. 3–18. Springer.

Gonzalezm, C., P. Melin, J. Castro, O. Mendoza, and O. Castillo. 2017b. Edge detection method based on general type-2 fuzzy logic applied to color images. *Information Sciences* 1–15.

Hoffmann, L.F., F.C. Parquet Bizarria, and J.W. Parquet Bizarria. 2020. Detection of liner surface defects in solid rocket motors using multilayer perceptron neural networks. *Polymeric Test*. https://doi.org/10.1016/j.polymertesting.2020.106559.

Jerome, T., R. Connor, and M. Douglas. 1994. Recurrent neural networks and robust time series prediction. *IEEE Transactions on Neural Networks* 5(2): 240–254.

Melin, P., O. Castillo, S. Gonzalez, J. Cota, W. Trujillo, and P. Osuna. 2007. Design of modular neural networks with fuzzy integration applied to time series prediction, 265–273. Berlin/Heidelberg: Springer.

Melin, P., J. Soto, O. Castillo, and J. Soria. 2011. A new approach for time series prediction using ensembles of ANFIS Models. *Expert Systems with Applications*.

Mendel, J. 2001. Uncertain rule-based fuzzy logic systems. Introduction an new directions, 213–231. Prentice-Hall, Inc.

Mikolov, T., S. Kombrink, L. Burget, J. Černocký, and S. Khudanpur. 2011. Extensions of recurrent neural network language model. In 2011 IEEE International Conference on Acoustics, Speech and Signal Processing (ICASSP), 5528–5531.

Min, H., X. Jianhui, X. Shiguo, and Y. Fu-Liang. 2004. Prediction of chaotic time series based on the recurrent predictor neural network. *IEEE* 3409–3416.

Ontiveros-Robles, E., and P. Melin, O. Castillo. 2018. Comparative analysis of noise robustness of type 2 fuzzy logic controllers. *Kybernetika* 54: 175–201.

Petnehazi, G. 2019. Recurrent neural networks for time series forecasting, University of Debrecen, 1–22.

Petroleum Database. 2020. https://mx.investing.com/commodities/crude-oil-historical-data. Accessed 5 June 2020.

Pulido, M., A. Mancilla, and P. Melin. 2009. An ensemble neural network architecture with fuzzy response integration for complex time series prediction, evolutionary design of intelligent systems in modeling, simulation and control, vol. 257/2009, 85–110. Berlin/Heidelberg: Springer.

Ramirez, E., P. Melin, G. Prado Arechiga. 2020. Hybrid model based on neural networks and fuzzy logic for 2-lead cardiac arrhythmia classification. *Hybrid Intelligent Systems in Control, Pattern Recognition and Medicine* 193–217.

Rohitash, C., and Z. Mengjie. 2012. Cooperative coevolution of Elman recurrent neural networks for chaotic time series prediction. *Neurocomputing* 86: 116–123.

Sanchez, M.A., O. Castillo, J.R. Castro, P. Melin. 2014. Fuzzy granular gravitational clustering algorithm for multivariate data. *Information Sciences* 279: 498–511.

Sánchez, D., and P. Melin. 2014. Optimization of modular granular neural networks using hierarchical genetic algorithms for human recognition using the ear biometric measure. *Engineering Applications of Artificial Intelligence* 27: 41–56.

Sánchez, D., P. Melin, and O. Castillo. 2017. Optimization of modular granular neural networks using a firefly algorithm for human recognition. *Engineering Applications of Artificial Intelligence* 64: 172–186.

Sharkey, A. 1996. One combining artificial of neural nets, department of computer science. University of Sheffield, U.K.

Sharkey, A. 1999. Combining artificial neural nets: Ensemble and modular multi-net systems. London: Springer-Verlag.

Sollich, P., and A. Krogh A. 1996. Learning with ensembles: How over-fitting can be useful. In *Advances in neural information processing systems*, eds. Touretzky, D.S., M.C. Mozer, M.E. Hasselmo, vol. 8, 190–196. Cambridge, MA: Denver, CO, MIT Press.

Soto, J., M. Patricia, and O. Castillo. 2014. Time series prediction using ensembles of ANFIS model with genetic optimization of interval type-2 and type-1 fuzzy integrators. *International Journal of Hybrid Intelligent system* 11: 211–226.

Sudipto, S., and G. Raghava. 2006. Prediction of continuous B-cell epitopes in an antigen using recurrent neural network. *National Library of Medecine* 40–48.

Turabee, G., M.R. Khowja, P. Giangrande, V. Madonna, G. Cosma, G. Vakil, C. Gerada, and M. Galea. 2020. The role of neural networks in predicting the thermal life of electrical machines. *IEEE Access* 8: 40283–40297.

Walid, M.M., and I.U. Alamsyah. 2016. Recurrent neural network for forecasting time series with long memory pattern. *Journal of Physics: Conference Series* 824: 1–8.

Yao, Q., S. Dongjin, C. Haifeng, C. Wei, J. Guofei, and C. Garrison. 2017. A dual-stage attention-based recurrent neural network for time series prediction. *Computer* science, 1–7. Cornell University.

Zhang, J., and K.F. Man. 1998. Time series prediction using recurrent neural network in multi-dimension embedding phase space. *IEEE International Conference on Systems, Man and Cybernetics* 2: 11–14.

Zhang, J., and X. Xio. 2000. Predicting chaotic time series using recurrent neural network. *IOP Science* 88–90.

Zhang, G., H. Zhou, C. Wang, H. Xue, J. Wang, and H. Wan. 2020. Forecasting time series albedo using NARnet based on EEMD decomposition. *IEEE Transactions on Geoscience and Remote Sensing* 58: 3544–3557. https://doi.org/10.1109/tgrs.2019.2958048.

Genetic Optimization of Ensemble Neural Network Architectures for Prediction of COVID-19 Confirmed and Death Cases

Julio C. Mónica, Patricia Melin, and Daniela Sánchez

Abstract In this work a genetic algorithm for ensemble neural network architecture optimization applied to COVID-19 time series prediction is proposed. The main objective of this paper is to show the results of the optimized number of neurons in two hidden layers of an ensemble artificial neural network used for time series prediction using a real genetic algorithm. The time series dataset used in this work is the confirmed and death cases of COVID-19 of 12 states of Mexico (and information about the whole country). Being the COVID-19 the pandemic that has been affecting many lives in Mexico, for this reason, this work seeks to find a prediction for confirmed and death cases in this country.

Keywords Genetic algorithm · Time series prediction · Artificial neural network · Fuzzy logic · NAR · Optimization

1 Introduction

The artificial neural networks (ANNs) and the genetic algorithms (GAs) are two techniques that have their own strengths and weaknesses. The ANNs have the learning but their main problem is how to establish their architecture, this problem can be resolved using an optimization technique such as the GAs. Nowadays, hybrid systems have been of vital importance to solve problems in a better way, it is a valuable application in technological advances, covering a large number of different areas, such as, medicine, finances, industry, among many others (Mahajan 2013).

J. C. Mónica · P. Melin (✉) · D. Sánchez
Tijuana Institute of Technology, Tijuana, Mexico
e-mail: pmelin@tectijuana.mx

J. C. Mónica
e-mail: julio.monica@tectijuana.edu.mx

D. Sánchez
e-mail: danielasanchez.itt@hotmail.com

O. Castillo and P. Melin (eds.), *Fuzzy Logic Hybrid Extensions of Neural and Optimization Algorithms: Theory and Applications*, Studies in Computational Intelligence 940,
https://doi.org/10.1007/978-3-030-68776-2_5

The proposed method is applied using genetic algorithms to optimize the numbers of neurons in the two hidden layers, having as objective function minimizing the errors of the 10 days ahead prediction of the ensemble artificial neural network. The ANNs have been used in other works, a prediction model was constructed to deal with the fatigue of polymer using artificial neural networks (Yang 2020). For the analysis of the spatial evolution of coronavirus pandemic around the world using Self Organizing Maps (Melin et al. 2020). The ANNs have also been used to identify aggression, hate speech, cyberbullying, etc. on tweets (Sadiq et al. 2020). To optimize a modular granular neural network using a firefly algorithm for human recognition (Sánchez et al. 2017), it was also used for series prediction of the Mexican Stock using Particle swarm optimization of ensemble neural networks with fuzzy aggregation (Pulido et al. 2014). The genetic algorithms have been used for feature selection for early time series classification (Ahn and Hur 2020), among elder drivers to predict the severity of fixed object crashes (Amiri et al. 2020). And have been combined with GA such as in predicting a solar space heating system performance (Jamali et al. 2019), for the electricity consumption of a real-world campus building (Luo et al. 2020), in a method designed for human body shape prediction (Cheng 2020), in a model for the vehicle insurance fraud identification (Yan et al. 2020), for the optimization of modular granular neural networks where a hierarchical genetic algorithm for human recognition was used using the ear biometric measure (Sánchez and Melin 2014) and for the prediction of soil wind erodibility (Kouchami-Sardoo et al. 2020) among many others.

This paper is organized as follows: Sect. 2 describes the concepts used in this work, the Sect. 3 contains the general architecture of the proposed method, Sect. 4 shows the experimental results and Sect. 5 presents the conclusion of this work.

2 Basic Concepts

In this document we present the basic concepts that were used in this work.

2.1 *Artificial Neural Networks*

Artificial neural networks (ANN) are computing systems that imitate the mechanism of the biological neural network. The ANN can be used in a supervised or unsupervised learning method. The components of the models are nodes, weights and layers. There are many different types of ANN. The FITNET is a commonly used Multi-Layer Perceptron (MLP) which is a class of feedforward artificial neural network containing an input layer, a hidden layer and an output layer (Ali and Shahbaz 2020; Cheshmberah et al. 2020). The general architecture of an ANN is shown in Fig. 1.

Where X_1, X_2 and X_n are the variables of the inputs layer, n is the number of neurons, W_{ij}, W_{kj}, W_{lj} are the weights and y is the output (Kim 2020).

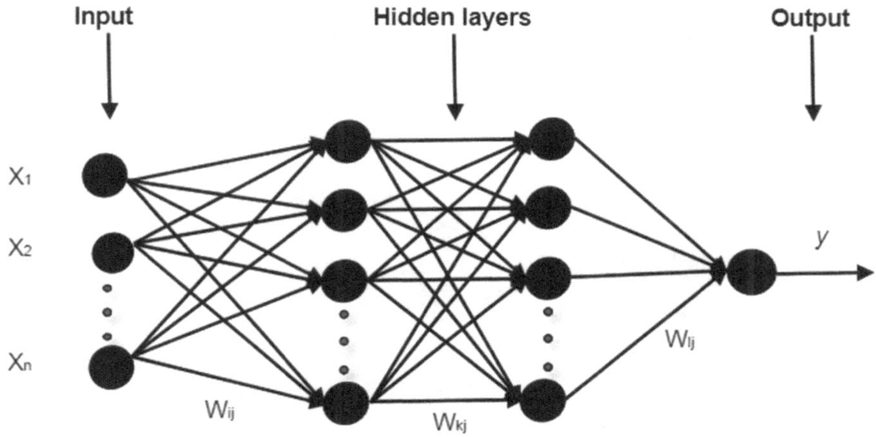

Fig. 1 The architecture of an Artificial Neural Network

2.2 *Nonlinear Autoregressive Neural Network*

The nonlinear autoregressive neural network (NAR) to estimate values uses past values of the time series. The NAR model consists of one input layer, one or more hidden layers and one output layer. NAR is a recurrent and dynamic network with feedback connections. In time series forecasting NAR is used in one-step-ahead or multi-step-ahead (Khan 2020; Ruiz et al. 2016). In the following Eq. (1) we can see how the NAR model is mathematically expressed.

$$y(t) = F(y(t-1), y(t-2), \ldots, y(t-d)) \tag{1}$$

where $y(t)$ is the output value of the considered time series y at time t, and d is the time delay and F denotes the transfer function (Benrhmach 2020; Pan and Pu 2019). In Fig. 2, the NAR neural network is shown.

An ensemble neural network (ENN) is a model for learning where the artificial neural networks are together to perform much better when solving robust problems (Chen et al. 2019; Zhou et al. 2002).

2.3 *Fuzzy Logic*

Fuzzy logic was introduced by Lotfi A. Zadeh in 1965 with the proposal of fuzzy set theory. In the concept of fuzzy logic, the truth values of each variable may fall in between the range of 0 and 1 unlike in the concept of Boolean logic where the variable values are either 0 (represents false) or 1 (represents true) (Chiueh 1991; Reddy et al. 2017). Fuzzy logic is a methodology that solves computational problems

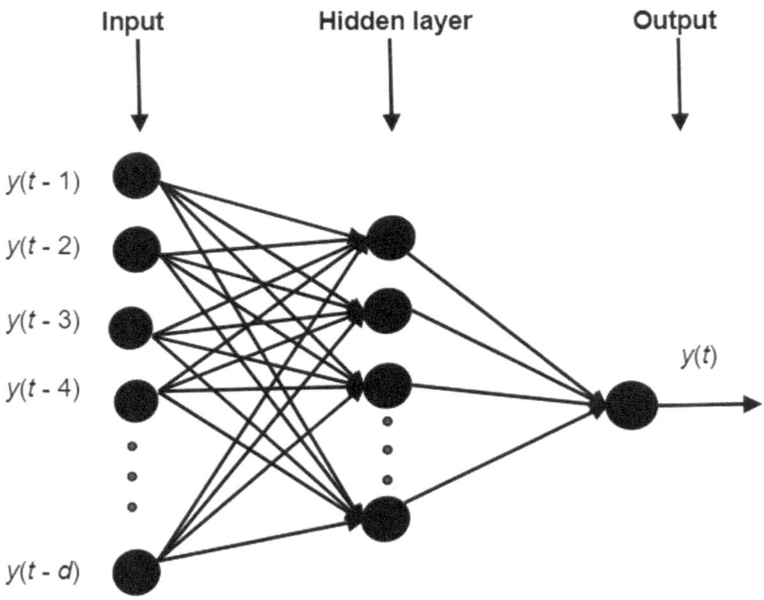

Fig. 2 The architecture of the NAR neural network

based on linguistic variables in an environment of imprecision or incompleteness of information. One of the fuzzy logic misconceptions is that fuzzy logic is fuzzy. Fuzzy logic is a precise conceptual system of reasoning and deduction where the analysis is associated with imperfect information. Fuzzy logic uses two remarkable human capabilities, the first one, the capability to communicate, reason, converse and make rational decisions. The second, the capability to perform a wide variety of physical and mental tasks (Zadeh 2009).

2.4 Genetic Algorithms

Genetic algorithm (GA), is one of the optimization algorithms. It is based on the Darwinian theory as their inspiration by natural evolution, the survival of the fittest, this is measured by an evaluation function called fitness function in computational terms (Sánchez 2017). Using the principle of evolution, the genetic algorithm techniques are based on many parameters that are required but the critical are, population size, selection, crossover and mutation rate to obtain the best solutions for optimization problems. A chromosome is a sequence of genes, in Fig. 3 a chromosome with genes is shown (Chiroma et al. 2017).

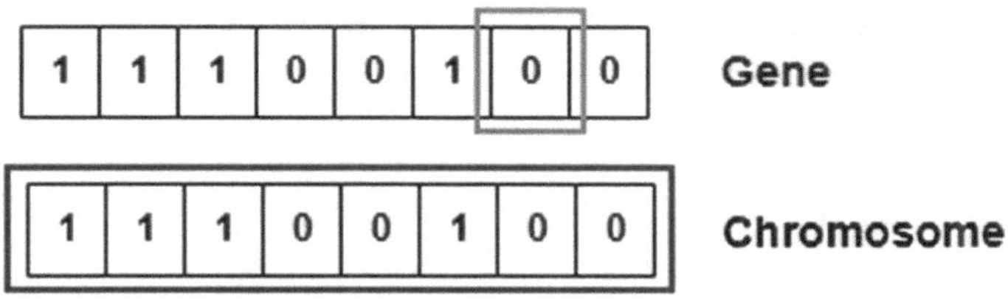

Fig. 3 The representation of a gene and a chromosome

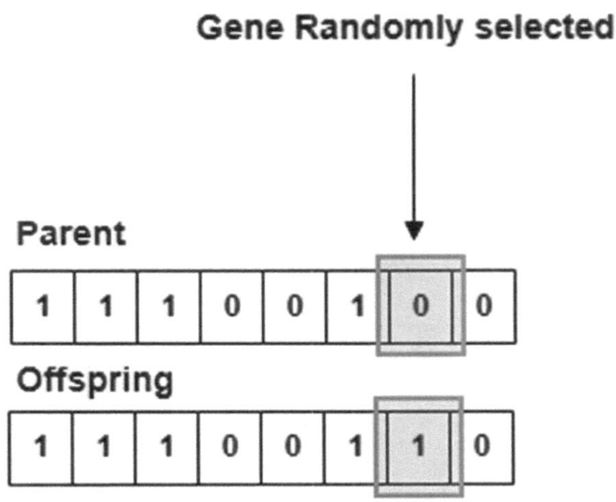

Fig. 4 Mutation process using single point

The population is a group of individuals, a chromosome is an individual. The reason for crossover in genetic algorithms is to help the reproduction of better chromosomes and the mutation of the strings is needed because genetic material may be lost in the crossover process, in order to fix the distortion of genetic information we do it through mutation. In Fig. 4 we show the mutation process using a single point (Chiroma et al. 2017).

3 Proposed Method

The proposed method consists of an ensemble neural network with 3 artificial neural networks, its optimal architecture is achieved using a real genetic algorithm technique. The main goal is to obtain the optimal number of neurons in the two hidden layers. We have a dataset of COVID-19 confirmed and death cases of 12 states of Mexico and the information of the whole country, the data of each state is learned from a Fitnet and two NAR neural networks, where each one obtains a prediction and normalized errors are calculated. The normalized errors are sent to a type-1 fuzzy logic integrator to finally give us a weight for each prediction and calculate a final prediction. The real genetic algorithm optimizes the numbers of neurons in the two hidden layers of each artificial neural network. This model is presented in Fig. 5. The values of the ANN are shown in Table 1. And for the genetic algorithm in Table 2. The training function used for Fitnet neural network is *trainlm (Levenberg-Marquardt backpropagation)* and for NAR is *trainlm* and *trainbr (Bayesian regularization backpropagation).*

For the fuzzy integrator system, where the inputs e_1, e_2 and e_3 are the normalized mean square errors of the values in the range between 0 and 1 of the 3 artificial neural networks being used to produce the weights w_1, w_2 and w_3 and p_1, p_2, p_3 are the predicted values of each module respectively. Then we combine the predictions to

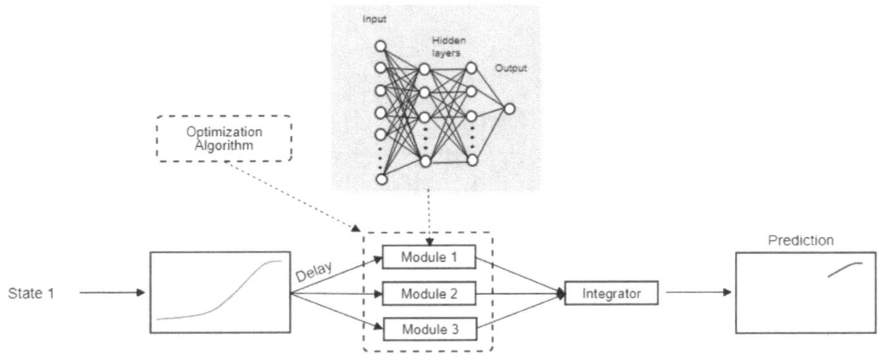

Fig. 5 Proposed model

Table 1 Values for the ENN

Parameters of the ENN	Values
Modules	3
Days to predict	10
Error goal	0.001
Training function	Trainlm, Trainbr
Hidden Layers	2
Neurons for each hidden layer	1–50

Table 2 Values for the genetic algorithm

Parameters of the GA	Values
Individuals	10
Generations	30
Selection	Tournament
Mutation Rate	0.3
Crossover	Uniform

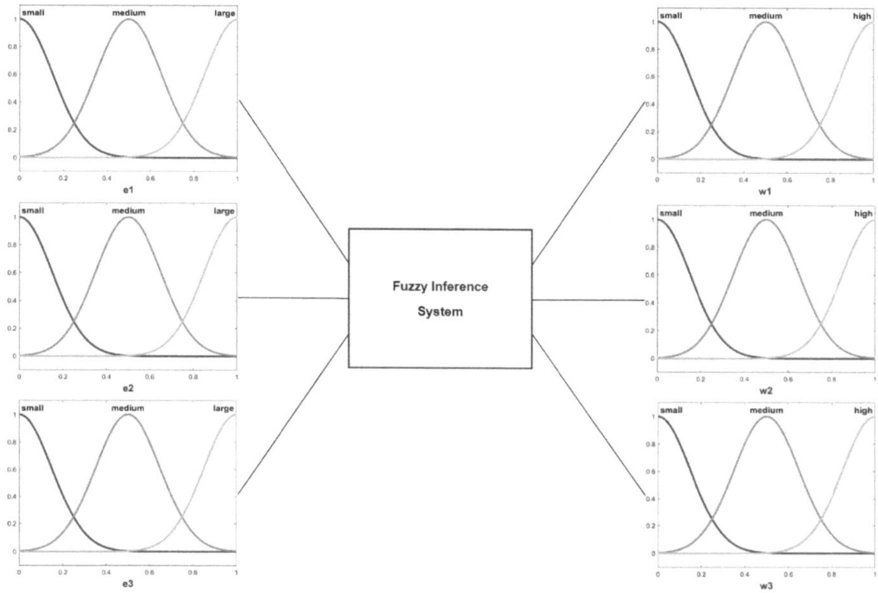

Fig. 6 The fuzzy inference system

obtain the final prediction PT using the expression in Eq. (2).

$$PT = \frac{w_1 p_1 + w_2 p_2 + w_3 p_3}{w_1 + w_2 + w_3} \tag{2}$$

The fuzzy inference system has 3 rules, 3 inputs e_1, e_2, e_3 and 3 outputs w_1, w_2, w_3 using Gaussian functions. In Fig. 6, the fuzzy inference system is shown.

The fuzzy system contains 3 rules which are the following:

1. If (e_1 is small) and (e_2 is medium) and (e_3 is large) then (w_1 is high) (w_2 is medium) (w_3 is small).
2. If (e_1 is large) and (e_2 is small) and (e_3 is medium) then (w_1 is small) (w_2 is high) (w_3 is medium).
3. If (e_1 is medium) and (e_2 is large) and (e_3 is small) then (w_1 is medium) (w_2 is small) (w_3 is high).

Based on their corresponding errors these rules express the knowledge of how to combine predictions.

The real genetic algorithm used in this work after the initial population has been created randomly at each generation a selection, crossover and mutation is performed to update the population of solutions and the tournament selection is used to select the individuals from the population to continue with the mating pool where the individuals known as parents are paired and crossover is applied to produce two offspring (Sawyerr et al. 2014). First, the whole population is evaluated based on

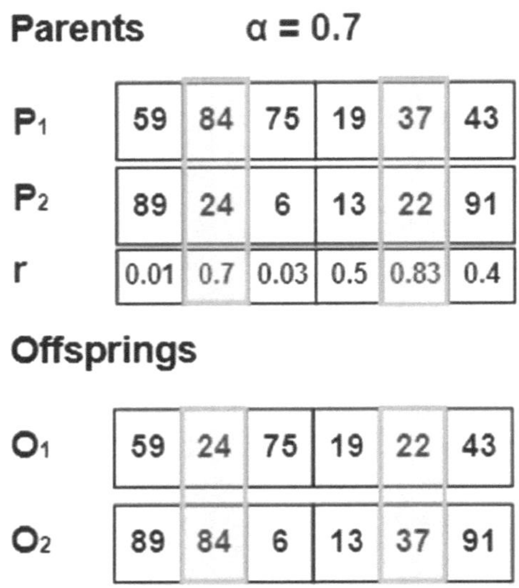

Fig. 6 Uniform crossover in a real genetic algorithm

the fitness function, then after the individual is ranked, the best fitness is saved to avoid modifications, process known as elitism. When the individuals from each generations of the population are selected for reproduction the uniform crossover combines the two chromosomes known as parents to produce new chromosomes known as offspring, the main goal is to attempt obtaining offspring better than their parents. In Fig. 7 a uniform crossover for a real genetic algorithm is shown.

Where P_1, P_2 are the chromosome parents and O_1, O_2 are the offspring obtained after the crossover. r are the random numbers generated to determine if the crossover occurs and α is the crossover rate.

In this work the fitness function used is the mean square error of the final prediction to obtain the optimization of the neurons in the two hidden layers of each artificial neural network.

4 Results of the Experiment

The optimized results using the real generic algorithm technique and all the numbers of neurons of the results are shown in this section.

4.1 Genetic Algorithms

The architectures obtained by the optimization technique using the genetic algorithm are shown in Table 3 of the COVID-19 confirmed cases, where the results show that

Table 3 The results of the COVID-19 confirmed cases

States	Type of ANN	Neurons	Final error
Baja California	FITNET	39,31	59.07
	NAR	34,42	
	NAR	24,32	
Ciudad de Mexico	FITNET	14,14	2718.96
	NAR	41,35	
	NAR	14,10	
Coahuila	FITNET	27,31	38.51
	NAR	40,14	
	NAR	26,15	
Estado de Mexico	FITNET	42,10	45.56
	NAR	40,24	
	NAR .3	50,37	
Jalisco	FITNET	47,7	2.62
	NAR	1,23	
	NAR	41,30	
Nuevo Leon	FITNET	19,6	37.92
	NAR	39,9	
	NAR	34,49	
Puebla	FITNET	9,50	17.57
	NAR	34,1	
	NAR	6,27	
Quintana Roo	FITNET	39,48	56.94
	NAR	18,13	
	NAR	26,13	
Sinaloa	FITNET	11,18	19.32
	NAR	24,27	
	NAR	21,38	
Tabasco	FITNET	11,18	85.98
	NAR	24,27	
	NAR	21,38	
Veracruz	FITNET	11,18	53.59
	NAR	25,26	
	NAR	21,38	
Yucatan	FITNET	32,8	17
	NAR	1,7	
	NAR	46,31	

(continued)

Table 3 (continued)

States	Type of ANN	Neurons	Final error
Nacional	FITNET	39,12	12190.84
	NAR	26,11	
	NAR	17,38	

the state with less final error is Jalisco.

In Table 4, the architecture obtained for the death cases are shown, where the results show that the states with less final error are Coahuila and Jalisco.

The comparison with non-optimized results presented (Pulido et al. 2014) for confirmed cases are shown in Table 5.

The comparison with non-optimized results presented (Melin et al. 2020) for death cases are shown in Table 6.

5 Conclusions

The proposed real generic algorithm used to find the optimal neural network ensemble architecture for the COVID-19 time series prediction of the death and confirmed cases of 12 states of Mexico and the total data of the country helped us find the optima number of neurons in the two hidden layers of the artificial neural networks (FITNET and NAR). The method used as an integration was a fuzzy interference system where the inputs received the prediction errors of the artificial neural networks to give us a final prediction. We obtained good results, as we can notice in the Tables 5 and 6 the optimized results of the final errors gave us a big improvement. One of the aspects of using the real genetic algorithm to consider is the time consuming, nowadays the hybrid systems need to get results in a faster way and that is a weakness while using a genetic algorithm in certain cases. For future work, different optimization techniques will be implemented to compare results and other application areas could be considered, like in (Castillo 1998; Castillo and Melin 2003; Sanchez et al. 2014).

Table 4 The results of the COVID-19 death cases

States	Type of ANN	Neurons	Final error
Baja California	FITNET	16,20	11.09
	NAR	43,1	
	NAR	14,33	
Ciudad de Mexico	FITNET	16,27	8.18
	NAR	16,10	
	NAR	27,14	
Coahuila	FITNET	41,40	0.11
	NAR	44,42	
	NAR	39,9	
Estado de Mexico	FITNET	18,14	7.32
	NAR	20,37	
	NAR .3	40,17	
Jalisco	FITNET	6,15	0.11
	NAR	25,3	
	NAR	10,28	
Nuevo Leon	FITNET	36,8	0.19
	NAR	37,36	
	NAR	28,23	
Puebla	FITNET	7,10	2.74
	NAR	45,49	
	NAR	13,15	
Quintana Roo	FITNET	28,8	1.15
	NAR	11,33	
	NAR	9,32	
Sinaloa	FITNET	37,8	2
	NAR	21,15	
	NAR	27,29	
Tabasco	FITNET	33,10	2.37
	NAR	27,3	
	NAR	12,39	
Veracruz	FITNET	17,50	2.19
	NAR	23,23	
	NAR	31,41	
Yucatan	FITNET	7,42	0.20
	NAR	38,31	
	NAR	48,5	

(continued)

Table 4 (continued)

States	Type of ANN	Neurons	Final error
Nacional	FITNET	31,19	115.65
	NAR	36,50	
	NAR	13,6	

Table 5 Comparison of the COVID-19 confirmed cases

State	Final error	
	Non-optimized	Optimized
Baja California	2529.07	59.07
Ciudad de Mexico	1263297.77	2718.96
Coahuila	109.83	38.51
Estado de Mexico	41570.11	45.56
Jalisco	1055.77	2.62
Nuevo Leon	131.86	37.92
Puebla	5516.75	17.57
Quintana Roo	7513.09	56.94
Sinaloa	74.22	19.32
Tabasco	56.71	85.98
Veracruz	9528.70	53.59
Yucatan	3811.82	17
National	2415010.109	12190.84

Table 6 Comparison of the COVID-19 death cases

State	Final error	
	Non-optimized	Optimized
Baja California	1119.39	11.09
Ciudad de Mexico	202.10	8.18
Coahuila	22.50	0.11
Estado de Mexico	2578.22	7.32
Jalisco	24.95	0.11
Nuevo Leon	2.69	0.19
Puebla	20.73	2.74
Quintana Roo	254.18	1.15
Sinaloa	168.05	2
Tabasco	61.23	2.37
Veracruz	241.81	2.19
Yucatan	80.21	0.20
National	28901.55	115.65

References

Ahn, G., and S. Hur. 2020. Efficient genetic algorithm for feature selection for early time series classification. *Computers & Industrial Engineering* 142: 106345.

Ali, S., and M. Shahbaz. 2020. Streamflow forecasting by modeling the rainfall–streamflow relationship using artificial neural networks. *Modeling Earth Systems and Environment* 6 (3): 1645–1656.

Amiri, A., A. Sadri, N. Nadimi, and M. Shams. 2020. A comparison between artificial neural network and hybrid intelligent genetic algorithm in predicting the severity of fixed object crashes among elderly drivers. *Accident Analysis and Prevention* 138: 105468.

Benrhmach, G., K. Namir, A. Namir, and J. Bouyaghroumni. (2020). Nonlinear autoregressive neural network and extended Kalman filters for prediction of financial time series. *Journal of Applied Mathematics.*

Castillo, O., and P. Melin. 1998. A new fuzzy-fractal-genetic method for automated mathematical modelling and simulation of robotic dynamic systems. *1998 IEEE International Conference on Fuzzy Systems (FUZZ-IEEE 1998) Proceedings*. 2, 1182–1187.

Castillo, O., and P. Melin. 2003. Intelligent adaptive model-based control of robotic dynamic systems with a hybrid fuzzy-neural approach. *Applied Soft Computing* 3 (4): 363–378.

Chen, Y., H. Chang, J. Meng, and D. Zhang. 2019. Ensemble Neural Networks (ENN): A gradient-free stochastic method. *Neural Networks* 110: 170–185.

Cheng, P., D. Chen, and J. Wang, (2020). Clustering of the body shape of the adult male by using principal component analysis and genetic algorithm–BP neural network. *Soft Computing*, 1–19.

Cheshmberah, F., H. Fathizad, G. Parad, and S. Shojaeifar. 2020. Comparison of RBF and MLP neural network performance and regression analysis to estimate carbon seques-tration. *International Journal of Environmental Science and Technology* 17 (9): 3891–3900.

Chiroma, H., S. Abdulkareem, A. Abubakar, and T. Herawan. 2017. Neural networks optimization through genetic algorithm searches: a review. *Applied Mathematics & Information* 11 (6): 1543–1564.

Chiueh, T. C. (1991, January). Optimization of fuzzy logic implementation. In *1991 Proceedings of the Twenty-First International Symposium on Multiple-Valued Logic* (pp. 348–349).

Jamali, B., M. Rasekh, F. Jamadi, R. Gandomkar, and F. Makiabadi. 2019. Using PSO-GA algorithm for training artificial neural net-work to forecast solar space heating system parameters. *Applied Thermal Engineering* 147: 647–660.

Khan, F.M., and R. Gupta. (2020). ARIMA and NAR based prediction model for time series analysis of COVID-19 cases in India. *Journal of Safety Science and Resilience.*

Kim, M.K., Y.S. Kim, and J. Srebric. (2020). Predictions of electricity consumption in a campus building using occupant rates and weather elements with sensitivity analysis: Artificial neural network vs. linear regression. Sustainable Cities and Society, 102385.

Kouchami-Sardoo, I., H. Shirani, I. Esfandiarpour-Boroujeni, A. Besalatpour, and M. Hajabbasi. 2020. Prediction of soil wind erodibility using a hybrid Genetic algorithm—Artificial neural network method. *CATENA* 187: 104315.

Luo, X., L. Oyedele, A. Ajayi, O. Akinade, J. Delgado, H. Owolabi, and A. Ahmed. 2020. Genetic algorithm-determined deep feedforward neural network architecture for predicting electricity consumption in real buildings. *Energy and AI* 2: 100015.

Mahajan, R., and G. Kaur. 2013. Neural networks using genetic algorithms. *International Journal of Computer Applications*, 77(14).

Melin, P., J. Monica, D. Sanchez, and O. Castillo. 2020a. Multiple ensemble neural network models with fuzzy response aggregation for predicting COVID-19 time series: The case of Mexico. *Healthcare* 8 (2): 181.

Melin, P., J. Monica, D. Sanchez, and O. Castillo. 2020b. Analysis of spatial spread relationships of coronavirus (COVID-19) pandemic in the world using self organizing maps. *Chaos, Solitons & Fractals* 138: 109917.

Pan, Y., L. Chen, J. Wang, H. Ma, S. Cai, S. Pu,... and E. Li. 2019. Research on deformation prediction of tunnel surrounding rock using the model combining firefly algorithm and nonlinear auto-regressive dynamic neural network. *Engineering with Computers*, 1–11.

Pulido, M., P. Melin, and O. Castillo. 2014. Particle swarm optimization of ensemble neural networks with fuzzy aggregation for time series prediction of the Mexican Stock Exchange. *Information Sciences* 280: 188–204.

Reddy, G.P., Y. Deepika, K.S. Prasad, and G.K. Kumar. 2017. Fuzzy Logics associated with neural networks in the real time for better world. *Materials Today: Proceedings* 4 (8): 8507–8516.

Ruiz, L.G.B., M.P. Cuéllar, M.D. Calvo-Flores, and M.D.C.P. Jiménez. 2016. An application of non-linear autoregressive neural networks to predict energy consumption in public buildings. *Energies* 9 (9): 684.

Sadiq, S., A. Mehmood, S. Ullah, M. Ahmad, G. Choi, and B. On. 2020. Aggression detection through deep neural model on Twitter. *Future Generation Computer Systems* 114: 120–129.

Sánchez, D., P. Melin, J. Carpio, and H. Puga. 2017. Comparison of optimization techniques for modular neural networks applied to human recognition. In *Nature-inspired design of hybrid intelligent systems* (pp. 225–241). Springer, Cham.

Sánchez, D., and P. Melin. 2014. Optimization of modular granular neural networks using hierarchical genetic algorithms for human recognition using the ear biometric measure. *Engineering Applications of Artificial Intelligence* 27: 41–56.

Sanchez, M.A., O. Castillo, J.R. Castro, and P. Melin. 2014. Fuzzy granular gravitational clustering algorithm for multivariate data. *Information Sciences* 279: 498–511.

Sánchez, D., P. Melin, and O. Castillo. 2017b. Optimization of modular granular neural networks using a firefly algorithm for human recognition. *Engineering Applications of Artificial Intelligence* 64: 172–186.

Sawyerr, B., A. Adewumi, and M. Ali. 2014. Real-coded genetic algorithm with uniform random local search. *Applied Mathematics and Computation* 228: 589–597.

Yan, C., M. Li, W. Liu, and M. Qi. 2020. Improved adaptive genetic algorithm for the vehicle insurance fraud identification model based on a bp neural network. *Theoretical Computer Science* 817: 12–23.

Yang, J., G. Kang, Y. Liu, K. Chen, and Q. Kan. 2020. Life prediction for rate-dependent low-cycle fatigue of PA6 polymer considering ratchetting: Semi-empirical model and neural network based approach. *International Journal of Fatigue*, 105619.

Zadeh, L.A. 2009. Toward extended fuzzy logic—A first step. *Fuzzy Sets and Systems* 160 (21): 3175–3181.

Zhou, Z.H., J. Wu, and W. Tang. 2002. Ensembling neural networks: many could be better than all. *Artificial Intelligence* 137 (1–2): 239–263.

Optimization of Modular Neural Networks for the Diagnosis of Cardiovascular Risk

Ivette Miramontes, Patricia Melin, Oscar Carvajal, and German Prado-Arechiga

Abstract These days, with the current situation we are experiencing worldwide due to the pandemic, it is of utmost importance to know our state of health, speaking more specifically of our cardiovascular health. Soft computing can be used by medical experts as a powerful tool to help and facilitate providing a diagnosis of our state of health. The objective of this work is to create a modular neural network to obtain the risk diagnosis that a patient has in developing a cardiovascular event in a period of 10 years likewise, find the heart age. In order to provide this information, a series of risk factors will be given as input to each of the modules, such as age, gender, body mass index, systolic pressure, if the patient is diabetic, if the patient smokes, if he/she is under hypertension treatment. Each module is optimized with two bio-inspired algorithms to test its performance and thereby obtain the best results to provide an accurate diagnosis.

Keywords Cardiovascular event · Hypertension · Optimization · Flower pollination algorithm (FPA) · Bird swarm algorithm (BSA)

1 Introduction

One of the biggest health problems we are currently facing is the risk of cardiovascular disease (CVD), the World Health Organization (WHO) (World Healt Organization 2017) states that CVD is the leading cause of death worldwide. It is estimated that in 2015 around 17.7 million people died, which represents 31% of the total deaths registered worldwide (World Healt Organization 2017). Soft computing has proven to be efficient in diagnosing different diseases such as glaucoma (Raghavendra et al. 2018), hepatitis (Sotudian et al. 2016), diabetes (Abu-Naser. 2018), lung cancer (Nasser and Abu-Naser 2019), thyroid (Ma et al. 2019), among others, this inspires us to implement a modular neural network to diagnose the risk of developing a cardiovascular event over 10 years. Modular neural networks have shown great efficiency in solving

I. Miramontes · P. Melin (✉) · O. Carvajal · G. Prado-Arechiga
Tijuana Institute of Technology, Tijuana, BC, Mexico
e-mail: pmelin@tectijuana.mx

O. Castillo and P. Melin (eds.), *Fuzzy Logic Hybrid Extensions of Neural and Optimization Algorithms: Theory and Applications*, Studies in Computational Intelligence 940,
https://doi.org/10.1007/978-3-030-68776-2_6

complex problems in an excellent way (Sánchez 2017; Sánchez et al. 2020). Another important part when talking about soft computing is optimization using bio-inspired algorithms. Optimization can be defined as a mathematical process to find the best solution to a problem (Haupt and Haupt 2004), some problems where optimization has been exposed to help to a great extent are in control (Lagunes 2020), industrial systems (Teo et al. 2014), pattern recognition (Poma et al. 2020), intrusion detection (Balasaraswathi et al. 2017), energy (González-Briones et al. 2018), etc.

In this work, and to give a more accurate diagnosis of the cardiovascular risk of patients, the architecture of each of the modules of the modular neural network is optimized, aimed at finding best number of layers and number of neurons per layer, where the Bird Swarm Algorithm (BSA) and Flower Pollination Algorithm (FPA) are used to achieve such optimization and, in addition to observing their performance. To experiment with the proposed approach a database of 250 patients is used to train each module in the modular neural network.

The main contribution of this work is to provide a diagnosis of the risk that a patient has in having a cardiovascular event, that is why the optimization of each module of the modular neural network is performed, to find the best architecture and provide a more accurate diagnosis.

This paper has been organized as follows: in Sect. 2 the literature review is presented, in Sect. 3 the proposed method is presented, in Sect. 4 the results are presented and, in Sect. 5 the conclusions obtained of this work are presented.

2 Literature Review

This section describes important concepts that help clarify the work done.

2.1 Flower Pollination Algorithm

The Flower Pollination Algorithm (Yang et al. 2014), proposed by Xin-She Yang in 2012, it's an algorithm that it is inspired by the process of plant pollination. The method uses local and global search computationally, which can be represented by both biotic and abiotic pollination, respectively. Likewise, the algorithm uses other parts of the pollination process which helps to find the best solution to the problem it is solving.

2.2 Bird Swarm Algorithm

The Bird Swarm Algorithm (Meng et al. 2016), proposed by Meng in 2015 to solve optimization problems. This is based on swarm intelligence drawn from social behaviors and social interactions in bird swarms. Birds mainly have three types of behaviors: foraging behavior, watchful behavior, and flight behavior. Birds can forage for food and escape predators through social interaction for a high probability of survival.

2.3 Blood Pressure and Hypertension

To focus and know the problem to be solved, it is important to know the concept of blood pressure and hypertension. Blood pressure is the force the blood exerts against the walls of the blood vessel walls, and it will be normal when it is below 120/80 mm of mercury (mmHg) (Papademetriou et al. 2019; Paul et al. 2019). As you can see, blood pressure has two measures, the first and highest is the systolic pressure, which is when the heart contracts. Diastolic pressure is when the heart relaxes and fills with blood (Bakris et al. 2018).

The European Guidelines for management of hypertension (Zanchetti et al. 2018) defines hypertension as the sustained elevation of blood pressure above the normal limits, which are above 140/90 mmHg.

2.4 Cardiovascular Disease and Heart Age

The cardiovascular disease is different from existing conditions that affect the heart or circulation, this condition is also known as heart and circulation disease (American Heart Association 2015).

Some of the conditions are heart attacks, heart failure, arrhythmias, stroke, among others (Carretero and Oparil 2000).

Another important concept is the age of the heart, since it can determine if a patient is prone to having a heart attack or stroke, and different risk factors are used to calculate it, such as age, blood pressure, if the patient smokes, etc. (British Heart Fundation n.d.).

2.5 Framingham Heart Study

The Framingham Heart Study (Framingham Heart Study 2019), it is a study that has been carried out for 72 years, which was born in the city of Framingham in Massachusetts and is directed by the National Cardiac Institute, which has different calculators to diagnose the risk of different diseases such as diabetes and hypertension.

In 2008, D'Agostino et al. (2008) proposed a calculator for the diagnosis of the risk of a cardiovascular event in 10 years, which uses patient risk factors to provide said diagnosis.

3 Proposed Method

A neuro-fuzzy hybrid model is designed (Miramontes 2018; Melin et al. 2019) to diagnose the risk of developing hypertension in people. Observing the current needs and to provide complete diagnosis, a modular neural network is integrated into this model, with which the risk that a patient has in developing a cardiovascular event in 10 years will be diagnosed, as well as it also happens to know the age of the heart, since this is of a sum to be able to specify the risk diagnosis.

The modular neural network consists of three modules, the first to obtain the risk of developing a cardiovascular event in 10 years, the second is to know the age of the heart when the patient is under treatment for hypertension and the last module it is to know the age of the heart when the patient is without treatment, this separation is made since when it is carrying out tests we notice that for the neural network it was more difficult to learn these behaviors together, so it was decided to do the training in independent modules. As input to each of the modules, different risk factors will be given, such as age, sex, systolic pressure, if the patient is being treated for hypertension, if he is a smoker, if he has diabetes and the rate of body mass, the modular neural network will learn the behavior of the data to obtain the aforementioned diagnoses as output.

To find the appropriate architecture in each module, their optimization is performed. To carry out this work, two bio-inspired algorithms are used to compare their performance, the first is the FPA and the second is the BSA, which searches for the number of layers and the number of neurons for each one. Due to the nature of the problem and previous experience, the layers are limited to two and the neurons per layer to thirty, since it has been observed that they are sufficient to obtain satisfactory results. Each of the modules is trained with 250 patients and in each iteration, it is tested with 50 patients. The representation of the optimization of the modular neural network is illustrated in Fig. 1

To find out which of the architectures provides a better result, the mean square error is used as the objective value (MSE), which is found in (1):

$$\mathrm{MSE} = \frac{1}{n}\sum_{i=1}^{n}\left(\widehat{Y} - Y_i\right)^2 \tag{1}$$

where:

n = number of data points,

$\hat{Y}$ = output value by the model,

Y = the real value for data point i.

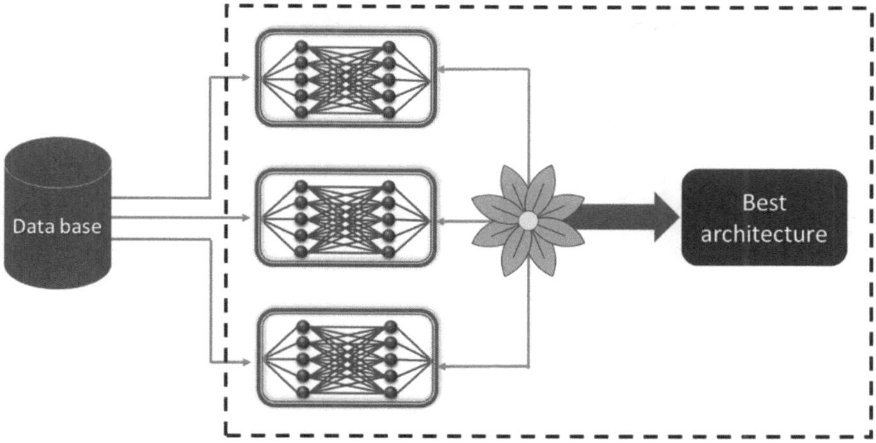

Fig. 1 Optimization of modular neural network

4 Results

To analyze the behavior of each algorithm and observe its performance, 30 experiments are performed. In each experiment, the parameters used in the algorithms are varied to find which is the combination that provides optimal results in the problem to be solved. Tables 1 and 2 present the parameters used in each experiment by the FPA and BSA algorithms, respectively.

The parameters associated with the FPA algorithm are:

n: number of individuals

No. Iter: number of iterations

Dim: Dimensions

P: Probability of change, this parameter is the one that controls the exploration and exploitation of the algorithm.

The parameters associated with the BSA algorithm are:

Pop: Bird population.

M: Total iterations in each experiment.

a1 and a2: Cognitive and social acceleration coefficient respectively.

c1 and c2: Parameter related to surveillance behavior.

CF: Flight frequency.

In Table 3, the results obtained by the FPA algorithm in each module are presented, where the number of layers and neurons per layer are listed by experiments, in addition to the error obtained. The error is obtained from testing each experiment with 50 patients, the best workouts can be seen in bold letters, and the worst experiments in italics.

The best architectures found for each module are presented in Table 4.

It is worth mentioning that 250 epochs were used to carry out the training, as well as the learning algorithm used is the Levenberg-Marquard.

Table 1 Parameters used in FPA for each experiment

No	n	No. Iter	Dim	P
1	15	62	3	0.9
2	17	55	3	0.3
3	19	49	3	0.5
4	21	44	3	0.7
5	23	40	3	0.3
6	25	37	3	0.2
7	27	34	3	0.8
8	29	32	3	0.6
9	31	30	3	0.9
10	33	28	3	0.1
11	35	27	3	0.4
12	37	25	3	0.8
13	39	24	3	0.2
14	41	23	3	0.7
15	43	22	3	0.9
16	45	21	3	0.5
17	47	20	3	0.3
18	49	19	3	0.2
19	51	18	3	0.5
20	53	18	3	0.9
21	55	17	3	0.3
22	57	16	3	0.6
23	59	16	3	0.8
24	61	15	3	0.7
25	63	15	3	0.9
26	65	14	3	0.4
27	67	14	3	0.3
28	69	13	3	0.9
29	71	13	3	0.5
30	73	13	3	0.7

In Table 5, the results obtained by the BSA algorithm for each of the modules are presented, where the number of layers and neurons per layer are listed by experiments, in addition to the error obtained. In this case, 50 patients are also used to test the architectures generated by the algorithm, the highlighted letters represent the best workouts and the worst experiments in italics.

Table 2 Parameters used in BSA for each experiment

No	Pop	M	a1	a2	FQ	c1	c2	dim
1	16	58	0.2	0.2	3	0.4	0.4	3
2	18	52	1.33	1.33	8	0.6	0.6	3
3	20	47	0.9	0.9	1	1.4	1.4	3
4	22	42	0.4	0.4	10	1.8	1.8	3
5	24	39	0.1	0.1	12	0.2	0.2	3
6	26	36	1.2	1.2	7	1.9	1.9	3
7	28	33	1	1	6	2	2	3
8	30	31	1.42	1.42	2	3.48	3.48	3
9	32	29	0.75	0.75	10	1.22	1.22	3
10	34	27	1.25	1.25	9	3.2	3.2	3
11	36	26	1.95	1.95	15	3.15	3.15	3
12	38	24	0.6	0.6	5	0.8	0.8	3
13	40	23	1.3	1.3	12	2.84	2.84	3
14	42	22	0.9	0.9	14	0.5	0.5	3
15	44	21	1.6	1.6	7	1.44	1.44	3
16	46	20	1.4	1.4	5	4	4	3
17	48	19	0.9	0.9	15	0.9	0.9	3
18	50	19	0.6	0.6	11	1.18	1.18	3
19	52	18	2	2	14	2.4	2.4	3
20	54	17	1	1	10	3	3	3
21	56	17	0.4	0.4	11	0.9	0.9	3
22	58	16	0.9	0.9	11	2	2	3
23	60	16	1.6	1.6	5	1.88	1.88	3
24	62	15	0.4	0.4	6	3.12	3.12	3
25	64	15	1.2	1.2	7	2.5	2.5	3
26	66	14	1.8	1.8	2	3.44	3.44	3
27	68	14	0.7	0.7	5	3.7	3.7	3
28	70	13	0.5	0.5	11	2.83	2.83	3
29	72	13	2	2	6	1.56	1.56	3
30	74	13	1.7	1.7	14	4	4	3

The best architectures found for each module are presented in Table 6.

It is worth mentioning that 250 epochs were used to carry out the training, as well as the learning algorithm used is the Levenberg-Marquard.

Once having the best architectures, the best training are tested with 15 patients, where the result provided by each module is compared with the result provided by the Framingham study, In Table 7, the list of patients with their risk factors is presented.

Table 3 Results obtained by the FPA per module

No	Module 1				Module 2				Module 3			
	Lyr	Neurons		Error	Lyr	Neurons		Error	Lyr	Neurons		Error
		L1	L2			L1	L2			L1	L2	
1	2	18	2	2.71E-03	2	2	30	1.06E-04	1	18		6.19E-05
2	2	17	2	2.05E-03	1	17		1.69E-04	1	16		6.09E-05
3	**2**	**13**	**2**	**1.76E-03**	*1*	*12*		*3.77E-04*	**2**	**2**	**18**	**5.00E-06**
4	2	20	2	2.43E-03	**2**	**3**	**22**	**2.07E-05**	2	2	13	3.84E-05
5	2	18	3	2.59E-03	1	11		2.74E-04	1	14		2.03E-04
6	2	14	2	2.85E-03	2	2	24	9.70E-05	2	2	17	1.24E-04
7	2	22	2	2.82E-03	1	9		1.74E-04	1	2		3.47E-04
8	2	12	5	3.72E-03	1	9		1.57E-04	1	2		1.11E-04
9	2	11	2	3.07E-03	1	13		2.35E-05	2	2	24	1.87E-05
10	2	13	2	3.54E-03	1	14		6.53E-05	1	2		1.58E-05
11	2	10	2	3.55E-03	2	3	28	1.30E-04	*2*	*2*	*11*	*3.21E-04*
12	2	11	2	3.70E-03	1	13		1.55E-04	2	2	25	1.80E-04
13	2	20	2	3.88E-03	2	2	26	2.61E-04	2	2	29	2.01E-04
14	2	19	2	3.49E-03	2	4	13	2.69E-04	2	2	30	3.05E-05
15	2	22	2	5.59E-03	1	5		2.43E-04	2	3	15	2.28E-04
16	2	12	3	3.50E-03	2	2	21	3.08E-04	1	7		1.94E-04
17	2	9	4	4.00E-03	1	18		2.99E-04	2	2	17	8.47E-05
18	2	16	3	6.44E-03	1	12		1.77E-04	1	3		2.61E-04
19	2	16	2	3.57E-03	1	12		1.77E-04	2	2	13	8.80E-05
20	*2*	*15*	*2*	*6.85E-03*	1	10		2.82E-04	2	2	28	9.82E-06
21	2	15	3	1.82E-03	1	22		1.04E-04	1	2		5.81E-05
22	2	12	3	4.53E-03	2	2	16	8.06E-05	2	2	14	8.18E-05
23	2	14	2	3.27E-03	2	2	25	2.45E-04	1	2		3.13E-05
24	2	21	2	2.22E-03	2	2	30	9.01E-05	1	2		1.16E-05
25	2	15	2	3.87E-03	1	3		2.65E-04	1	4		1.49E-05
26	2	9	5	5.87E-03	2	2	21	1.11E-04	1	4		6.52E-05
27	2	13	5	4.54E-03	2	2	26	2.64E-04	2	2	10	2.07E-05
28	2	12	2	3.48E-03	1	21		2.60E-04	1	7		2.25E-04
29	2	14	2	2.85E-03	2	5	30	2.04E-04	1	3		9.18E-05
30	2	20	2	2.78E-03	1	2		3.08E-04	2	2	27	1.53E-05

Table 4 Results obtained by the FPA per module

	Module 1	Module 2	Module 3
Number of layers	2	2	2
Number of neurons per layer	13, 2	3,22	2,18

Table 5 Results obtained by the BSA per module

No	Module 1				Module 2				Module 3			
	Lyr	Neurons		Error	Lyr	Neurons		Error	Lyr	Neurons		Error
		L1	L2			L1	L2			L1	L2	
1	2	15	2	1.99E-03	1	11		2.67E-04	1	1		1.26E-05
2	2	13	3	2.16E-03	1	11		2.49E-05	1	3		3.56E-04
3	2	13	2	2.64E-03	2	2	25	2.72E-04	1	2		1.27E-05
4	2	6	19	1.58E-02	1	13		2.17E-04	1	4		2.13E-04
5	2	14	3	2.99E-03	*1*	*6*		*9.77E-04*	1	2		1.50E-05
6	2	15	3	2.49E-03	1	6		5.55E-05	1	5		2.86E-04
7	2	14	4	3.00E-03	1	7		1.87E-04	2	5	29	5.92E-05
8	2	10	3	2.07E-03	1	16		1.47E-04	*1*	*6*		*1.02E-03*
9	2	15	3	3.32E-03	1	6		9.54E-04	1	12		1.62E-04
10	**2**	**11**	**3**	**1.92E-03**	2	3	21	6.04E-04	2	2	14	5.30E-06
11	2	13	3	2.65E-03	2	7	30	1.25E-04	2	1	29	7.79E-05
12	2	12	4	2.15E-03	2	2	12	2.73E-04	1	4		7.80E-05
13	2	15	3	3.04E-03	1	6		8.92E-05	1	8		2.29E-05
14	2	7	9	5.85E-03	1	5		1.82E-04	1	11		2.50E-04
15	2	14	4	3.42E-03	**1**	**11**		**1.16E-06**	1	4		1.37E-05
16	2	14	3	2.91E-03	1	18		5.08E-05	1	4		3.78E-05
17	2	12	4	3.45E-03	1	9		6.94E-05	1	3		1.23E-04
18	2	9	3	1.24E-02	1	13		9.30E-05	1	2		2.87E-05
19	2	13	4	2.51E-03	1	15		3.43E-04	1	26		4.27E-04
20	2	12	3	4.57E-03	1	5		1.03E-04	**2**	**2**	**27**	**1.31E-06**
21	*2*	*8*	*9*	*7.02E-03*	1	11		5.08E-04	1	3		5.10E-06
22	2	14	3	3.11E-03	1	11		1.24E-04	2	2	30	4.86E-05
23	2	19	3	3.73E-03	2	8	29	4.90E-05	2	1	9	2.59E-05
24	2	18	3	4.15E-03	1	9		5.52E-05	2	2	25	3.58E-05
25	2	12	2	2.38E-03	1	6		1.19E-04	1	1		1.96E-05
26	2	19	2	1.95E-03	2	2	5	4.04E-04	2	1	28	1.09E-04
27	2	10	5	2.13E-03	1	14		3.23E-04	1	1		1.60E-05
28	2	9	4	4.35E-03	1	8		1.05E-04	1	5		4.39E-05
29	2	15	3	2.38E-03	1	8		6.15E-05	1	8		2.25E-05
30	2	20	1	4.05E-03	2	4	30	1.72E-04	2	1	30	3.60E-05

Table 6 Results obtained by the BSA per module

	Module 1	Module 2	Module 3
Number of layers	2	1	2
Number of neurons per layer	11, 3	11	2, b27

Table 7 List of patients with risk factors

No.	Risk factors						
	Age	Sex	BMI	Syst	Diab	Smoke	Treatment
1	27	W	24.3	122	No	No	No
2	28	W	23.36	120	No	No	No
3	28	M	29.76	123	No	No	No
4	25	W	24.4	114	No	No	No
5	45	W	24.9	116	No	No	No
6	31	M	35.26	95	No	No	No
7	33	M	25.26	130	No	No	No
8	32	W	29.98	123	No	No	No
9	25	W	21.7	108	No	No	No
10	30	W	30.3	123	No	No	No
11	30	M	21.55	107	No	No	No
12	32	W	24.49	112	No	No	No
13	31	M	30.07	112	No	No	No
14	29	M	21.5	99	No	No	No
15	31	M	23.4	106	No	No	No

In Table 8, the results generated by the modular neural network already optimized by the different algorithms are presented. It can be observed that the modular neural network optimized with the FPA algorithm provides an accuracy in the information provided.

Table 9 presents a comparison of the percentage of success provided by the training performed with the best parameters provided by each algorithm, it is observed that when comparing both algorithms with the training of the non-optimized modular neural network, both algorithms improve said percentage, but FPA shows a slight improvement when compared to the results provided by the BSA algorithm.

5 Conclusions and Future Work

In this work, a modular neural network for the diagnosis of the risk of developing a cardiovascular event in 10 years is proposed. To obtain the diagnosis, different risk factors are given as input and 2 outputs are obtained, the first is the risk of developing a cardiovascular event and the second is the patient's heart age, a result that helps the cardiologist to have a broader picture of the health of the patient's heart.

Each architecture of these modules was optimized with two bio-inspired algorithms, which are the FPA and BSA to observe their performance and find out which

Table 8 List of patients with risk factors

No.	FHS		FPA			BSA		
	Risk %	Heart age	Risk %	Heart age	ABS	Risk %	Heart age	ABS
1	0.8	27	0.8	27	0	0.8	27	0
2	0.8	28	0.8	28	0	0.8	28	0
3	1.7	28	1.7	28	0	1.8	28	0.1
4	0.5	25	0.5	25	0	0.5	25	0
5	2.8	45	2.8	45	0	2.8	45	0
6	1.7	31	1.7	31	0	1.7	31	0
7	1.7	33	1.7	33	0	1.6	33	0.1
8	1.4	32	1.4	32	0	1.4	32	0
9	0.4	25	0.4	25	0	0.4	25	0
10	1.2	30	1.2	30	0	1.2	30	0
11	1.3	30	1.3	30	0	1.3	30	0
12	1	32	1	32	0	1	32	0
13	2	31	2	31	0	2.1	31	0.1
14	1	29	1	29	0	1.1	29	0.1
15	1.5	31	1.5	31	0	1.5	31	0

Table 9 Percent of success

Non-optimized MNN	Optimized MNN	
	FPA	BSA
85%	100%	96%

of the architectures found with these algorithms gave us a more accurate diagnosis. For this, 30 experiments are carried out varying the parameters of the algorithms. Once having the best architectures given by each algorithm, they are tested with a set of patients, with the result that the architecture given by the FPA algorithm provides us with 100% success in diagnosing patients compared to 85% of success when the modular neural network is not optimized. It is concluded in this study that, although optimizing with the two methods improves the success rate, the FPA provides better results compared to the results obtained by the BSA algorithm, so it can say that the architecture provided by the FPA algorithm gives us a better diagnosis on this problem. As future work, multiple objective optimization could also be used, as in Sánchez and Melin (2014), Sánchez et al. (2017), Sánchez (2017) or other kinds of applications like in Castillo (1998), Castillo and Melin (2003), Sanchez et al. (2014).

References

American Heart Association. 2015. [Online]. Available: http://www.heart.org/HEARTORG/Conditions/HighBloodPressure/High-Blood-Pressure-or-Hypertension_UCM_002020_SubHomePage.jsp. Accessed 15 Oct 2018.

Bakris, G.L., and M.J.B.T.-H.A.C. to B.H.D. (Third E. Sorrentino, Eds., "Braunwald's Heart Disease Family of Books," Elsevier, 2018, pp. xv–xviii.

Balasaraswathi, V.R., M. Sugumaran, and Y. Hamid. 2017. Feature Selection Techniques for Intrusion Detection using Non-Bio-Inspired and Bio-Inspired Optimization Algorithms. *Journal of Communications and Information Networks* 2 (4): 107–119.

Carretero, O.A., and S. Oparil. 2000. Essential Hypertension. *Circulation* 101 (3): 329–335.

Castillo, O., P. Melin. 1998. A new fuzzy-fractal-genetic method for automated mathematical modelling and simulation of robotic dynamic systems. *1998 IEEE International Conference on Fuzzy Systems (FUZZ-IEEE 1998) Proceedings*. Volume 2, 1182–1187.

Castillo, O., and P. Melin. 2003. Intelligent adaptive model-based control of robotic dynamic systems with a hybrid fuzzy-neural approach. *Applied Soft Computing* 3 (4): 363–378.

D'Agostino, R.B., et al. 2008. General cardiovascular risk profile for use in primary care: the framingham heart study. *Circulation* 117: 743–754.

Framingham Heart Study. 2019. [Online]. Available: https://www.framinghamheartstudy.org/risk-functions/hypertension/index.php. Accessed 15 Jul 2019.

González-Briones, L., J. Prieto, F. De La Prieta, E. Herrera-Viedma, and J.M. Corchado. 2018. Energy optimization using a case-based reasoning strategy. *Sensors* 18 (3): 1–27.

Haupt, R.L., and S.E. Haupt. 2004. *Practical genetic algorithms, Second*. New Jersey: A Wiley-Interscience publication.

Lagunes, M.L., O. Castillo, F. Valdez, and J. Soria. 2020. Comparison of fuzzy controller optimization with dynamic parameter adjustment based on of Type-1 and Type-2 Fuzzy Logic BT—hybrid intelligent systems in control, pattern recognition and medicine. O. Castillo and P. Melin, Eds. Cham: Springer International Publishing, pp. 47–56.

Ma, L., C. Ma, Y. Liu, and X. Wang. 2019. Thyroid diagnosis from SPECT images using convolutional neural network with optimization. *Computational Intelligence and Neuroscience* 2019: 6212759.

Melin, P., G. Prado-Arechiga, I. Miramontes, and M. Medina-Hernandez. 2016. Hybrid intelligent model based on modular neural network and fuzzy logic for hypertension risk diagnosis. *Journal of Hypertension* 34: e153.

Melin, P., I. Miramontes, and G. Prado-Arechiga. 2018. A hybrid model based on modular neural networks and fuzzy systems for classification of blood pressure and hypertension risk diagnosis. *Expert Systems with Applications* 107: 146–164.

Melin, P., G. Prado-Arechiga, I. Miramontes, and J.C. Guzmán. 2019. Hypertension diagnosis with a soft computing model using a graphical user interface. *Journal of Hypertension* 37: e233.

Meng, X.-B., X.Z. Gao, L. Lu, Y. Liu, and H. Zhang. 2016. A new bio-inspired optimisation algorithm: Bird Swarm Algorithm. *Journal of Experimental & Theoretical Artificial Intelligence* 28 (4): 673–687.

Miramontes, I., C.J. Guzman, P. Melin, and G. Prado-Arechiga. 2018. Optimal design of interval Type-2 fuzzy heart rate level classification systems using the bird swarm algorithm. *Algorithms*, 11(12).

Miramontes, I., P. Melin, and G. Prado-Arechiga. 2020. Comparative study of bio-inspired algorithms applied in the optimization of fuzzy systems. BT—Hybrid Intelligent Systems in Control, Pattern Recognition and Medicine, pp. 219–231.

Nasser, I.M., and S.S. Abu-Naser. 2019. Lung cancer detection using artificial neural network. *International Journal of Engineering and Information Systems* 3 (3): 17–23.

Papademetriou, V., E.A. Andreadis, and C. Geladari. 2019. *Management of hypertension*. Cham: Springer International Publishing AG.

Paul, M., et al. 2019. Measurement of blood pressure in humans: A scientific statement from the american heart association. *Hypertension* 73 (5): e35–e66.

Poma, Y., P. Melin, C. González, and G. Martinez. 2020. Optimization of Convolutional Neural Networks Using the Fuzzy Gravitational Search Algorithm. *Journal of Automation, Mobile Robotics and Intelligent Systems* 14: 109–120.

Raghavendra, U., H. Fujita, S.V. Bhandary, A. Gudigar, J.H. Tan, and U.R. Acharya. 2018. Deep convolution neural network for accurate diagnosis of glaucoma using digital fundus images. *Information Science (Ny)* 441: 41–49.

N.S. El_Jerjawi and S. S. Abu-Naser. 2018. Diabetes prediction using artificial neural network. *International Journal* of *Advanced Science* and *Technology*, vol. 121.

Sánchez, D., P. Melin, J.M.C. Valadez, and H. Puga. 2017. Comparison of optimization techniques for modular neural networks applied to human recognition. In *Nature-Inspired Design of Hybrid Intelligent Systems*, pp. 225–241.

Sánchez, D., P. Melin, J. Carpio, and H. Puga. 2017. Comparison of optimization techniques for modular neural networks applied to human recognition. In *Nature-Inspired Design of Hybrid Intelligent Systems* (pp. 225–241). Springer, Cham.

Sánchez, D., and P. Melin. 2014. Optimization of modular granular neural networks using hierarchical genetic algorithms for human recognition using the ear biometric measure. *Engineering Applications of Artificial Intelligence* 27: 41–56.

Sanchez, M.A., O. Castillo, J.R. Castro, and P. Melin. 2014. Fuzzy granular gravitational clustering algorithm for multivariate data. *Information Sciences* 279: 498–511.

Sánchez, D., P. Melin, and O. Castillo. 2017c. Optimization of modular granular neural networks using a firefly algorithm for human recognition. *Engineering Applications of Artificial Intelligence* 64: 172–186.

Sánchez, D., P. Melin, and O. Castillo. 2020. Comparison of particle swarm optimization variants with fuzzy dynamic parameter adaptation for modular granular neural networks for human recognition. *Journal of Intelligent & Fuzzy Systems* 38 (3): 3229–3252.

Sotudian, S., M.H.F. Zarandi, and I.B. Turksen. 2016. From Type-I to Type-II fuzzy system modeling for diagnosis of hepatitis. *Journal of Computing and Information* 10 (7): 1280–1288.

Teo, K.L., H.-F. Wang, and C. Wu. 2014. Optimization in Industrial Systems. *Mathematical Problems in Engineering* 2014: 824539.

World Healt Organization. 2017. [Online]. Available: https://www.who.int/news-room/fact-sheets/detail/cardiovascular-diseases-(cvds).

British Heart Fundation. [Online]. Available: https://www.bhf.org.uk/informationsupport/risk-factors/check-your-heart-age.

Yang, X.S., M. Karamanoglu, and X. He. 2014. Flower pollination algorithm: A novel approach for multiobjective optimization. *Optimization and Engineering* 46 (9): 1222–1237.

Zanchetti, A., et al. 2018. 2018 ESC/ESH Guidelines for the management of arterial hypertension. *European Heart Journal* 39 (33): 3021–3104.

A Review on the Cuckoo Search Algorithm

Maribel Guerrero-Luis, Fevrier Valdez, and Oscar Castillo

Abstract A review of the Cuckoo Search (CS) algorithm, which is within the classification of bio-inspired algorithms and is based on the breeding parasitism of a bird species called cuckoo, is presented. The CS contains a set of rules, which simplifies the process of reproduction of these birds in real life and is translated into a computer algorithm in order to solve optimization problems. With these rules, new solutions are discovered and which go into the next iteration, this process helps to find the best quality solutions. There are different variants of the cuckoo search algorithm, which have been compared with other meta-heuristics in order to find better results. The purpose of this paper is to review the literature on the contributions that have been made to CS throughout history, likewise identify the proposed variants that the algorithm and the applications in various intelligent computing applications. As science and technology advance, the complexity of the problems also increases, therefore the importance of continuing to improve the CS algorithm and adapting it various engineering problems.

Keywords Cuckoo search · Levy flights · Optimization and applications

M. Guerrero-Luis · F. Valdez · O. Castillo (✉)
Tijuana Institute of Technology, Tijuana, B.C., Mexico, Calzada Tecnológico S/N Fracc. Tomas Aquino, 22379 Tijuana, Mexico
e-mail: ocastillo@tectijuana.mx

M. Guerrero-Luis
e-mail: maribel.guerrero@tectijuana.edu.mx

F. Valdez
e-mail: fevrier@tectijuana.mx

O. Castillo and P. Melin (eds.), *Fuzzy Logic Hybrid Extensions of Neural and Optimization Algorithms: Theory and Applications*, Studies in Computational Intelligence 940,
https://doi.org/10.1007/978-3-030-68776-2_7

1 Introduction

The Science and technology is constantly changing, it seeks to adapt to various complex problems applied to the real world, which can be expressed mathematically, such problems have an increasing level of complexity, implicit variables and dimensions, therefore the importance of optimization.

The exact methods require a lot of mathematical calculations and time to find a solution or sometimes it is very complicated to arrive at the result. his implies investing a lot of time of calculation and computer power, there are problems of the real world applied to problems of combinatory optimization of engineering that is difficult to find solution that is, they have a great computational complexity of Hard-NP such as: to find the shortest way between N points, to minimize the costs of delivery of services to clients, optimal sequence of processes in production lines, optimal resource allocation, etc.

There are heuristic optimization algorithms, which are used in artificial intelligence that have to do with aspects of employing expert knowledge in specific tasks, these algorithms are the basis for the construction of metaheuristic algorithms.

The metaheuristics are model-based, have greater complexity than heuristic algorithms and are also more complex. It consists of iterative processes that guide a heuristic in an intelligent way to explore and exploit search spaces that lead to more promising solutions.

In diverse AI problems, we will see implicit the models associated to optimization problems and search of better solutions, the metaheuristics are approximate algorithms, not exact and in addition, they are probabilistic, to find good solutions it is required to have a balance between exploration and exploitation of the space.

There are search meta-heuristics: in neural networks, PSO, ant colony, based on birds, among others.

The nature has been a fundamental source of inspiration, able to provide efficient, simple and adaptable solutions to the search for better results, there are several meta-heuristics which have been used to solve some engineering and optimization problems.

In this paper, it is intended to make a review of the implementations, variants, advances that have had the Cuckoo Search algorithm and the potential of using Levy flights in the algorithm itself as in others.

In 2009 Xin-She Yang and Suash Deb developed the Cuckoo Search algorithm in Bacanin (2011). CS is based on the parasitism of offspring of some bird species called cuckoo. The CS algorithm implements the so-called Lévy flights.

This algorithm has had a variety of applications, variants such as: adjust the parameters with the intention of improving convergence, implementation of Levy flights to avoid stagnation of solutions that are not promising, these proposals help to leave a local minimum and analyze the performance to problems of N dimensions, because one of the limitations of the method is not able to converge to the optimal solution when dealing with complex problems of high dimension, there are currently many search methods, however there is no perfect optimization algorithm that fits any real world problem.

We will start with the analogy with the nature of cuckoo birds, the CS algorithm, rules, Levy flights, algorithms that use Levy flights to improve results, variants of the Cuckoo Search method and its applications in the real world.

In this paper, the importance of Levy flights is described from the bio-inspired perspective, some examples of algorithms that implement this type of strategy to avoid being trapped in a local minimum are mentioned, the behaviour of the algorithm is explained by means of a flow chart, and some variants and applications that the algorithm has had recently are specified.

2 An Analogy with Nature

The Cuckoo birds are fascinating and emit beautiful sounds, they have an aggressive breeding strategy. Cuckoo species such as Ani and Guira lay their eggs in communal nests, although they may remove the eggs of others to increase the likelihood of their eggs hatching (Berrazouane 2014; Al-Saati 2013).

Many bird species have the same approach to motherhood (Bacanin 2011).

The birds form quickly and lay an egg, usually covered with two protective layers to prevent cracking, because cuckoo birds drop their egg into a host nest. Cuckoo birds seek the best nest to place their young, because the egg has a high nutritional level and care must be taken with predators (Rajabioun 2011), in order to maximize the survival rate of the egg and achieve the best society of the habitat (Walton 2013).

Some cuckoo species such as Tapera evolved to such a degree that they specialize in mimicking the color and pattern of the eggs of some host species. This reduces the probability of host bird discovery of eggs and therefore increases their reproductivity (Ghosh 2014).

A particular characteristic of cuckoo birds is that they choose the nests where the host birds have just laid their eggs. The cuckoo egg is the first to hatch and its first instinctive action is to eliminate the other eggs in the nest, in order to keep all the food brought by the host birds.

The studies also show that a cuckoo chick can also imitate the host's chick call to gain access to more feeding opportunities (Yang 2009).

2.1 *Cuckoo Search Algorithm*

Cuckoo Search (CS), is a metaheuristic algorithm inspired by the nature of a bird family, and has been used to solve some engineering designs and optimization problems with promising results (Walton 2013; Joshi et al. 2017).

The placement of their eggs in host nests (often other species) and the breeding has been the motivation for the development of this new bio-inspired optimization algorithm. Additionally, this algorithm is reinforced by the so-called Lévy flights (Yang 2013). CS is a population-based search procedure that is used as an optimization tool, in the solution of complex, non-linear and non-convex optimization problems (Marichelvam et al. 2014).

2.2 Algorithm Rules

The Cuckoo search algorithm includes the following rules (Chandrasekaran and Simon 2012; Jamil and Zepernick 2013; Patwardhan et al. 2014):

(1) Each cuckoo bird has only one egg at a time and throws its egg into a nest which is selected through so-called Lévy flights.
(2) Nests containing the best quality eggs are transferred to the following iterations.

The host nests in the algorithm are fixed, the intruder egg is discovered by the host bird with a probability $p_a \in [0,1]$. When this happens, the host bird can:

(1) Dispose of the egg
(2) Leave the nest and build a new nest

In rule number 2, the quality of a solution or fitness is known as a value proportional to the objective function, and fitness is similar to the fitness function in a genetic algorithm (Ong and Zainuddin 2013).

The following criteria should be considered during the implementation of the algorithm (Yang 2010):

(1) In a nest, each egg is a solution
(2) Each cuckoo bird can only lay one egg (i.e. a single solution).

The aim of this type of aggressive breeding strategy is to use the new and most potentially promising solutions (Cuckoo birds) and to forget all the solutions that are not promising.

In several metaheuristic algorithms a matrix is used, to place several random solutions that are related to the problem of interest. With this, information is exchanged and the best results are obtained. In this matrix, each line represents the location of the artificial nest (Yang 2010).

The CS algorithm uses a balanced combination of local random search and global random scan search, controlled by a parameter p_a (Yang 2013).

2.3 Levy Flights

So-called Lévy flights are required in the CS algorithm to perform global and local searches in the solution domain (Manikandan 2012). The representation of a Lévy flight in a search space is called a random walk (i.e. a trajectory consisting of successive random steps) and its peculiarity is due to the successive sudden jumps it makes, which are chosen from a probability density function that has a power-law tail, such behavior is also performed by flies and various insects (Molga and Smutnicki 2005).

Lévy flights have had great success in several bio-inspired algorithms, because they contain a percentage of new movements which are based on the best solution found until the iteration in which the algorithm goes, this generates an acceleration in the process of finding the best solution, while ensuring that the remaining percentage is generated at random and is far from the current solution this generates an efficient exploration and not avoid being trapped in local minimums (Manikandan 2012).

2.4 Mathematical Formulas

The formula for generating a new position is described below, for example to generate a cuckoo i is defined by Eq. 1:

$$x_i^{(t+1)} = x_i^t + \alpha \cdot randn(size(D)) \oplus Le'vy(\beta) \oplus s_i \tag{1}$$

where:

$\boldsymbol{x}_i^{(t+1)}$: It is a vector that contains the new position of the nest.
$\boldsymbol{t}$: Is the number of the current iteration (Time).
$\boldsymbol{x}_i^t$: Is a vector containing the current position of the nest.
α: It is a constant, it is related to the scale of the problem of interest. >0.
randn: Random scalar value.
The product $\oplus$ is a vector multiplication.
Lévy (β): It is a flight of Lévy.
D: Dimensions.

$$S_i = (x_i^t - \boldsymbol{x}_{best}) \tag{2}$$

The Eq. 2 is described below:
$\boldsymbol{x}_i^t$: It's the vector at the position of the nest we're evaluating.
Xbest: It is a vector containing the best position found so far.
If $\boldsymbol{x}_i^t = \boldsymbol{Xbest}$, the solution remains.

$$Le'vy(\beta) = \frac{\sigma(\beta) \cdot randn}{|randn|^{\frac{1}{\beta}}}, \tag{3}$$

Described the Eq. 3:

β : It's a constant ($1 \leq \beta \leq 2$).

The sigma standard deviation is calculated with the following Eq. 4.

$$\sigma(\beta) = \left\{ \frac{\Gamma(1+\beta)\sin\left(\frac{\pi\beta}{2}\right)}{\Gamma\left[\frac{(1+\beta)}{2}\right]\beta 2^{\left(\frac{\beta-1}{2}\right)}} \right\}^{1/\beta}, \quad \sigma(\beta) = 1 \tag{4}$$

The Gamma function is defined by the integral Eq. 5:

$$\boldsymbol{\Gamma}(z) = \int_0^\infty t^{z-1} e^{-t} dt \tag{5}$$

The formulas described above are used by the cuckoo search algorithm to find high quality solutions.

2.5 Flowchart CS

CS implements a strategy to control the balance between intensive local search and more efficient exploration throughout the search space. One of the advantages of this algorithm is the reduced number of parameters, which makes CS a less complex and therefore potentially more generic algorithm. In principle, the population is fixed and the variable Pa allows to control essentially the elitism and balance in the randomization and local search, however other parameters have been studied that are implicit in the selection of the best value in which the alpha and beta values can also help to improve the solutions. The reduced number of parameters makes the algorithm less complex and therefore potentially more generic. All this is explained by the ease of tuning both CS parameters while remaining robust to solve a wide range of optimization problems, even qualified problems are NP-hard (Yang and Suash 2009; Molga and Smutnicki 2005).

Figure 1 shows the cuckoo search algorithm in which it is important to define the target function F(x), we generate an initial population called n, number of nests Xi, number of iterations, Pa, Beta, Alpha, Dimensions and the lower and upper limits.

This flow chart shows the process of finding the best nests with good quality solutions throughout the iterations.

3 Implementation of Levy Flights in Other Algorithms

Below are some papers that implement Levy walks with trajectories that take successive random steps that suddenly change trajectories helping to improve results.

Artificial Bee Colony (ABC) is a population-based probabilistic approach considered a simple and newly created algorithm to find the global optimum. The ABC

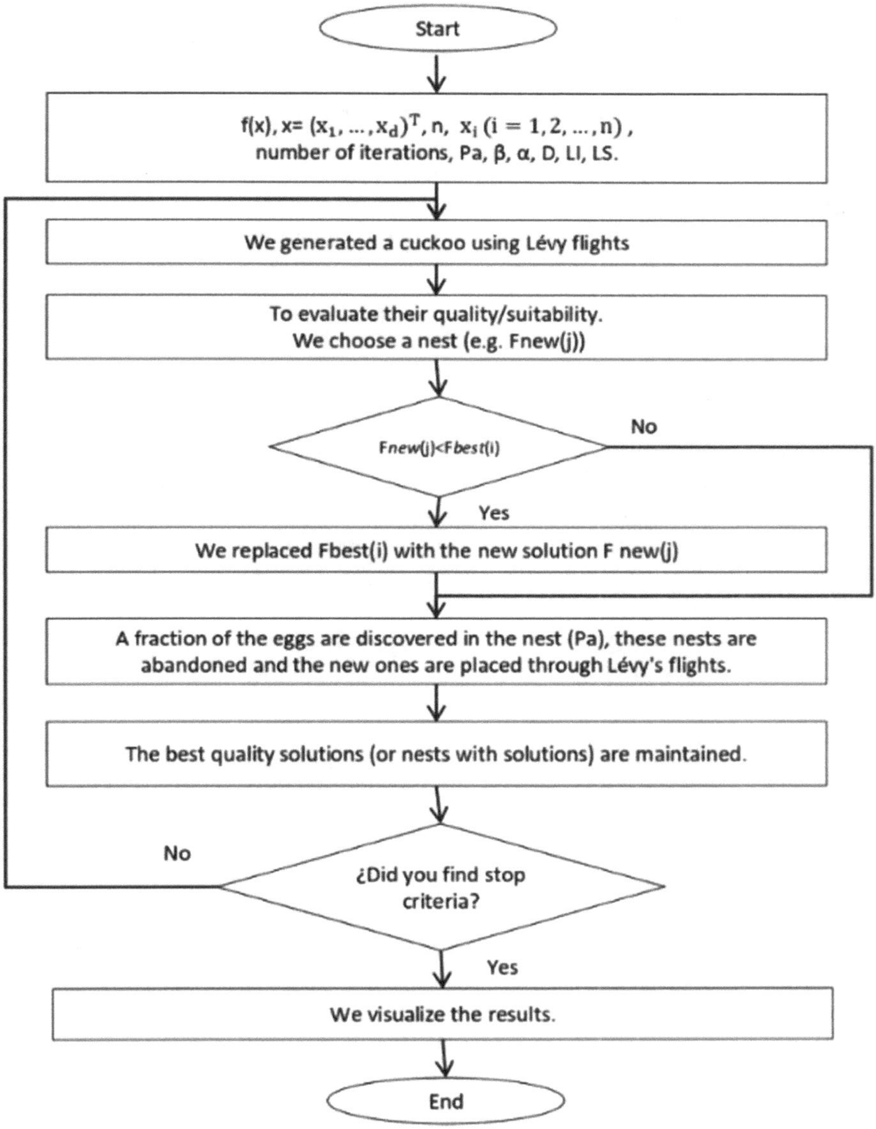

Fig. 1 Flowchart the cuckoo search algorithm (Guerrero et al. 2015)

algorithm is computationally expensive and slow processing. Therefore, in this pa the Opposition Based on Lévy Flight ABC (OBLFABC) is proposed. This is an algorithm that implements a local search strategy inspired by the Lévy Flight random walk and called Lévy Flight Local Search (LFLS) when implementing Lévy Flights, the simulation results show that OBLFABC is faster between ABC, Gbest guided ABC (GABC), best ABC (BSFABC), modified ABC (MABC) (Sharma et al. 2013).

Flower Pollination Algorithm (FPA), perform tests and compare their performance. The simulation results show that the flower algorithm is more efficient than GA and PSO (Yang 2012).

Particle Swarm Optimization and Firefly Algorithm are among the most powerful algorithms for optimization. Numerical studies and results suggest that the L'evy-Flight Firefly Algorithm is efficient in finding global optimums with higher success rates (Yang 2010).

4 Variants of the Cuckoo Search Algorithm

We reviewed some papers related to cuckoo search algorithms, and found some variants or strategies implemented in such algorithm to improve results.

OBL It is a learning strategy based on opposition and improved local search to help the normal CS algorithm. It consists of forgetting the previous solutions of the original swarm, then selecting three cuckoos from the swarm at random to produce a new individual. It replaces the previous solutions as long as the new one is better. The objective is to find a potential optimal solution locally, around the current global optimal solution. This strategy seeks to improve the global search capacity, the speed with which the algorithm converges and the convergence precision of the algorithm (Zefan 2017).

In order to eliminate slow search and low accuracy of cuckoo search for optimization function, testing shows that the improved algorithm has better performance and can work well in high dimensional optimization (Jia 2013).

Firefly-Cuckoo Search Algorithm (FCSA) this algorithm was designed to solve random antenna array (RAA) beam pattern optimization problems. To achieve improved performance of the firefly algorithm, the algorithm is implemented by implementing the Levy flight mechanism of the cuckoo search algorithm as the global search operator to improve the performance of the conventional firefly algorithm. In the various simulations the results show that FCSA has a better performance in terms of accuracy and convergence rate to reduce the maximum sidelobe level (SLL) of the RAA (Luo 2018).

New approach to ultrasonic echo pattern estimation using a hybrid of Cuckoo Search (CS) and Adaptive Cuckoo Search (ACS).

Backscattered ultrasonic echo modeling is represented by superimposed Gaussian echoes corrupted by noise.

It is represented by a non-linear function to the Gaussian echo with the following set of parameters: center frequency, echo bandwidth, echo amplitude, arrival time and phase.

The parameter estimation is formulated as a non-linear optimization problem. The CS algorithm converges to the best solution with less time than ACS. However, the ACS algorithm outperforms CS (Chibane 2018).

Modification of the Cuckoo Optimization Algorithm (COA), Unequal Limit COA (ULCOA), which is intended to find the optimal calibration parameters. This algorithm was evaluated by several non-linear reference functions that indicate superior performance of the ULCOA compared to particle swarm optimization and genetic algorithms in terms of accuracy and robustness. In addition, the ULCOA-based field calibration for the MEMS triaxial accelerometer is analyzed. Finally, experimental results are provided (Shokri-Ghaleh et al. 2020).

5 Applications

To be able to visualize the performance of this algorithm, it is important to be able to make an investigation of applications that it has in some problems of engineering in which the algorithm of CS has been implemented.

In Table 1 w e can find some applications that have been made with the cuckoo search algorithm in real life, then described.

A medical image always needs to be a reliable and precise data to be able to detect if an object is known or unknown in the human body, the complex in this area is the lack of techniques that have an adequate optimization, dedicated hardware that is an accurate diagnostic complex. for this reason, a version of the modified algorithm GW (gray wolf) is implemented with the help of the cuckoo search algorithm which has the purpose of finding an optimization process for the detection of objects in the human body with high precision, but also allows physicians to take measurements with greater precision in real time and reliably (Dutta 2020).

In this article he addresses the issue of random key combination with cuckoo search to solve combinatorial problems (Ouaarab 2020).

Table 1 Applications of the cuckoo search algorithm

Authors and Year	Application
Dutta (2020)	Medical imaging
Ouaarab (2020)	The random-key encoding scheme
Troxler (Troxler 2020)_ENREF_28 2020	Nest implements the multi-threaded
Oruc (Oruc 2020)	Parallel communication strategy
Wu (2019)	Fuzzy Logic control optimized by the cuckoo search algorithm
Tran-Ngoc et al. (2019)	Damage detection in bridges and beam-like structures
Sanchez et al. (2015)	Technique for forming information granules is shown with CS and Type-2 Fuzzy Logic
Gonzalez (2015)	Applied to edge detection for digital images
	Durgun and Yildiz (2012) Structural design optimization of vehicle

The multi-threaded cuckoo search algorithm with n number of eggs per nest implements the multi-threaded variant using Java, to run multiple CPU cores (Troxler 2020).

New model that adopts the CS algorithm for fuel flow in the ascent stage of flight (Oruc 2020).

The cuckoo search algorithm has been implemented to solve the problem of control of traffic signals in real time, the control system is implemented with fuzzy logic in an intersection with four lanes depending on the number of vehicles, in addition the time of the green light is given by the CS algorithm (Wu 2019).

CS has been implemented in the optimal structural design of vehicle and improve cost, fuel efficiency, profitability and lightweight design (Tran-Ngoc et al. 2019).

The cuckoo search algorithm is used in this proposal where granules that represent the information will be formed using the technique of fuzzy logic type-2 by intervals (Sanchez et al. 2015).

Implementation of the CS algorithm for the optimization of interval type-2 fuzzy logic parameters applied to edge detection for digital images (Gonzalez 2015)

The CS and Artificial Neural Network (ANN) algorithm is used, the backward propagation algorithm based on the descent of the gradient is implemented, neural networks are implemented to improve the parameters of the algorithm and it is implemented in the location and quantification of structural damage (Durgun and Yildiz 2012).

6 Conclusions

In the present paper, for simplicity, we have assumed that each cuckoo has one egg at a time and is ringed to a random nest, this greatly simplifies implementation. However, there are other multi-objective algorithms that store the results in matrices and having n quantity of eggs in a nest increases its advantages for some applications. This may form a topic for future research. The cuckoo algorithm has few parameters compared to other meta-heuristics, such parameters are: n, Pa, alpha and beta, in previous paper we have already discussed what are the ideal values for alpha and beta.

It is initialized with an initial population called n of $xi(i = 1.2, \ldots n)$, number of iterations, a value of Pa, α,β and number of dimensions.

Another possible extension is to use artificial intelligence techniques to help improve convergence and get out of local minima where the algorithm could get stuck and try increasing the dimensions. In addition, it is possible to analyze in detail the levy flights and if there are any techniques that avoid going out of the search space and find the balance of parameter adjustment that allows the algorithm to have a good balance and further improve the results.

Independently that it is considered a relatively new algorithm, in this article we could observe that there are many applications in diverse engineering with the intention of optimizing and finding the best solution to any problem of the real world.

This algorithm has been very popular in the last years due to its simplicity and promising results.

The future work to be done is to implement artificial intelligence techniques to help improve the results of the cuckoo search algorithm.

References

Al-Saati, N., and M. Alabajee. 2013. The use of cuckoo search in estimating the parameters of software reliability growth models. *International Journal of Computer Science and Information Security*, 11.

Bacanin, N. 2011. *An object-oriented software implementation of a novel cuckoo search algorithm*, in *Proceedings of the 5th European conference on European computing conference*. 2011, World Scientific and Engineering Academy and Society (WSEAS): Paris, France. p. 245–250.

Berrazouane, S. 2014. Parameter optimization via cuckoo optimization algorithm of fuzzy controller for energy management of a hybrid power system. *Energy Conversion and Management*, 78, 652-660-2014 v.78.

Chandrasekaran, K., and S.P. Simon. 2012. Multi-objective scheduling problem: Hybrid approach using fuzzy assisted cuckoo search algorithm. *Swarm and Evolutionary Computation* 5: 1–16.

Chibane, F., A. Benammar, and R. Drai. 2018. Parameters Estimation of Ultrasonics Echoes using the Cuckoo Search and Adaptive Cuckoo Search Algorithms. In *2018 26th European signal processing conference (EUSIPCO)*. IEEE.

Durgun, İ., and A.R. Yildiz. 2012. Structural design optimization of vehicle components using cuckoo search algorithm. *Materials Testing* 54 (3): 185–188.

Dutta, S., and A. Banerjee. 2020. Optimal image fusion algorithm using modified grey wolf optimization amalgamed with cuckoo search, levy fly and Mantegna algorithm. In *2020 2nd International Conference on Innovative Mechanisms for Industry Applications (ICIMIA)*. IEEE.

Ghosh, S., et al. 2014. Gray level image enhancement using cuckoo search algorithm. In *SIRS*.

Gonzalez, C.I., et al. 2015. Cuckoo search algorithm for the optimization of type-2 fuzzy image edge detection systems. In *2015 IEEE Congress on Evolutionary Computation (CEC)*. IEEE.

Guerrero, M., O. Castillo, and M. García, Cuckoo search algorithm via Lévy flight with dynamic adaptation of parameter using fuzzy logic for benchmark mathematical functions. In *Design of Intelligent Systems Based on Fuzzy Logic*, Neural Networks and Nature-Inspired Optimization. 2015, Springer. p. 555–571.

Jamil, M., and H.-J. Zepernick. 2013. Multimodal function optimisation with cuckoo search algorithm. *The International Journal of Bio-Inspired Computing* 5 (2): 73–83.

Jia, R.-M., and D.-X. He. 2013. Complex valued cuckoo search with local search. In *2013 Ninth International Conference on Natural Computation (ICNC)*. IEEE.

Joshi, A.S., et al. 2017. Cuckoo search optimization- a review. *Materials Today: Proceedings* 4 (8): 7262–7269.

Luo, X., Y.-J. Yang, and Q.-Y. Zhou. 2018. A Firefly-Cuckoo Search Algorithm for Optimizing the Beam Patterns of the Random Antenna Arrays. In *2018 4th Annual International Conference on Network and Information Systems for Computers (ICNISC)*. IEEE.

Manikandan, P., and S. Selvarajan. 2012. Data clustering using cuckoo search algorithm (CSA). In *SocProS*.

Marichelvam, M.K., T. Prabaharan, and X.S. Yang. 2014. Improved cuckoo search algorithm for hybrid flow shop scheduling problems to minimize makespan. *Applied Soft Computing* 19: 93–101.

Molga, M., and C. Smutnicki. 2005. Test functions for optimization needs. *Test Functions for Optimization Needs* 101: 48.

Ong, P., and Z. Zainuddin. 2013. An efficient cuckoo search algorithm for numerical function optimization. *AIP Conference Proceedings* 1522 (1): 1378–1384.

Oruc, R., and T. Baklacioglu. 2020. Modelling of fuel flow-rate of commercial aircraft for the climbing flight using cuckoo search algorithm. *Aircraft Engineering and Aerospace Technology*.

Ouaarab, A. 2020. Random-key cuckoo search (RKCS) applications. In *Discrete cuckoo search for combinatorial optimization*. Springer, pp. 71–86.

Patwardhan, A.P., R. Patidar, and N.V. George. 2014. On a cuckoo search optimization approach towards feedback system identification. *Digital Signal Processing* 32: 156–163.

Rajabioun, R. 2011. Cuckoo optimization algorithm. *Applied Soft Computing* 11 (8): 5508–5518.

Sanchez, M.A., O. Castillo, and J.R. Castro. 2015. Information granule formation via the concept of uncertainty-based information with interval type-2 fuzzy sets representation and Takagi–Sugeno–Kang consequents optimized with Cuckoo search. *Applied Soft Computing* 27: 602–609.

Sharma, H., J.C. Bansal, and K. Arya. 2013. Opposition based lévy flight artificial bee colony. *Memetic Computing* 5 (3): 213–227.

Shokri-Ghaleh, H., et al. 2020. Unequal limit cuckoo optimization algorithm applied for optimal design of nonlinear field calibration problem of a triaxial accelerometer. *Measurement* 164: 107963.

Tran-Ngoc, H., et al. 2019. An efficient artificial neural network for damage detection in bridges and beam-like structures by improving training parameters using cuckoo search algorithm. *Engineering Structures* 199: 109637.

Troxler, D., T. Hanne, and R. Dornberger. 2020. A multi-threaded cuckoo search algorithm for the capacitated vehicle routing problem. In *Proceedings of the 2020 4th international conference on intelligent systems, Metaheuristics & Swarm Intelligence*.

Walton, S., et al. 2013. Comment on Cuckoo search: A new nature-inspired optimization method for phase equilibrium calculations by V. Bhargava, S. Fateen, A. Bonilla-Petriciolet.

Wu, S., et al. 2019. Adaptive fuzzy logic traffic signal control based on cuckoo search algorithm. In *International symposium for intelligent transportation and smart city*. Springer.

Yang, X., and D. Suash. 2009. *Cuckoo search via lévy flights*. in *2009 world congress on nature & biologically inspired computing (NaBIC)*.

Yang, X.-S. 2010. Nature-Inspired Metaheuristic Algorithms: Second Edition. Luniver Press.

Yang, X.-S. 2010. Firefly algorithm, Levy flights and global optimization. In *Research and development in intelligent systems XXVI*. 2010, Springer. p. 209–218.

Yang, X.-S. 2012. Flower pollination algorithm for global optimization. In *International conference on unconventional computing and natural computation*. Springer.

Yang, X.-S. 2013. *Cuckoo search and firefly algorithm: Theory and applications*. Incorporated: Springer Publishing Company.

Zefan, C., and Y. Xiaodong. 2017. Cuckoo search algorithm with deep search. In *2017 3rd IEEE International Conference on Computer and Communications (ICCC)*. IEEE.

An Improved Convolutional Neural Network Based on a Parameter Modification of the Convolution Layer

Ruth Rodriguez, Claudia I. Gonzalez, Gabriela E. Martinez, and Patricia Melin

Abstract Convolutional Neural Network models represent a powerful tool that can be applied in image classification tasks. Convolutional Neural Networks have a wide variety of parameters and these can produce different results for the same tasks depending on the particular parameter settings. This paper presents an improved Convolutional Neural Networks model based on changing the number of layers and convolution filters weights with the aim of analyzing the impact in the image classification and recognition systems. The Handwritten digits (MNIST), the American Sign Language (ASL MNIST) and, the Mexican Sign Language (MSL) databases were selected to perform this study. The proposed approach achieved greater precision than the accuracy obtained without varying the parameters; therefore, the modification of the parameters in the convolution layer represents a significant impact on the results.

Keywords Kernel initializer · Deep learning · Pattern recognition · Convolutional neural network

1 Introduction

Convolutional Neural Networks are considered a powerful classification tool and can be very useful for solving diverse kinds of problems. It is a supervised learning method inspired by the vision system, which manages to process the information captured by the eye and interpret it, in order to identify elements of an environment. Convolutional Neural Networks (CNNs) have produced good results in different fields, such as Computer Vision, Medicine and Robotics. Recent works have shown that they can obtain more precise results in solving specific problems, compared to other methods used. For this reason, its use has increased in the field of industry, leading to the development of architectures that make it possible to maximize its

R. Rodriguez · C. I. Gonzalez · G. E. Martinez · P. Melin (✉)
Tijuana Institute of Technology, Tijuana, Mexico
e-mail: pmelin@tectijuana.mx

O. Castillo and P. Melin (eds.), *Fuzzy Logic Hybrid Extensions of Neural and Optimization Algorithms: Theory and Applications*, Studies in Computational Intelligence 940,
https://doi.org/10.1007/978-3-030-68776-2_8

potential in various applications, making it essential to know the functionality of each one of the parameters that compose it.

Parameter modification within Convolutional Networks is usually done based on experimentation. Research is currently underway to explore different architectures that will allow for better results. CNNs have a wide variety of parameters and can produce different classification accuracy for the same tasks depending on the different parameters; several studies have been carried out since the emergence of popular architectures such as AlexNet (Krizhevsky et al. 2017) and GoogleNet (Szegedy 2015) that allow finding better results.

In Poma et al. (2020a, b, c), a Convolutional Neural Network model is proposed that optimizes the parameters of the Convolutional Neural Network by using the FGSA (Fuzzy Gravitational Search Algorithm) in which batch size, number of convolution filters and filters size are optimized. The results show an improvement with optimized parameters compared to other methods used.

In Ozbulak (2018), a Gabor filter is used to solve weights initialization weights problem and transfer learning of pre-trained models using an architecture where the number of layers, number of filters, filters size and filter initialization are modified; in this research, very good results are obtained, even better than with the pre-trained models.

In Sinha et al. (2017) demonstrated how varying the parameters of a Convolutional Neural Network has a high accuracy result, experimental analysis shown the change the number of layers and the input window size has a significant impact on the recognition accuracy the result are compared against AlexNet (Krizhevsky et al. 2017).

In other work such as Mendoza et al. (2009), Gonzalez (2017a, b), Gonzalez (n.d.), edge detectors are used for image processing, although the methods for pattern recognition that are used are different from those discussed in this paper, the use of Edges has produced very good results for images processing. Edges include one of the most important information in the image, and can provide the information of the position of the object. Edge detection is an important tool in image processing, used in feature detection and texture analysis (El-Sayed 1998). Previous works also use CNN for pattern recognition focused on static images from sign language (Bin 2019; Xie 2018; Bantupalli et al. 2018).

The parameter modifications have an impact on the CNN accuracy in image classification tasks that are investigated and analyzed in this study. A new methodology is presented that incorporates the modification of the parameters in the CNN, such as the number of convolution layers and the initialization of the filter weights in the convolution layer using the Sobel operator, Prewitt Operator, and Laplacian Operator, in order to find the impact that it generates in image recognition and classification systems.

The experiments were conducted by varying different network parameters and recording the accuracy. Experimental analysis has shown that changing the number of layers, the appropriate initialization, and the weight of the convolutional filters have a significant impact on classification accuracy. The MNIST Handwritten Digit and the American Sign Language (ASL MNIST) databases were selected; also we have created a Mexican Sign Language database (MSL). Officially, the MSL database has not been published, according to the National Institute of Statistics and Geography, it is estimated that in Mexico there are around 800 thousand people with hearing and speech problems (Discapacidad 2020); for this reason, we believe it is important to contribute in the development of technologies that can help to the well-being, development and, quality of life of people affected by this condition. The proposed approach achieved greater precision than the accuracy obtained without varying parameters.

This paper was developed with the following structure: In Sect. 2, some basic concepts related to the subject of study are briefly explained. Sequentially, in Sect. 3, the methodology used is described, as well as the architecture used in the CNN model. Section 4 shows the results obtained from experimentation and, finally, Sect. 5 offers some conclusions.

2 Background and Basic Concepts

This Section presents the basic concepts and the process that a Convolutional Neural Network performs to extract the features of an image, and also explains the main ideas of an edge detector in image processing.

2.1 Convolutional Neural Network Concepts

Convolutional neuronal networks are a powerful tool to solve classification problems. It is a type of Neural Network architecture in the Deep Learning area, not only consists of a Neural Network that has many hidden layers, but is a computational structure that mimics how the visual cortex of the brain processes and recognizes images. Its objective is to learn the characteristics of the data, using convolutions that improve the images recognition, when the properties are extracting, it can learn in such a way that it can be converted into an adaptive system (Kim 2017).

A Convolutional Neural Network is based on a mathematical linear operation between matrices called convolution, and the network architecture is formed by 4 main layers, a convolutional layer, a non-linearity layer, a Pooling layer and finally a layer fully-connected (Albawi 2017).

The Convolution is a mathematical operation to merge two sets of information; in this case, the convolution is applied to the input data, the grayscale images can be represented in a matrix of m columns and n rows, where each pixel has a value from 0 to 255; using a Kernel each value from the input matrix is multiplied to produce a feature map. This operation is illustrated in Fig. 1, where the matrix shown on the left is the entry image and the filter is superimposed like a mask and slides from left to right multiplying and adding each of the elements, on the right is the resulting feature map. The convolution formula is defined in Eq. (1), (Heaton 2017).

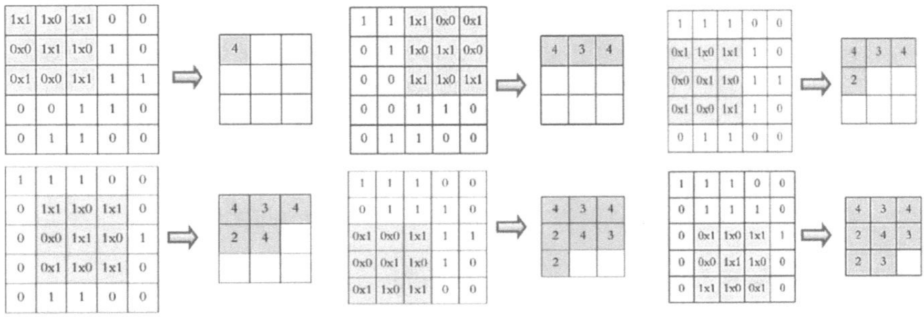

Fig. 1 Convolution operation

$$net(x, y) = \sum_{m=1}^{k} \sum_{x=1}^{P} I(m, n)w(x - m, y - n) \quad (1)$$

where:

net: feature matrix (map).
x: row of the feature matrix (map).
y: column of the feature matrix (map).
I: input image.
w: filter matrix.
1 m: filter matrix row.
n: filter matrix column.
k: filter row size.
p: filter column size.

When adding more layers, it is possible that each layer can be associated with different filters. Therefore, we can extract different features from the given image. Each layer has its own filter and therefore extracts different features from the input, as is shown in Fig. 2, the matrix can be set to detect edges in the image (Kokkinos n.d.). These matrices are also called a filter or kernel because they act like the classic filters in image processing. However, on Neural Networks these filters are initialized, followed by the training procedure shape filters, which are more suitable for the given task. Usually, the weights used in kernels of the convolutional layer are initialized with random functions (Albawi 2017). Therefore, it is intended that, after several iterations, the weights are adjusted until the desired result is reached (Ozbulak 2018).

2.2 *Edge Detectors*

Edge detectors are an important tool for image processing and have been very useful for several areas, such as extraction and feature detection tasks. There are many approaches to perform edge detection; one of the most common can be grouped into

Operation	Filter	Convolved image
Identity	$\begin{bmatrix} 0 & 0 & 0 \\ 0 & 1 & 0 \\ 0 & 0 & 0 \end{bmatrix}$	
Edge detection	$\begin{bmatrix} -1 & 0 & 1 \\ -2 & 0 & 2 \\ -1 & 0 & 1 \end{bmatrix}$	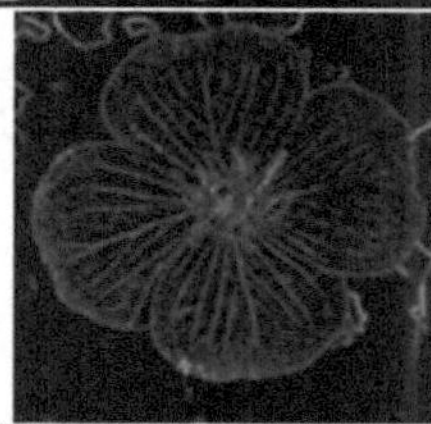
	$\begin{bmatrix} -1 & 0 & 1 \\ -1 & 0 & 1 \\ -1 & 0 & 1 \end{bmatrix}$	
	$\begin{bmatrix} 0 & 1 & 0 \\ 1 & -4 & 1 \\ 0 & 1 & 0 \end{bmatrix}$	
Blur	$\frac{1}{9}\begin{bmatrix} 1 & 1 & 1 \\ 1 & 1 & 1 \\ 1 & 1 & 1 \end{bmatrix}$	
Sharpen	$\frac{1}{10}\begin{bmatrix} 0 & -1 & 0 \\ -1 & 5 & -1 \\ 0 & -1 & 0 \end{bmatrix}$	

Fig. 2 Effects of using a different convolution matrix

two categories, gradient and Laplacian methods. The gradient based method detects edges looking for the maximum and the minimum in the first image derivative, which consists of emphasizing with gray tones pixels that are different from their neighbors, as in Sobel (1970) and Prewitt (1970). Laplacian-based methods look for a zero crossing of the second derivative of the image, in areas where the image has constant intensities, the operator response will be zero, such as the Laplacian operator (Wang 2007). In this section, some of these methods are explained as they are used for this study.

2.3 Sobel Operator

The Sobel operator is a very popular algorithm in gradient-based techniques that use neighborhood scanning, which in gray-scale images works by defining two kernels, usually, 3 × 3, which convolute the image producing a feature map. Kernels are defined as *Sobelx* and *Sobely* in their horizontal Eq. (2) and vertical positions in Eq. (3) (Gonzalez 2017).

$$Sobelx = \begin{bmatrix} -1 & -2 & -1 \\ 0 & 0 & 0 \\ 1 & 2 & 1 \end{bmatrix} \tag{2}$$

$$Sobely = \begin{bmatrix} -1 & 0 & 1 \\ -2 & 0 & 2 \\ -1 & 0 & 1 \end{bmatrix} \tag{3}$$

Consequently, the filtering calculation carried out is defined by the following equation, based on a 3 × 3 filter as shown in Eqs. (4) and (5).

$$gx = \sum_{i=1}^{\mathrm{i}=3} \sum_{j=1}^{\mathrm{j}=3} Sobelx_{ij} \cdot f(x + i - 2, y + j - 2) \tag{4}$$

$$gy = \sum_{i=1}^{\mathrm{i}=3} \sum_{j=1}^{\mathrm{j}=3} Sobely_{ij} \cdot f(x + i - 2, y + j - 2) \tag{5}$$

2.4 Prewitt Operator

In addition to locating or improving the pixels that belong to the edges, the Prewitt operator performs a smoothing operation, which makes the filter more robust in relation to the image noise. In general, it uses a matrix filter with 3×3 coefficients, which facilitates the possibility of configuring it in such a manner that the filter is not so vulnerable to image noise (Prewitt 1970).

This operator uses two kernels defined in Eqs. (6) and (7) in horizontal and vertical positions respectively (Gonzalez 2017).

$$Prewittx = \begin{bmatrix} -1 & -1 & -1 \\ 0 & 0 & 0 \\ 1 & 1 & 1 \end{bmatrix} \tag{6}$$

$$Prewitty = \begin{bmatrix} -1 & 0 & 1 \\ -1 & 0 & 1 \\ -1 & 0 & 1 \end{bmatrix} \tag{7}$$

2.5 Laplacian Operator

This type of filter is based on a derivative operator; the result of this operation accentuates the areas that have a large discontinuity in the image, highlighting mainly edges, discontinuities, and fine details. The derivative function depends on two variables, the derivative of the second order with respect to the variable x and the derivative regarding the variable y. The Laplacian operator (Wang 2007) is defined by Eq. (8); therefore the previous equation is equivalent to applying a kernel defined in Eq. (9).

$$g(x, y) = \sum_{s=-a}^{a} \sum_{t=-b}^{b} w(s, t) \cdot f(x + s, y + t) \tag{8}$$

$$Laplacian = \begin{bmatrix} 0 & 1 & 0 \\ 1 & -4 & 1 \\ 0 & 1 & 0 \end{bmatrix} \tag{9}$$

3 Proposed Approach

This study explores the impact of the correct initialization of the weights and the variation of the number of convolution layers in the classification results of a CNN, in order to help the network to achieve an optimal convergence. This Section presents the architectures, parameters, and the processes that are used for this approach.

3.1 Proposed Architecture

The proposed approach for the analysis and modification of parameters of a Convolutional Neural Network consists of modifying the following parameters: number of convolution layers, the initialization of the convolution kernel weights (Sobel operator, Prewitt operator and Laplacian Filter), and the number of filters. The implemented architectures are represented in Table 1, in which the variation of the convolution layers numbers can be observed. A total of six architectures are proposed, the general architecture implemented can be found in Fig. 3, and the parameters used are defined in Table 2.

3.2 Convolution Kernel Initialization

LeCun et al. (1998a), proposed a training method of deep neural network using gradient based optimization, specifically, in stochastic gradient descent it requires that the weights of the network be initialized to small random values close to zero [0.0, 0.1]. Randomness is also used in the search process for the combination of the training data set (Sandve 2013). The training algorithms are commonly performed iteratively, so they require the user to specify some starting point from which to start the iterations, training the Deep Learning models represents a difficult task and most algorithms are strongly affected by initialization, the aim in our approach is to initialize weights statically without that randomness (Xu 2016).

The initialization of the weights is divided into 5 approaches to observe the behavior of the data from after applied applying the proposal.

- The first is to initialize the architecture with the Sobel or Prewitt kernels in their horizontal position (one architecture for Sobel another for Prewitt). The initialization of the weights is illustrated in Fig. 4.
- The second approach consists of initializing the architecture with the kernels in the horizontal direction and subsequently, the next Convolution layer is initialized with the filters in the vertical direction, as illustrated in Figs. 4 and 5.
- The third approach consists in using the complete Sobel and Prewitt operator in the vertical and horizontal direction contained in a convolution layer.
- In the fourth approach, the architectures initialize the weights with the Laplacian kernel which is shown in Fig. 6.
- Finally, the proposal is also performed by using a "Truncated Normal" random function (Gulli 2017) that randomly initializes the weights.

Table 1 Six proposals of architectures for CNN

1 Layer	2 Layers	3 Layers	4 Layers	5 Layers	6 Layers
1 Conv. + ReLu 1 Pooling layer 1 Fully-Connected layer	2 Conv. + ReLu 1 Pooling layer 1 Fully-Connected layer	3 Conv. + ReLu 1 Pooling layer 1 Conv. + ReLu 1 Pooling layer 1 Fully-Connected layer	3 Conv. + ReLu 1 Pooling layer 1 Conv.+ ReLu 1 Pooling layer 1 Fully-Connected layer	3 Conv. + ReLu 1Pooling layer 2 Conv.+ ReLu 1 Pooling layer 1 Fully-Connected layer	3 Conv. + ReLu 1 Pooling layer 3 Conv.+ ReLu 1 Pooling layer 1 Fully-Connected layer

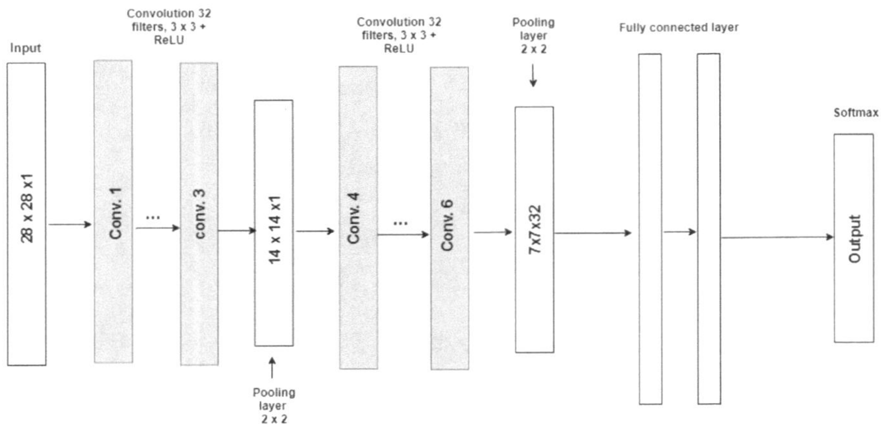

Fig. 3 General architecture of CNN with 6 convolution layers

Table 2 Parameters used in CNN architectures

Learning function	Filters numbers	Output function	Convolutional layer number	Kernel type	Batch size	Kernel size	Pooling layer size	Epoch
Adadelta	32	Softmax	1–6	Sobel Prewitt Laplacian	128	3 × 3	2 × 2	10–100

Horizontal Sobel kernel

1	2	1
0	0	0
-1	-2	-1

Horizontal Prewitt kernel

1	1	1
0	0	0
-1	-1	-1

Fig. 4 Sobel and Prewitt kernels in the horizontal direction

Vertical Sobel kernel

-1	0	1
-2	0	2
-1	0	1

Vertical Prewitt kernel

-1	0	1
-1	0	1
-1	0	1

Fig. 5 Sobel and Prewitt kernels in the vertical direction

Fig. 6 Laplacian kernel

0	1	0
1	-4	1
0	1	0

4 Experiments

This Section presents the results obtained by the experiments applied to all the CNN models described in Sect. 3, explaining what each of the experiments consisted and how they were developed.

4.1 Case Study MNIST Handwritten Digits

In the first case study, the tests were performed with the kernels of Fig. 4 and using the parameters presented in Table 2. For this experiment, the MNIST Handwritten Digits database (LeCun et al. 1998b) was selected, which contains a total of 70,000 images, 45,000 for training, 10,000 for testing and, 15,000 for validation. An example of this database is illustrated in Fig. 7.

The experiments were performed 15 times for each type of architecture described in Table 3, after obtaining 24 Tables of simulation results; these are summarized in Table 6. Tables 4 and 5 present the mean and best results after applying the Sobel, Prewitt in their horizontal positions, the Laplacian kernel and, the Random filter. According to Table 6, the mean reached by the Prewitt kernel was better than Sobel with a mean value of 91.03%; in the other hand, Laplacian kernel improves the results obtained by Prewitt and Sobel with a mean of 94.53%; but after applying the random weights, the best classification accuracy was reached with a mean of 98.83%.

The following experiments were performed using both Sobel and Prewitt kernels in the horizontal and vertical directions; in addition, the general parameters of the CNN were established with the values described in Table 2. In this test, one convolutional layer is initialized with the horizontal kernels followed by a layer with the vertical kernels as shown in Table 7.

Table 8 shows the results obtained with the 5 different architectures with each of the filters, it is observed that the results improve considerably using both kernels instead of using only one. The results are good with the Sobel and Prewitt operators; however, they do not exceed the results compared with the random filter.

Once the experimentation has been carried out, the results are summarized in Table 8, where we can notice that the best architecture was obtained with the Random filter applied in 6 convolution layers with a mean value of 98.83%.

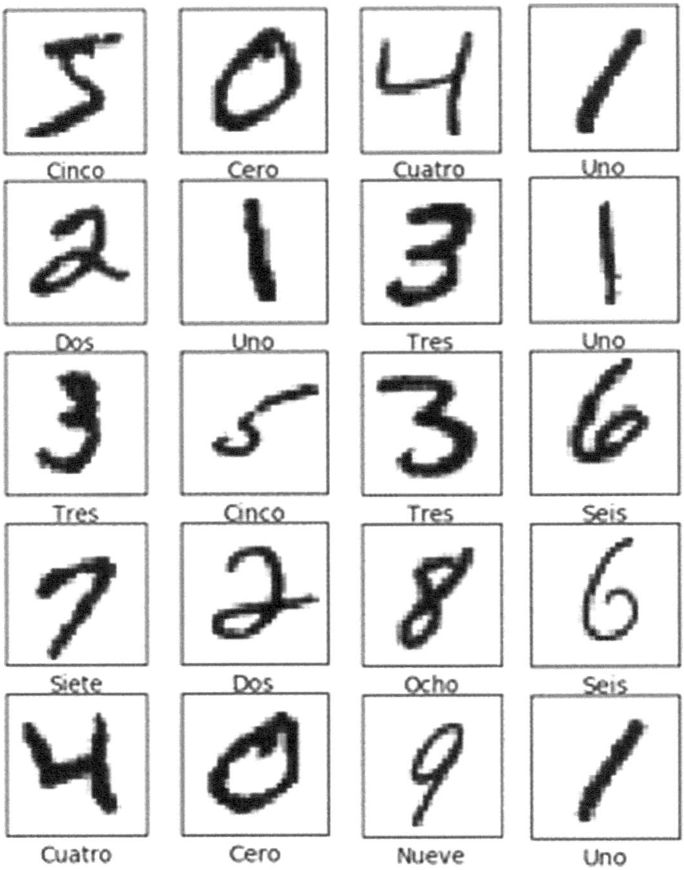

Fig. 7 MNIST handwritten Digits database

In another test, these parameters are taken to create a new architecture, but now the number of filters increases, and the proposed architecture is illustrated in Fig. 8. According to the results presented in Table 9, we can observe that the mean value is 99.27% and this improves the accuracy presented in Table 6 with a value reached of 98.83%.

4.2 *Case Study MNIST American Sign Language*

In a second case study, the MNIST American Sign Language Database was selected, previous works have already used this database (Zhao 2018; Rathi 2018), which has a total of 34,627 images; 27,455 of these were used for training, 7,172 for

Table 3 Architecture used with the Laplacian filter, Sobel and Prewitt in the horizontal direction

Number of Convolutional Layers				
2 Layers	3 Layers	4 Layers	5 Layers	6 Layers
1 Conv. Random filters	1 Conv random filters	1 Conv. Random filters	1 Conv. Random filters	1 Conv. Random filters
1 Conv. Laplacian/Sobel/Prewitt Horizontal direction	1 Conv. Laplacian/Sobel/Prewitt Horizontal direction	1 Conv. Laplacian/Sobel/Prewitt Horizontal direction	1 Conv. Laplacian/Sobel/Prewitt Horizontal direction	1 Conv. Laplacian/Sobel/Prewitt Horizontal direction
1 Capa Pooling	1 Conv. Laplacian/Sobel/Prewitt Horizontal direction Vertical	1 Conv. Laplacian/Sobel/Prewitt Horizontal direction	1 Conv. Laplacian/Sobel/Prewitt Horizontal direction	1 Conv. Laplacian/Sobel/Prewitt Horizontal direction
1 Fully-connect layer	1 Pooling Layer	1 Pooling Layer	1 Pooling Layer	1 Pooling Layer
	1 Fully-connect layer	1 Conv. Laplacian/Sobel/Prewitt Horizontal direction	1 Conv. Laplacian/Sobel/Prewitt Horizontal direction	1 Conv. Laplacian/Sobel/Prewitt Horizontal direction
		1 Pooling Layer	1 Conv. Laplacian/Sobel/Prewitt Horizontal direction	1 Conv. Laplacian/Sobel/Prewitt Horizontal direction
		1 Fully-connect layer	1 Pooling Layer	1 Conv. Laplacian/Sobel/Prewitt Horizontal direction
			1 Fully-connect layer	1 Pooling Layer
				1 Fully-connect layer

Table 4 Results obtained with the Sobel and Prewitt kernel in their horizontal position

Training number	% Accuracy	
	Sobel (%)	Prewitt (%)
1	94.24	94.78
2	84.4	94.15
3	94.12	94.44
4	93.45	86.62
5	90.12	94.06
6	85.23	91.75
7	88.90	94.99
8	84.44	95.12
9	94.20	86.0
10	94.06	87.45
11	89.56	88.69
12	85.23	90.50
13	93.50	93.63
14	90.16	84.12
15	92.54	89.14
Mean	**90.28**	**91.03**
Best	**94.24**	**94.78**

Table 5 Results obtained using the Laplacian kernel and the Random filter

Training number	% Accuracy	
	Laplacian (%)	Random filter (%)
1	94.96	97.52
2	94.11	97.72
3	94.04	97.69
4	94.34	97.75
5	94.96	97.50
6	94.97	97.57
7	94.88	97.29
8	95.42	97.61
9	94.30	98.04
10	94.12	97.62
11	93.94	97.52
12	93.70	97.66
13	95.41	97.02
14	94.27	97.70
15	94.35	97.61
Mean	**94.53**	**97.59**
Best	**95.41**	**98.04**

Table 6 Summary of 15 experiments applied for each kernel and architectures

Kernel Types	Number of convolution layers											
	1 Layer		2 Layers		3 Layers		4 Layers		5 Layers		6 Layers	
	% Mean	% Best	% Mean	% Best	% Mean	% Best	% Mean	% Best	% Mean	% Best	% Mean	% Best
Sobel	90.28	94.24	76.27	85.55	37.68	51.5	16.37	21.68	11.52	15.9	9.93	19.79
Prewitt	91.03	94.78	86.33	92.78	54.4	70.2	30.78	38.80	17.46	23.07	15.77	21.66
Laplacian	94.53	95.41	84.23	90.32	71.36	86.1	24.72	32.20	13.05	18.20	10.26	13.06
Random	97.59	98.04	97.81	98.21	98.64	98.80	98.40	98.91	98.58	98.80	**98.83**	**99.00**

Table 7 Architectures used for experiments with Sobel, Prewitt, Laplacian and Random filters

2 Convolutional layers	3 Convolutional layers	4 Convolutional layers	5 Convolutional layers	6 Convolutional layers
1 Conv Layer + Relu	1 Conv Layer + Relu	1 Conv Layer + Relu	1 Conv Layer + Relu	1 Conv Layer + Relu
1 Laplacian Filter Layer/Horizontal Sobel/Prewitt	1 Laplacian Filter Layer/Horizontal Sobel/Prewitt	1 Laplacian Filter Layer/Horizontal Sobel/Prewitt	1 Laplacian Filter Layer/Horizontal Sobel/Prewitt	1 Laplacian Filter Layer/Horizontal Sobel/Prewitt
1 Pooling Layer	1 Pooling Layer	1 Laplaciano Filter Layer/Vertical Sobel/Prewitt	1 Pooling Layer	1 Laplacian Filter Layer/Vertical Sobel/Prewitt
1 Fully-connect Layer	1 Laplacian Filter Layer/Vertical Sobel/Prewitt	1 Pooling Layer	1 Laplacian Filter Layer/Vertical Sobel/Prewitt	1 Pooling Layer
	1 Fully-connect Layer	1 Laplacian Filter Layer/Horizontal Sobel/Prewitt	1 Laplacian Filter Layer/Horizontal Sobel/Prewitt	1 Laplacian Filter Layer/Horizontal Sobel/Prewitt
		1 Pooling Layer	1 Laplacian Filter Layer/Vertical Sobel/Prewitt	1 Laplacian Filter Layer/Vertical Sobel/Prewitt
		1 Fully-connect Layer	1 Pooling Layer	1 Laplacian Filter layer/Horizontal Sobel/Prewitt
			1 Fully-connect layer	1 Pooling Layer
				1 Fully-connected layer

Table 8 The summary of experiments applied to the different architectures and filter types

Filter type	Number of convolution layer									
	2 Layers		3 Layers		4 Layers		5 Layers		6 Layers	
	% Mean	% Best	% Mean	% Best	% Mean	% Best	% Mean	% Best	% Mean	% Best
Sobel	96.62	97.17	96.64	97.28	83.79	96.59	87.43	94.59	32.82	92.65
Prewitt	93.24	94.27	97.30	94.70	94.18	95.01	93.35	94.63	87.51	94.48
Laplacian	92.70	94.11	76.44	93.87	10.60	11.39	10.59	11.35	10.21	11.35
Random	97.81	98.29	98.64	98.80	98.40	98.91	98.58	98.83	**98.83**	**99.00**

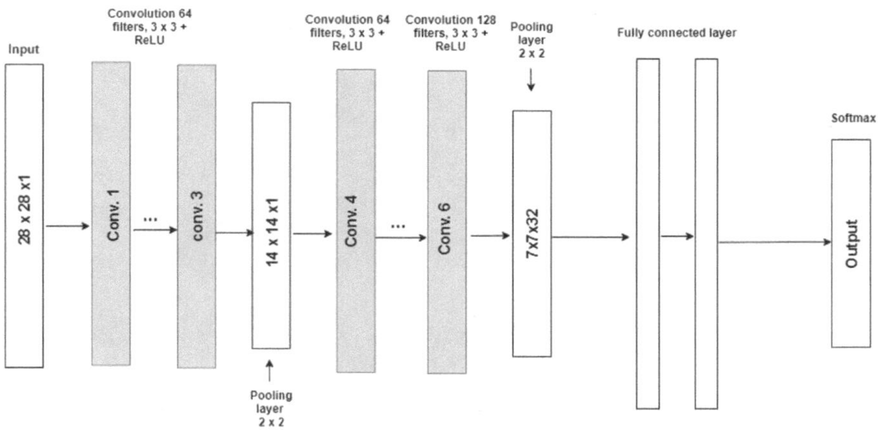

Fig. 8 Model of a CNN using the Random filter, 64 filters in the convolution layers (1–6) and 128 in the last convolution layer

Table 9 The results of the best architecture increasing the filters number

Training number	% Accuracy (%)
1	99.22
2	99.07
3	99.23
4	99.18
5	99.32
6	99.19
7	99.23
8	99.37
9	99.40
10	98.90
11	99.46
12	99.40
13	99.30
14	99.34
15	99.44
Mean	**99.27**
Best	**99.44**

validation, and 7,172 for tests, an example of this database is illustrated in Fig. 9 The experimental results were obtained with the same architectures shown in Table 3 and with the same parameters presented in Table 2. As shown in Table 10, the results obtained with the MNIST ASL (American Sign Language) database are good but we can notice that for the filters Sobel and Prewitt the results decrease when the

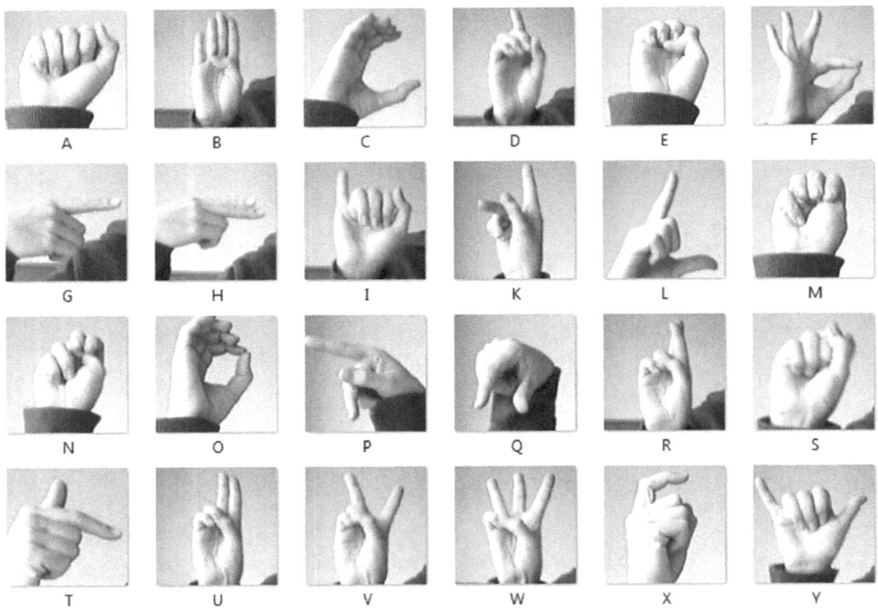

Fig. 9 MNIST American sign language database

Table 10 Simulation results using six different architectures and the filters Sobel and Prewitt with both kernels (horizontal and vertical)

Filter type	Number of Convolutional layers									
	2 Layers		3 Layers		4 Layers		5 Layers		6 Layers	
	% Mean	% Best	% Mean	% Best	% Mean	% Mean	% Best	% Mean	% Best	% Mean
Sobel	75.89	82.17	66.51	69.00	68.80	73.74	58.13	66.18	56.71	63.20
Prewitt	75.74	82.93	64.18	67.49	66.01	69.67	59.08	61.16	57.60	63.74
Random	79.21	83.95	76.96	81.56	71.09	80.20	85.10	86.32	**89.25**	**90.35**

convolutional layer is increasing; in the other hand, we can observe that the behavior of the random filter is very good increasing more layers.

Finally, the following experiments were performed using a complete layer of Prewitt and Sobel with both kernels (horizontal and vertical) as can be seen in Eqs. (2–3) and (6, 7). In this test, the Laplacian filter has been discarded due to better results with the other filters types are obtained; the Sobel and Prewitt filters are the first layer of the architecture and successively it is combined with the random filters to give greater robustness. Based on previous experimentation, we can observe that adding more pooling layers can contribute to a significant impact on recognition, which is why 10 different architectures were designed and these are presented in Tables 11 and 12, where we can notice that the number of convolution and pooling layers

Table 11 Part 1 of proposal architectures increasing the number of poo ling layers

2 layers, 2 Pooling	2 layers, 3 Pooling	3 layers, 2 Pooling	3 layers, 3 Pooling	4 layers, 2 Pooling
1 Sobel/Prewitt Layer	1 Sobel/Prewitt Layer	1 Sobel/Prewitt Layer	1 Sobel/Prewitt Layer	1 Sobel/Prewitt Layer
1. Pooling Layer	1 Pooling Layer	1 conv. +Relu layer	1 Pooling layer	1Conv +Relu layer
1 Conv +Relu	1 Conv + Relu	1 Pooling layer	1 Conv + Relu	1 Pooling layer
1 Pooling layer	1 Pooling layer	1 Conv +Relu	1 Pooling layer	1 Conv + Relu
1.Fully-connect layer	1 Pooling Layer	1 Pooling Layer	1 Conv +Relu	1 Conv + Relu
	1 Fully-connected layer	1 Fully-connected layer	1 Pooling layer	1 Pooling layer
			1 Fully-connected layer	Fully-connected layer

Table 12 Part 2 of proposal architectures increasing the number of poo ling layers

4 layers, 3Pooling	5 layers, 3 Pooling	5 layers, 4 Pooling	6 layers, 3 Pooling	6 layers, 4 Pooling
1 Sobel/Prewitt Layer	1 Sobel/Prewitt Layer	1 Sobel/Prewitt Layer	1 Sobel/Prewitt Layer	1 Sobel/Prewitt Layer
1 Pooling layer	1 Conv +Relu	1 Conv +Relu	1 Conv +Relu	1 Conv +Relu
1 Conv +Relu	1 Pooling layer	1 Pooling layer	1 Pooling layer	1 Pooling layer
1 Pooling layer	1 Conv +Relu	1 Conv +Relu	1 Conv +Relu	1 Conv +Relu
1 Conv + Relu	1 Conv + Relu	1 Pooling layer	1 Conv +Relu	1 Conv +Relu
1 Pooling layer	1 Pooling layer	1 Conv + Relu	1 Pooling layer	1 Pooling layer
1 Conv + Relu	1 Conv + Relu	1 Pooling layer	1 Conv + Relu	1 Conv + Relu
1 Pooling layer	1 Capa Pooling	1 Capa conv + Relu	1 Conv + Relu	1 Pooling layer
1 Fully-connect layer	1 Fully-connect layer	1 Pooling layer	1 Pooling layer	1 Conv + Relu
		1 Fully-connect layer	1 Fully-connect layer	1 Pooling layer
				1 Fully-connect layer

are increased. After implementing the 10 architectures, the experimental results are presented in Table 13 and we can notice that Sobel achieved better classification accuracy than Prewitt with a mean value of 94.23% for the architecture of 4-convolutional layers and 3-Pooling layers and for the same architecture, the random filter obtained a mean value of 95.26%, which was better than Sobel and Prewitt. Moreover, by adding more layers the random filter reached better classification results with a mean value of 95.66%, which increases the accuracy, but the architecture requires more computational times.

Table 13 Results obtained by applying Sobel and Prewitt as a first convolution layer

Filter type	Architectures									
	2 Conv. 2 Pooling		3 Conv. 2 Pooling		4 Conv. 3 Pooling		5 Conv. 3 Pooling		6 Conv. 3 Pooling	
	% Mean	% Best	% Mean	% Best	% Mean	% Best	% Mean	% Best	% Mean	% Best
Sobel	93.41	93.94	93.12	93.52	94.23	95.56	90.53	91.29	93.25	94.11
Random	93.31	94.01	94.61	95.10	95.26	96.00	95.10	95.73	**95.66**	**96.29**
Prewitt	92.00	92.5	90.87	91.54	93.93	94.36	93.74	94.52	93.48	94.13

4.3 Case Study Mexican Sign Language Database

The Mexican Sign Language (MS) Database was created from images of MSL students and teachers, this database has 10 images per letter of the alphabet, in total there are 21 characters without movement, as shown in Fig. 10 and as a result 210 images were obtained per person, 18 different people were photographed, in total 3,780 images were obtained for the database. The data is split in 80% for training and 20% for testing, having a total of 3,024 images for training and 756 for testing. Images include noise by slightly rotating the position of the hand sign. The format for the database is.CSV file to speed up the computation process.

The final experiments were performed using an architecture of 4-convolution layers and 3-pooling layers which is described in Table 12, this architecture was selected because good results were obtained and it does not need too many computational resources, the difference in results is minimal compared with the architecture of 6-convolution layers and 3-pooling layers. The experiments were carried out by increasing the number of epochs to 50, using the same parameters of the Table 2. Finally, the results were compared against other neural network techniques, as can

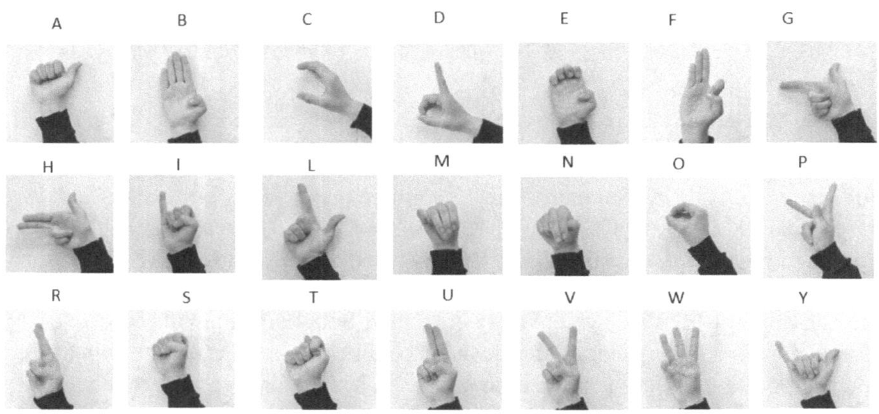

Fig. 10 Mexican sign language database

Table 14 Best architecture obtained versus traditional neural network methods

Best Architecture obtained versus others Neural network techniques	Validation accuracy % (average)
4-Convolution layers (random filter) + 3-Pooling layer	99.08
4-Convolución layers + 3-Pooling layer using Cross-validation	97.53
Sobel layer + 4-Convolution Layers + 3-Pooling layers	99.10
Sobel layer + Traditional Neural Network methods + 4 hidden layers con 128 neurons c/u	95.04

be seen in Table 14, the best results were obtained using the Sobel architecture layer following by 4-Convolutional layers and 3-Pooling layers with a 99.10% of classification. This means that for this study case, the Sobel operator helps to the CNN to better convergence. Furthermore, in Table 14 we can observe that the proposed approach was better than using a traditional neural network, in whose architecture the Sobel operator was also implemented as a preprocessing phase for all the input images.

5 Conclusions

In summary, according to the results presented using the Sobel and Prewitt as static filters in the implemented convolution layer, the CNN models have very good results. However, when compared with the random filters, which initialize the weights randomly and not act statically, these last methods helps to obtain better results because it represents a powerful method for the extraction of characteristics in images.

According to the results of Tables 9 and 13, the use of random filters reached a mean of 99.27% in the classification of the MNIST Handwritten Digits database and a mean of 95.66% for the MNIST ASL data set; then we can infer that the initialization of the weights in the convolution greatly influences the results obtained.

In fact, the main contribution of this paper is to demonstrate the impact of changing the parameters of the Convolutional Neural Network on the accuracy of the classification. The presented study reveals the importance of choosing the filter number, the convolution layer number, and the appropriate kernel initializer when a CNN is training.

In general, we can conclude that using static filters such as Sobel, Prewitt, and Laplacian in the convolution layers, good results are obtained; however, the method is affected by increasing the number of layers. Random filters provide the possibility of exploring a greater characteristics number; nevertheless, for the study case of the MSL database (Table 14) combining two types of filters, we can obtain good results. In this specific case, when the Sobel filter was used with the traditional method of Convolutional Neural Networks, the results were good, achieving a classification accuracy of 99.10%. As future work, we could consider applying multiple objective optimization to the neural networks, like in Sánchez and Melin (2014), Sánchez et al. (2017), Sánchez (2017), or other kinds of applications (Castillo 1998; Castillo and Melin 2003; Sanchez et al. 2014).

Acknowledgements We thank the Tijuana Institute of Technology, and the financial support provided by our sponsor CONACYT contract grant number: 701173.

References

Albawi, S., T. Mohammed, and S. Al-Zawi. 2017. Understanding of a convolutional neural network. In *International Conference on Engineering and Technology (ICET).*

Bantupalli, k., and Y. Xie. 2018. American sign language recognition using deep learning and computer vision. *2018 IEEE International Conference on Big Data (Big Data)*, Seattle, WA, USA, 2018, pp. 4896–4899. https://doi.org/10.1109/bigdata.2018.8622141.

Bin, L.Y., G.Y. Huann, and L.K. Yun. 2019. Study of convolutional neural network in recognizing static american sign language. In *2019 IEEE international conference on signal and image processing applications (ICSIPA)*, pp. 41–45.

Castillo, O., and P. Melin. 1998. A new fuzzy-fractal-genetic method for automated mathematical modelling and simulation of robotic dynamic systems. *1998 IEEE International Conference on Fuzzy Systems (FUZZ-IEEE 1998) Proceedings*. Volume 2, 1182–1187.

Castillo, O., and P. Melin. 2003. Intelligent adaptive model-based control of robotic dynamic systems with a hybrid fuzzy-neural approach. *Applied Soft Computing* 3 (4): 363–378.

El-Sayed, A., A.Y. Estaitia, and M. Khafagy. 1998. Automated edge detection using convolutional neural network. In *International Journal of Advanced Computer Science and Applications*, vol. 86, issue 11.

Gonzalez, C.I., P. Melin, J. Castro, and O. Castillo. 2017a. Edge detection methods based on generalized type-2 fuzzy logic. In *SpringerBriefs in Applied Sciences and Technology.*

Gonzalez, C.I., P. Melin, J.R. Castro, and O. Castillo. 2017b. Edge detection methods and filters used on digital image processing. *Edge Detection Methods Based on Generalized Type-2 Fuzzy Logic*, Published by Springer, pp. 11–16.

Gonzalez, C., P. Melin, J. Castro, O. Mendoza, and O. Castillo. General type-2 fuzzy edge detection in the preprocessing of a face recognition system. In *Applied soft computing*, vol. 9, issue 4, pp. 3–18.

Gulli, A., and S. Pal. 2017. Deep Learning with Keras, Published by Packt Publishing Ltd.

Heaton, I., J.I. Goodfellow, Y. Bengio, and A. Courville. 2017. Deep learning. *Genetic Programming and Evolvable Machines*, 19(1–2): 305–307.

INEGI, "Discapacidad", Inegi.org.mx, 2020. [Online]. Available: https://www.inegi.org.mx/temas/discapacidad/. Accessed 03 Aug 2020.

Kim, P. 2017. *MATLAB deep learning: With machine learning*. Seoul, Korea: Neural Networks and Artificial Intelligence. Apress.

Krizhevsky, A., I. Sutskever, and G. Hinton. 2017. ImageNet classification with deep convolutional neural networks. *Communications of the ACM* 60 (6): 84–90.

Kokkinos, I., E.C. Paris, and G. Group, Introduction to deep learning convolutional networks. *Dropout, Maxout* 1, pp. 1–70.

LeCun, Y., L. Bottou, Y. Bengio, and P. Haffner. 1998a. Gradient-based learning applied to document recognition. *Proceedings of the IEEE* 86: 2278–2324.

LeCun, Y., L. Bottou, Y. Bengio, and P. Haffner. 1998b. Gradient-based learning applied to document recognition. *Proceedings of the IEEE* 86 (11): 2278–2324.

Mendoza, O., P. Melín, and O. Castillo. 2009. Interval type-2 fuzzy logic and modular neural networks for face recognition applications. *Applied Soft Computing* 9 (4): 1377–1387.

Ozbulak, G., and H. Ekenel. 2018. Initialization of convolutional neural networks by Gabor filters. In *2018 26th Signal Processing and Communications Applications Conference (SIU).*

Poma, Y., P. Melin, C. González, and G.E. Martinez. 2020a. Optimal recognition model based on convolutional neural networks and fuzzy gravitational search algorithm method. *Hybrid intelligent systems in control, pattern recognition and medicine 2020*, vol. 827, pp. 71–81.

Poma, Y., P. Melin, C.I. González, and G.E. Martinez. 2020b. Filter size optimization on a convolutional neural network using FGSA. *Intuitionistic and Type-2 Fuzzy Logic Enhancements in Neural and Optimization Algorithms 2020*, pp. 391–403.

Poma, Y., P.Melin, C.I. González, and G.E. Martinez. 2020c. Optimization of convolutional neural networks using the fuzzy gravitational search algorithm. *Journal of Automation, Mobile Robotics & Intelligent Systems* 14(1).

Prewitt, J.M.S. 1970. Object enhancement and extraction. In *Picture analysis and psychopictorics*, ed. B.S. Lipkin and A. Rosenfeld, 75–149. New York, NY: Academic Press.

Rathi, D. 2018. Optimization of transfer learning for sign language recognition targeting mobile platform. *International Journal on Recent and Innovation Trends in Computing and Communication*, 6.

Sánchez, D., P. Melin, J. Carpio, and H. Puga. 2017. Comparison of optimization techniques for modular neural networks applied to human recognition. In *Nature-inspired design of hybrid intelligent systems*, pp. 225–241. Springer, Cham.

Sánchez, D., and P. Melin. 2014. Optimization of modular granular neural networks using hierarchical genetic algorithms for human recognition using the ear biometric measure. *Engineering Applications of Artificial Intelligence* 27: 41–56.

Sanchez, M.A., O. Castillo, J.R. Castro, and P. Melin. 2014. Fuzzy granular gravitational clustering algorithm for multivariate data. *Information Sciences* 279: 498–511.

Sánchez, D., P. Melin, and O. Castillo. 2017b. Optimization of modular granular neural networks using a firefly algorithm for human recognition. *Engineering Applications of Artificial Intelligence* 64: 172–186.

Sandve, G.K., A. Nekrutenko, J. Taylor, and E. Hovig. 2013. Ten simple rules for reproducible computational research. *PLoS Comput Biol* 9(10).

Sinha, T., B. Verma, & A. Haidar. 2017. Optimization of convolutional neural network parameters for image classification. In *IEEE Symposium Series on Computational Intelligence (SSCI).*

Sobel, I. 1970. Camera models and perception. Ph.D. thesis, Stanford University, Stanford, CA.

Szegedy, C., W. Liu, Y. Jia, P. Sermanet, S. Reed, D. Anguelov, D. Erhan, V. Vanhoucke, and A. Rabinovich. 2015. Going deeper with convolutions. In *The IEEE conference on computer vision and pattern recognition (CVPR)*, pp. 1–9.

Wang, X. 2007. Laplacian operator-based edge detectors. *IEEE Transactions on Pattern Analysis and Machine Intelligence* 29 (5): 886–890.

Xie, B., X. He, and Y. Li. 2018. GB-D static gesture recognition based on convolutional neural network. *The Journal of Engineering* 2018(16): 1515–1520. https://doi.org/10.1049/joe.2018.8327.

Xu, X., H. Ge, and S. Li. 2016. An improvement on recurrent neural network by combining convolution neural network and a simple initialization of the weights. *2016 IEEE International Conference of Online Analysis and Computing Science (ICOACS), Chongqing*, pp. 150–154. https://doi.org/10.1109/ICOACS.2016.7563068.

Zhao, Y., and L. Wang. 2018. The application of convolution neural networks in sign language recognition. *2018 ninth international conference on intelligent control and information processing (ICICIP)*, Wanzhou, pp. 269–272, https://doi.org/10.1109/icicip.2018.8606707.

Parameter Optimization of a Convolutional Neural Network Using Particle Swarm Optimization

Jonathan Fregoso, Claudia I. Gonzalez, and Gabriela E. Martinez

Abstract In this paper, parameter optimization of Convolutional Neural Network architectures is proposed by applying the Particle Swarm Optimization algorithm, where the parameters that are optimized are: the number of layers, number of convolution filters, the filter size, and the batch size. The optimized architecture is applied in two sign language databases, the American Sign Language Alphabet and American Sign Language MNIST. This research aims to analyze the performance of the proposed architecture, focusing on obtaining a better recognition rate of the signs to achieve an increase in the recognition and computational performance of the Convolutional Neural Network architecture and subsequently propose tools for assisted communication for the deaf community. The achieved results indicate that the proposed method is effective since a recognition rate above 99% was obtained in both study cases, achieving a reduction in processing times and computational cost.

Keywords Convolutional neural networks · Particle swarm optimization · Image recognition · Sign language recognition

1 Introduction

Convolutional Neural Networks (CNN) are a type of Artificial Neural Network architecture based on supervised learning, they have proven to be a very powerful tool of Deep Learning because they are an interesting adaptive method for the processing, classification, and recognition of patterns in images or videos. Besides this, they can be used in different application areas thanks to the implementation of the artificial

J. Fregoso · C. I. Gonzalez (✉) · G. E. Martinez
Tijuana Institute of Technology, Tijuana, BC, Mexico
e-mail: cgonzalez@tectijuana.edu.mx

J. Fregoso
e-mail: jonathan.fregoso@tectijuana.edu.mx

G. E. Martinez
e-mail: gmartinez@tectijuana.edu.mx

O. Castillo and P. Melin (eds.), *Fuzzy Logic Hybrid Extensions of Neural and Optimization Algorithms: Theory and Applications*, Studies in Computational Intelligence 940,
https://doi.org/10.1007/978-3-030-68776-2_9

vision for solving different problems in areas such as natural language processing, aerospace, robotics, and medicine; the latter being the preferred area to propose the solution to different health and social security problems (Goodfellow et al. 2016). A CNN requires high computational processing costs for information processing; various researchers have been dedicated their investigations to the development of new computational techniques to increase CNN performance, successfully testing different hybridization methods that together with CNN, they managed to improve the performance and the percentage of recognition of the same. Therefore, if the problem to be solved requires high processing costs, it is recommended to hybridize techniques attached to intelligent computing. Otherwise, it is not recommended to apply these techniques because they will increase the difficulty when solving the problem (Hemanth et al. 2020).

Among the most relevant techniques that offer greater performance when optimizing the parameters of a CNN is the application of metaheuristic techniques derived from evolutionary computation and fuzzy logic (Gaxiola et al. 2016). In recent years the optimization of parameters through the application of the Particle Swarm Optimization (PSO) algorithm for neural networks has shown favorable results, which means that it is applied in the optimization of different types of architectures focused on the solution of different problems (Gonzalez et al. 2019; Varela-Santos and Melin 2020), and thanks to this versatility can find studies dedicated to the optimization of modular neural networks to obtain the blood pressure trend (Miramontes et al. 2020). As well as different monolithic neural network models applied to the forecast of electricity prices for energy exchange (Peter and Reglend 2017). Thanks to the wide performance offered by PSO optimization in current times, it is a fundamental part of CNN optimization, taking as a clear example various studies focused on the recognition and extraction of object characteristics where the optimization of parameters is conditioned by complexity (Sánchez et al. 2020). Another problem is where a dynamic adjustment is made in the search spaces depending on the percentage of recognition obtained (Fernandes and Yen 2019).

In the area of fuzzy logic, there are techniques and algorithms such as the Fuzzy Gravitational Search Algorithm (FGSA) that are being used by researchers as another alternative for optimizing CNN architectures. FGSA is based on the calculation of the law of Newton's gravity to calculate distances and positions of the agents of the algorithm and thus achieve the readjustment of parameters of the CNN architecture, although each problem is different, the techniques can be applied to the vast majority of problems taking into account some will offer better results than others depending on the application area (Poma et al. 2020a, b).

The main contribution of this paper is to optimize the parameters of the architecture of a Convolutional Neural Network through the application of the PSO algorithm. The parameters to optimize are the number of layers, number of convolution filters, the filter size and, the batch size. Designing parameters for a CNN is a difficult task; this is usually obtained based on an expert or after several tests. The aim of this paper is that the PSO helps to find a better parameter design and to improve the performance of a CNN. The optimized architecture is applied in two databases that include the American Sign Language Alphabet and American Sign Language MNIST.

The paper is structured as follows: the general theory and operation of CNN and PSO are introduced in Sect. 2 and Sect. 3 respectively. The general methodology to develop the optimization process is described and explained in Sect. 4. Section 5 presents the experiments performed and the analysis of results obtained after applying the optimized architecture for the two databases: the American Sign Language Alphabet and American Sign Language MNIST. Also in this section, a comparison results is presented against other works focused on image recognition using CNN. Finally, Sect. 6 explains the conclusions and some future works.

2 Convolutional Neural Network

A Convolutional Neural Network (CNN) is a type of Artificial Neural Network architecture based on supervised learning, they are specifically determined to work with image and/or video processing, aiming at the extraction of patterns and/or characteristics in the input data. This type of network imitates the biological functioning of the visual cortex of the human being (Brain and Eyes) (Hubel and Wiesel 1959), so it focuses on detecting characteristics such as lines, curved shades of color together with relevant patterns in the images to carry out an identification process and classification of objects (Kim 2017; Cheng et al. 2018).

The architecture is made up of deep hierarchical layers, which are the following: the input layer, convolution layer, non-linearity layer, pooling layer, and the classifying layer. With this, it is proposed that each one of the layers carry out a specific process within the learning process of the network. Figure 1 represents the basic architecture of a Convolutional Neural Network (Zou et al. 2016; Huang et al. 2019).

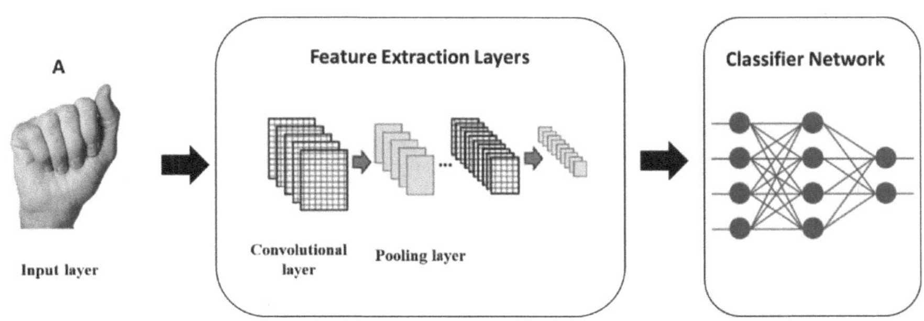

Fig. 1 The basic architecture of a convolutional neural network

2.1 Input Layer

This is the first layer that makes up the network architecture, in this layer the images or videos that will be used for the extraction of characteristics, training, and validation in the network are entered. In general, the information entered is two-dimensional matrices that represent the composition of the images to be processed. It is highly recommended that the images have a pre-processing this to reduce the computational cost of the algorithms and increase the effectiveness, so it is recommended to normalize the pixel values, segment the images, use gray-scale in the images and even perform the extraction of background to keep only the relevant information in the image (Xianwei and Mingzhou 2019).

2.2 Convolution Layer

It consists of the fundamental layer in this neural network architecture, in this layer convolution filters or kernels are applied to the input images to generate multiple images of the same called feature maps, these feature maps are different and they depend on the convolution filter applied to the image. There are different types of convolution filters that focus on extracting different characteristics from the image, they can be lines, curves, shades, colors, etc. The process consists of moving the filter from left to right throughout the image, operating the pixel-by-pixel values of the filter with those of the image, as shown in Fig. 2 (Xianwei et al. 2019).

Figure 3 shows the convolution process where the * inside the circle denotes the convolution operation and the symbol φ represents the activation function. While the gray-scale square icons between the operators indicate the filters applied. It is important to emphasize that the same amount of characteristic maps are generated according to the number of filters applied (Engelbrecht 2007a).

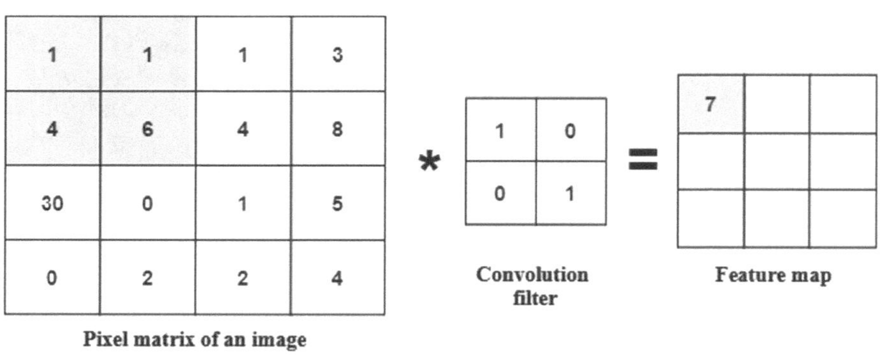

Fig. 2 Convolution process to generate feature maps

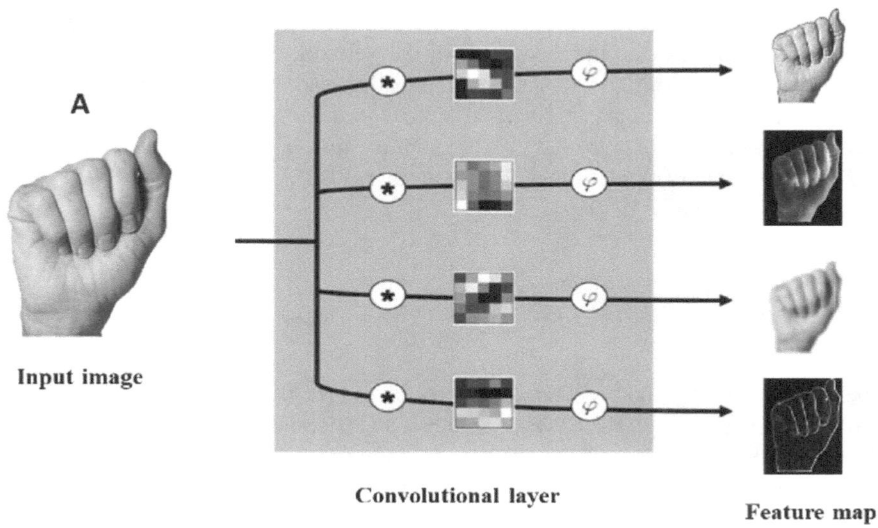

Fig. 3 Convolution process

The convolution process consists of a mathematical operation in which two sets of information are merged when operating on each other, as shown in Eq. (1) (Kim 2017).

$$S(x, y) = \sum_{m=1}^{k} \sum_{n=1}^{p} I(m, n)w(x - m, y - n) \quad (1)$$

where

S: is the feature map.
x: is the row of the feature map.
y: is the column of the feature map.
I: is the input image.
w: is the filter matrix.
m: is the row of the filter matrix.
n: is the column of the filter matrix.
k: is the size of the filter matrix row.
p: is the size of the filter matrix column.

2.3 Non-linearity Layer

It is in charge of activating the neurons within the network through the analysis of the feature map previously extracted in the convolution layer, where if the value in the map position is ≤ 0, this information is discarded from the image and, if it is >0 the information passes to the next layer of the network. The Rectified Linear Unit (ReLU) function is commonly used. The main advantage is computational efficiency thanks to its speed of convergence, while the significant disadvantage occurs when the values of the feature map tend to zero and this causes a decrease in the adjustment and training capacity of the model. Some variants of ReLU are LeakyReLU, PReLU, ELU, etc. (Fukushima 1980).

2.4 Pooling Layer

The operation performed in this layer aims to reduce the image dimension while maintaining the characteristics with greater relevance, thus reducing the processing load. To make the pooling, it is necessary to combine neighboring pixels in a single representative value, have to select the kernel size and the type of pooling that you want to carry out (Schmidhuber 2015).

Pooling types are as follows:

- Maximum (Selection of the pixel with the largest value).
- Average (Arithmetic mean of the values).
- Sum (Sum of values).

An example of how pooling is performed in a 4 (rows) $\times$ 4 (columns) feature map and a 2 $\times$ 2 kernel is represented in Fig. 4, whereby the input feature map is divided into 4 segments of 2 $\times$ 2 and each of these segments is obtained the pixel value according to the type of pooling selected. The kernel will be superimposed on

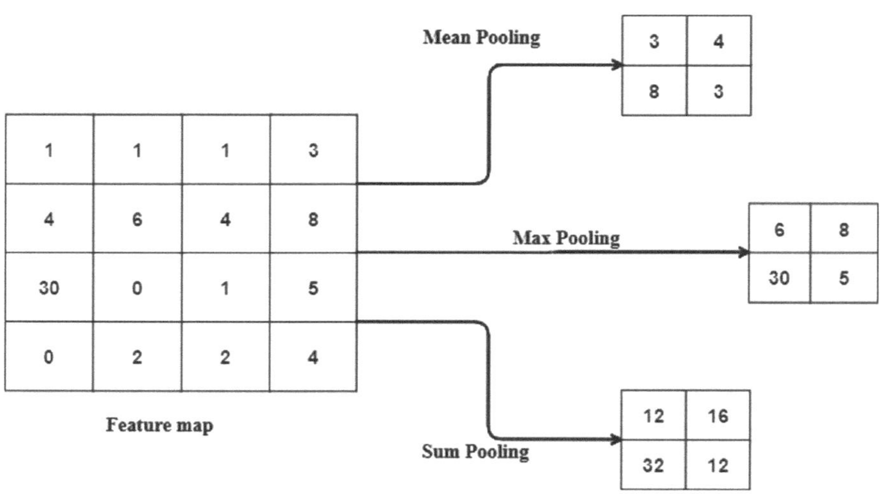

Fig. 4 Mean, max and sum pooling examples

the input data and then carry out a similar operation to the convolution layer it is possible to reduce the size of the image, managing to maintain the characteristics with greater relevance.

2.5 *Classifier Layer*

This is the layer in charge of classifying the images, it consists of the decision making of the network to determine to which class an image belongs employing the analysis of the result of the grouping layer (Pooling), the high-level characteristics are identified for the later map it to a particular class. For multi-class classification, the neural network has the same number of outputs as classes. The *softmax* activation function is used, which generates a probability distribution vector for the output values (Aggarwal 2018).

3 Particle Swarm Optimization

The population-based algorithm Particle Swarm Optimization (PSO) consists of a metaheuristic optimization method aimed at finding global minimums or maximums within objective functions. It is an operation inspired by the behavior of flocks of birds where the social and cognitive-communication factor directly influences the food search process when communicating with each other to find the best route to the search for food. In this algorithm each individual (bird) is represented by a direction, speed, and acceleration, the movement of individuals is the result of combining individual decisions of each with the behavior of the rest (Engelbrecht 2007b; Yannakakis and Togelius 2015).

PSO allows optimizing a problem from a population of candidate solutions called particles, which are transferred to hyper-dimensional search spaces in which the global minimum and maximum search is performed to achieve the optimal solution to the problem, the changes in the position of the particles is influenced by its best local position so far, as well as by the best global positions found by other particles, as it travels through a search space, aiming to achieve rapid convergence towards the best possible solutions. The representation of a swarm is similar to a population, while a particle is similar to an individual. The process is simple, the particles move through a multidimensional search space, where the position of each particle is adjusted according to their own experience and that of their neighbors in such a way that $x_i(t)$ is represented as the position of the particle i within the search space, in time step t; otherwise, t represents discrete time steps. The position of the particle is changed by adding a velocity $v_i(t)$, for the current position. Equation (2) represents the position of the particle (Xiaojing et al. 2019; Fielding and Zhang 2018).

$$x_i(t+1) = x_i(t) + v_i(t+1) \tag{2}$$

3.1 Global Best PSO

For the best global PSO solution, the entire swarm is taken as the sample where the social component of the speed update represents the information obtained from the total particles in the swarm, in this case, the social information represents the best position found by the swarm, and is described as $\hat{y}(t)$. For GBEST PSO, the velocity of particle i is calculated in Eq. (3).

$$v_{ij}(t+1) = v_{ij}(t) + c_1 r_{1j}(t)\left[y_{ij}(t) - x_{ij}(t)\right] + c_2 r_{2j}(t)\left[\hat{y}_j(t) - x_{ij}(t)\right] \quad (3)$$

where $v_{ij}(t)$ represents the velocity of particle i in dimension j at time step t, $x_{ij}(t)$ is the position of particle i in dimension j at time step t, c_1 and c_2 are positive acceleration constants used to scale the contribution of the cognitive and social components respectively and $r_{1j}(t)$, $r_{2j}(t)$ are random values in the range [0, 1], sampled from a uniform distributions (Engelbrecht 2007b; Adeleh et al. 2017).

3.2 Local Best PSO

For the best local PSO solution, smaller sample groups are defined for each particle. The social component reflects information exchanged within the vicinity of the particle, reflecting local knowledge of the environment (Engelbrecht 2007b; Giaquinto and Fornarelli 2009). Concerning the velocity equation, the social contribution to the velocity of the particle is proportional to the distance between a particle and the best position found by the vicinity of the particles. Speed is calculated in Eq. (4).

$$v_{ij}(t+1) = v_{ij}(t) + c_1 r_{1j}(t)\left[y_{ij}(t) - x_{ij}(t)\right] + c_2 r_{2j}(t)\left[\hat{y}_{ij}(t) - x_{ij}(t)\right] \quad (4)$$

where $\hat{y}_{ij}$ is the best position, found by the neighborhood of particle i in dimension j the local best particle position $\hat{y}_i$.

4 Proposed Method

The proposed method consists of optimizing the parameters of a CNN architecture by applying the PSO algorithm. From the technical point of view, the aim is to determine the parameters that have the greatest significance on the performance of the architecture, then apply the PSO method to optimize them, and find the ideal parameters that allow us to obtain the highest percentage of recognition in the analyzed images.

The general architecture of the proposed method is illustrated in Fig. 5, which

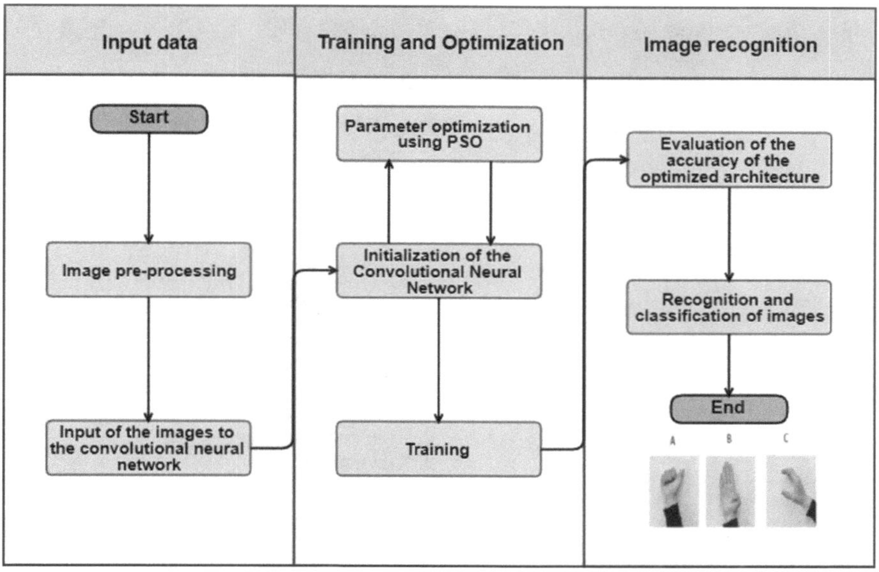

Fig. 5 CNN optimization using PSO algorithm

consists of three processes. The first process is the input data that will feed the Convolutional Neural Network, in the second, these images will reach the training and optimization process, this being the essential point when looking for the best parameters for the network using PSO and finally evaluating the accuracy of the optimized architecture.

4.1 Parameter Optimization of the CNN

The application of Convolutional Neural Networks to solve problems that require the use of artificial vision is a watershed in the innovation of processes in industries and everyday life, although it offers great benefits, its main disadvantage is that they require high processing computational costs, so it becomes necessary to use techniques to increase their performance. Therefore, an optimization of the architecture parameters is proposed in order to reduce computational costs and execution times, ensuring that the recognition percentage is optimal. A summary of some parameters of a CNN that can be optimized is represented in Table 1.

After carrying out an exploratory study (explained in Sect. 5.1), it was decided to optimize the following four parameters, since these represent a high impact on the final recognition rate.

1. The number of layers: Number of convolution and grouping layers.

Table 1 Parameters per layer of a CNN

Layer name	Parameter
Convolution	1. The number of filters per convolution layer 2. Size of the mask or filter 3. Lot size or batch size 4. The number of convolution layers 5. Type of filter to use 6. Type of filling 7. Stride or step size
Non-linearity	1. Type of activation function
Pooling	1. Size of the pooling window 2. Type of pooling 3. The number of pooling layers 4. Stride or step size
Classifier	1. The number of layers to use 2. The number of neurons per layer 3. The number of epochs 4. Type of activation function

2. The number of convolution filters: Number of filters used for the extraction of characteristics.
3. Filter Dimensions: Represents the filter size.
4. Batch size: Represents the number o f images per block that enter the network.

For the correct operation of the PSO algorithm, it is necessary to design the architecture of the particle used by the algorithm. Tables 2, 3, and 4 detail the proposed particle.

Table 2 presents the design of the proposed particle consisting of four positions, where each position represents a parameter to be optimized. Position 1 corresponds to the number of layers in the architecture. Position 2 and 3 represent the number of convolution filters and the dimension of the convolution filter respectively. Finally, position 4 indicates the batch size to be used.

Table 2 The proposed particle design used in the PSO

Position	1	2	3	4
Particle	No. layers	No. filters	Convolutional filter dimension	Batch size

Table 3 Details of the search spaces of the particle used

Particle	No. layers	No. filters	Convolutional filter dimension	Batch size
Min/Max value	[1–3]	[32–128]	[1–4]	[32–256]

Table 4 Convolutional filter dimensions

Filter dimension	[3, 3] [5, 5] [7, 7] [9, 9]

Table 3 represents the search spaces used for each parameter where a lower limit and an upper limit are specified, so each position of the particle must be kept within this range of values for experimentation.

Table 4 denotes the dimensions of the convolution filters, each filter size is associated with a position in the search space, so position 1 within the search space corresponds to the filter [3, 3], position 2 to the filter [5, 5], and so on.

4.2 CNN-PSO Optimization Process

The parameter optimization process consists of giving inputs to the image database to be used on CNN, afterward, the particle populations of the PSO algorithm are generated according to the design of the proposed particle (Table 2), the parameters of the CNN architecture are dynamic and are taken from the positions of the generated particles. The CNN architecture is initialized and the training and validation process begins, the process is repeated, evaluating all the particles until the stop criteria is found (in this case, it is the number of iterations). Finally, the algorithm selects the optimal solution that consists of the particle with which the highest recognition percentage was obtained. The optimization process is illustrated in Fig. 6.

5 Experiments and Results

In this section, we present the experiments carried out and the results obtained in the different case studies. To compare the performance and the ability of the method to adapt to solving different problems, two datasets were included in the tests: the American Sign Language Alphabet (ASL Alphabet) (American Sign Language dataset 2018), and the American Sign Language MNIST (ASL MNIST) (Sign Language MNIST 2017). The characteristics of the databases are found in Tables 5 and 6 respectively.

Figure 7 illustrates a sample of the elements that make up the ASL Alphabet database.

Figure 8 shows a sample of the elements that make up the ASL MNIST database.

5.1 Exploratory Experiment

An exploratory study was performed in the ASL Alphabet database to determine which parameters had greater relevance on the recognition percentage and at the same time analyzing the performance of the CNN architecture without optimization;

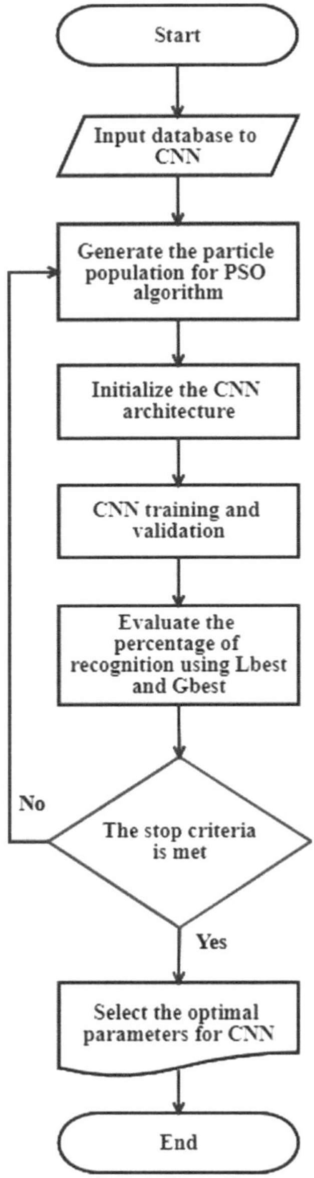

Fig. 6 Flowchart of the CNN-PSO optimization process

the results obtained were favorable and helped to establish the convenient parameters to optimize in the proposed method.

The study consisted of performing 10 experiments where the learning function, the nonlinearity activation function, and the activation function of the classifying

Table 5 American Sign Language Alphabet (ASL Alphabet) database description

Data base	
Name	American Sign Language Alphabet (ASL Alphabet)
Total images	87,000
Training images	82,650
Test images	4350
Image size	200 × 200
Data base format	JPGE
Representation	26 classes are for letters and 3 more for space, delete and nothing

Table 6 American Sign Language MNIST (ASL MNIST) database description

Data base	
Name	American Sign Language MNIST (ASL MNIST)
Total images	34,627
Training images	27,455
Test images	7172
Image size	28 × 28
Data base format	CSV
Representation	24 classes, sign letters without movement

layer were static parameters, as shown in Table 7. The manually adjusted parameters were the number of convolution layers, the number of filters, filter size, the batch size, and the training epochs. The experimental results are shown in Table 8, and we can appreciate that the best recognition percentage achieved has a value of 99.95%, the worst recognition with 88.53%, and the mean of 96.64%.

5.2 *American Sign Language Alphabet (ASL Alphabet) Experiment*

This study consisted of the execution of 30 experiments of the proposed optimized architecture to find the optimal configuration of parameters that would allow obtaining the best percentage of recognition possible. In this study, static parameters were handled in the CNN architecture, such as the learning function, the nonlinearity activation function, the activation function of the classifier layer, and the epochs, while in the PSO optimization algorithm the static parameters are, the quantity of particles, number of iterations, the weight of inertia, the cognitive and social constants. This is shown in Tables 9 and 10 respectively.

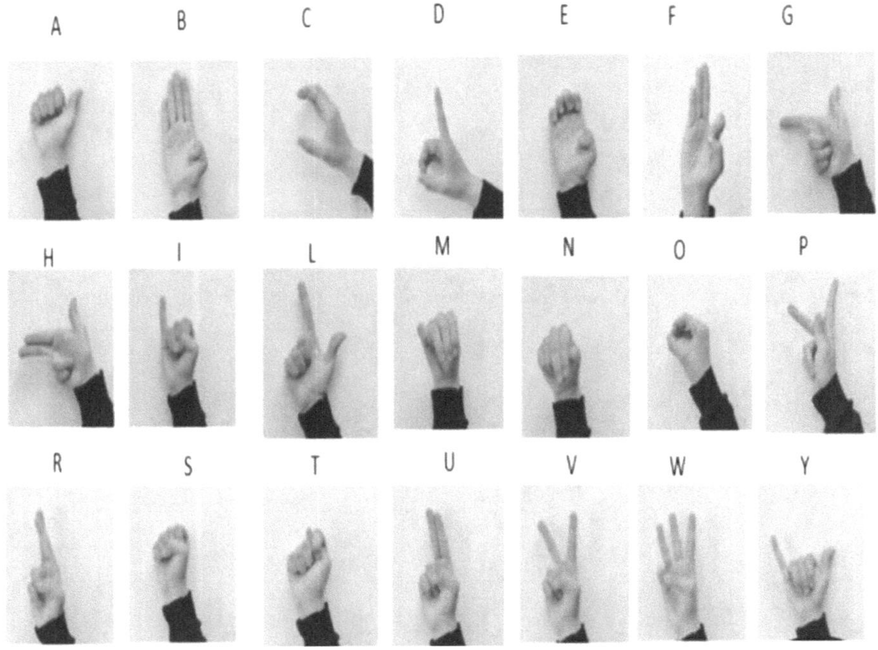

Fig. 7 Example of the ASL Alphabet database

Regarding the parameters optimized in this study, they are the number of convolution layers, number of filters, the filter size, and the batch size, as shown in Table 2. The analysis of the results obtained was performed, achieving the best recognition percentage of 99.87%, the worst with 98.18%, and maintaining an average of 99.58%. The results obtained are shown in Table 11.

5.3 American Sign Language MNIST Experiment

In this case study, 30 experiments were performed using the same static and dynamic parameters presented in Tables 9 and 10. According to the analysis of the results shown in Table 12, the best percentage of recognition is 99.98%; the worst is 98.82% and maintains an average of 99.53% of recognition.

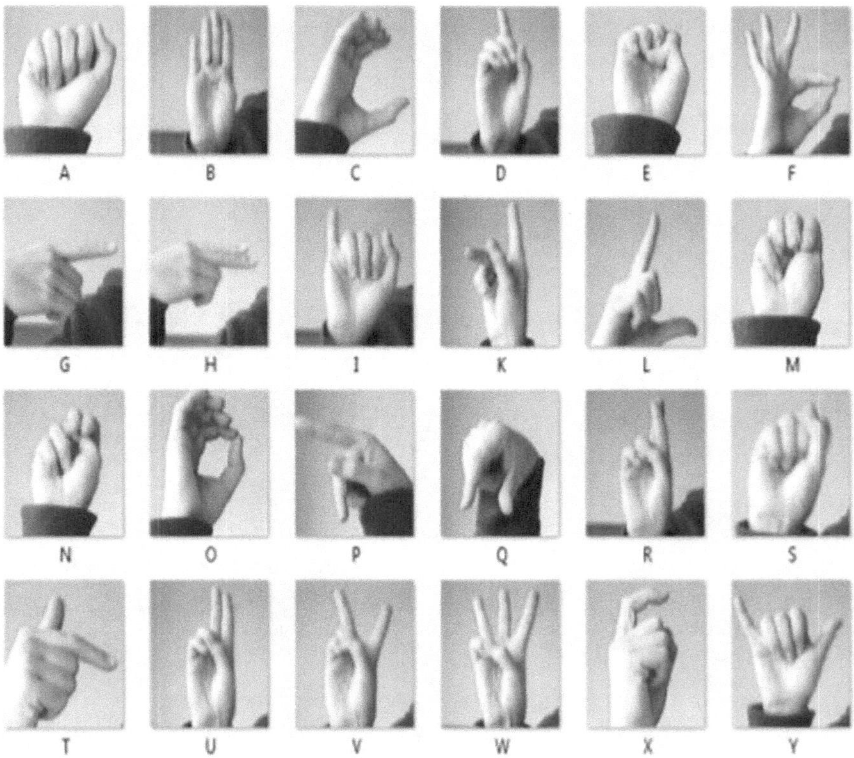

Fig. 8 Example of the ASL MNIST database

Table 7 Static parameters of the exploratory study

Learning function	Nonlinearity activation function	Activation function of the classifying layer	Filter size
Adam optimizer (RMSprop and stochastic gradient descent with momentum)	ReLU	Softmax	[3 × 3]

5.4 Analysis and Comparison of Results

The results obtained for the American Sign Language MNIST database were as follows: after 30 executions, an average recognition rate of 99.53% was achieved, while the best recognition rate was of 99.98% with an architecture consisting of 2-layers convolutions and pooling, 117 filters with a dimension of [7 × 7], and a value of 129 in the batch size. Otherwise, for the American Sign Language Alphabet database, the average recognition rate was 99.58% and the best result reached is

Table 8 Results of the exploratory study with manually adjusted parameters

No. experiment	Convolutional layers	No. filters	Batch size	Epochs	Recognition (%)
1	1	32	64	5	91.63
2	1	32	64	10	88.53
3	2	32	64	5	93.75
4	2	32	64	10	97.40
5	3	32	64	5	97.68
6	3	32	64	10	99.24
7	3	64	128	5	99.43
8	3	64	128	10	99.15
9	6	32	64	5	99.72
10	6	64	128	10	**99.95**
				Mean	96.64

Table 9 Static parameters of the CNN—ASL Alphabet experiment

Learning function	Nonlinearity activation function	Activation function of the classifying layer	Epoch
Adam optimizer (RMSprop and Stochastic Gradient Descent with momentum)	ReLU	Softmax	5

Table 10 Static parameters of the PSO algorithm—ASL Alphabet experiment

Particles	Iterations	Inertial weight (W)	Cognitive constant (W1)	Social constant (W2)
10	10	0.85	2	2

99.87% with an architecture of 3-layers convolutions and pooling, 128 filters with a dimension of [7 × 7], and a batch size of 256.

Table 13 shows a summary of the results obtained in the two case studies that were carried out, as well as a comparison with other previous research works that focus on the recognition of American Sign Language. In Bin et al. (2019), the database used was created by themselves and it resembles the characteristics of the ASL MNIST. The database consists of 4800 images with a size of 32 × 32 and this was taken with a mobile device. The architecture that they propose consists of four convolution layers, two pooling layers and a fully connected layer, the architecture consists of two blocks composed of two continuous convolution layers and one pooling layer, finally the fully connected layer. In the research, the authors only present the highest percentage of recognition achieved with a value of 95%; therefore, according to the results of the proposed optimization architecture, our approach was better with a value of 99.87% for the ASL Alphabet database and a value of 99.98 for the ASL MNIST.

Table 11 Results of the ASL Alphabet experiment with parameter optimization

No. experiment	Convolutional layers	No. filters	Filter size	Batch size	Recognition (%)
1	3	99	[7 × 7]	107	98.85
2	3	104	[9 × 9]	256	99.66
3	3	128	[9 × 9]	256	99.70
4	3	128	[7 × 7]	256	99.79
5	3	128	[9 × 9]	256	99.72
6	3	128	[7 × 7]	256	99.62
7	2	32	[7 × 7]	256	98.18
8	3	109	[7 × 7]	256	99.73
9	3	128	[7 × 7]	197	99.75
10	3	128	[7 × 7]	256	99.81
11	3	66	[7 × 7]	181	99.31
12	3	118	[7 × 7]	256	99.87
13	3	128	[9 × 9]	256	99.67
14	3	128	[7 × 7]	256	99.85
15	3	128	[9 × 9]	256	99.61
16	3	128	[9 × 9]	256	99.63
17	3	90	[9 × 9]	256	99.66
18	3	128	[7 × 7]	256	99.82
19	3	128	[7 × 7]	256	99.79
20	3	128	[7 × 7]	256	99.76
21	3	128	[9 × 9]	256	99.68
22	3	128	[9 × 9]	256	99.67
23	3	128	[7 × 7]	256	99.75
24	3	123	[7 × 7]	32	98.38
25	3	128	[9 × 9]	256	99.64
26	3	128	[7 × 7]	256	99.82
27	3	128	[9 × 9]	215	99.56
28	3	128	[7 × 7]	256	**99.87**
29	3	100	[9 × 9]	256	99.64
30	3	128	[7 × 7]	256	99.84
				Mean	99.58

6 Conclusion and Future Work

In this paper, we describe the parameters optimization of the CNN architecture through the implementation of the PSO algorithm for the recognition of American Sign Language. The obtained results indicate that the proposed method is effective since in both study cases carried out, the recognition percentages achieved are

Table 12 Results of the ASL MNIST experiment with parameter optimization

No. experiment	Convolutional layers	No. filters	Filter size	Batch size	Recognition (%)
1	3	128	[9 × 9]	137	99.27
2	2	128	[9 × 9]	218	99.54
3	2	128	[7 × 7]	205	99.52
4	3	128	[7 × 7]	136	99.33
5	2	128	[9 × 9]	232	99.59
6	3	96	[9 × 9]	107	98.82
7	2	118	[7 × 7]	189	99.36
8	2	128	[9 × 9]	256	99.59
9	2	112	[9 × 9]	256	99.49
10	2	128	[9 × 9]	256	99.60
11	2	128	[7 × 7]	256	99.59
12	2	128	[7 × 7]	256	99.61
13	2	128	[9 × 9]	220	99.67
14	2	128	[9 × 9]	256	99.57
15	2	128	[9 × 9]	256	99.51
16	2	128	[7 × 7]	237	99.55
17	2	128	[7 × 7]	256	99.61
18	2	128	[9 × 9]	256	99.58
19	2	128	[9 × 9]	256	99.53
20	2	128	[9 × 9]	256	99.65
21	2	128	[7 × 7]	148	99.42
22	2	128	[9 × 9]	256	99.51
23	2	128	[9 × 9]	215	99.53
24	2	128	[9 × 9]	255	99.56
25	2	128	[9 × 9]	256	99.65
26	2	128	[7 × 7]	256	99.57
27	2	128	[9 × 9]	256	99.53
28	2	117	[7 × 7]	129	**99.98**
29	3	128	[5 × 5]	242	99.87
30	2	128	[7 × 7]	256	99.55
				Mean	99.53

stable, providing a reduction in processing times and computational cost. In summary, for the American Sign Language MNIST database, the results were as follows: an average recognition rate of 99.53%, while the best recognition rate was 99.98%. For the American Sign Language Alphabet database, the average recognition rate was 99.58% and the best result achieved was 99.87%. In comparison with other works

Table 13 Results comparison

Authors	Description	Accuracy (%)
Bin and Huann (2019)	CNN on 24 ASL using phone camera	95
Proposal method	Parameter optimized of a CNN using PSO for ASL Alphabet database	99.87
Proposal method	Parameter optimized of a CNN using PSO for ASL MNIST database	99.98

focused on the recognition of signs, we can affirm that the optimization approach presented in this paper was better. The use of metaheuristics represents a good way to find the parameters of the CNN architecture.

In future work, we will continue with the implementation of the proposed method as a fundamental basis in the approach and development of human–computer iteration tools that allow improving the quality of life of the deaf community through the use of assisted communication tools, the application area is varied. On the one hand, we can start from a system that allows the recognition of signs so that in this way it is possible to establish communication between users and non-users of sign language. As well as in the development of a sign language translator that allows the translation of signs in different languages, managing to communicate the users of sign languages from different countries. In addition, we could consider the optimization with multiple objectives of the neural network, like in (Sánchez and Melin 2014, 2017a, b) or other kinds of applications (Castillo and Melin 1998; Castillo and Melin 2003; Sanchez et al. 2014).

Acknowledgments We thank the Tijuana Institute of Technology, and the financial support provided by our sponsor CONACYT with the scholarship number: 954950.

References

Adeleh, E., D. Vajiheh, B. Ali, & S. Vahid. 2017. Improved particle swarm optimization through orthogonal experimental design. In *2nd conference on swarm intelligence and evolutionary computation (CSIEC2017)*, 153–158.

Aggarwal, C.C. 2018. *Neural networks and deep learning*. New York: Springer.

Bin, L.Y., G.Y. Huann, and L.K. Yun. 2019. Study of convolutional neural network in recognizing static American sign language. In *2019 IEEE International Conference on Signal and Image Processing Applications (ICSIPA)*, 41–45.

Castillo, O., and P. Melin. 1998. A new fuzzy-fractal-genetic method for automated mathematical modelling and simulation of robotic dynamic systems. In *1998 IEEE International Conference on Fuzzy Systems (FUZZ-IEEE 1998) Proceedings*, vol. 2, 1182–1187.

Castillo, O., and P. Melin. 2003. Intelligent adaptive model-based control of robotic dynamic systems with a hybrid fuzzy-neural approach. *Applied Soft Computing* 3 (4): 363–378.

Cheng, J., P.-S. Wang, G. Li, Q.-H. Hu, and H.-Q. Lu. 2018. Recent advances in efficient computation of deep convolutional neural networks. *Frontiers of Information Technology & Electronic Engineering* 19: 64–77.

Engelbrecht, P. 2007a. *Computational intelligence: An introduction*, 2nd ed. Wiley: University of Pretoria, South Africa.
Engelbrecht, P. 2007b. *Computacional intelligence*. South Africa: WILEY.
Fernandes, F.E., and G.G. Yen. 2019. Particle swarm optimization of deep neural networks architectures for image classification. *Swarm and Evolutionary Computation* 49: 62–74.
Fielding, B., and L. Zhang. 2018. Evolving image classification architectures with enhanced particle swarm optimisation. *IEEE Access* 6: 68560–68575.
Fukushima, K. 1980. A self-organizing neural network model for a mechanism of pattern recognition unaffected by shift in position. *Biological Cybernetics* 36: 193–202.
Giaquinto, and G. Fornarelli. 2009. PSO-based cloning template design for CNN associative memories. IEEE Transactions on Neural Networks 20 (11): 1837–1841.
Gaxiola, F., P. Melin, F. Valdez, J.R. Castro, and O. Castillo. 2016. Optimization of type-2 fuzzy weights in backpropagation learning for neural networks using GAs and PSO. *Applied Soft Computing* 38: 860–871.
Gonzalez, B., P. Melin, and F. Valdez. 2019. Particle swarm algorithm for the optimization of modular neural networks in pattern recognition. *Hybrid Intelligent Systems in Control, Pattern Recognition and Medicine* 827: 59–69.
Goodfellow, I., Y. Bengio, and A. Courville. 2016. *Deep learning*. Cambridge: MIT Press.
Hemanth, J.D., O. Deperlioglu, and U. Kose. 2020. An enhanced diabetic retinopathy detection and classification approach using deep convolutional neural network. *Neural Computing and Applications* 32: 707–721.
Huang, J., W. Zhou, H. Li, and W. Li. 2019. Attention-based 3D-CNNs for large-vocabulary sign language recognition. *IEEE Transactions on Circuits and Systems for Video Technology* 29 (9): 2822–2832.
Hubel, D.H., and T.N. Wiesel. 1959. Receptive fields of single neurons in the cat's striate cortex. *The Journal of Physiology* 148: 574–591.
Kaggle. 2017. Sign language MNIST. https://www.kaggle.com/datamunge/sign-language-mnist. Accessed 8 Feb 2020.
Kaggle. 2018. American Sign Language dataset. https://www.kaggle.com/grassknoted/asl-alphabet. Accessed 10 Feb 2020.
Kim, P. 2017. *MATLAB deep learning*. Seoul: Apress.
Miramontes, I., P. Melin, and G. Prado-Arechiga. 2020. Particle swarm optimization of modular neural networks for obtaining the trend of blood pressure. In *Intuitionistic and type-2 fuzzy logic enhancements in neural and optimization algorithms: Theory and applications*, vol. 862, 225–236.
Peter, S.E., and I.J. Reglend. 2017. Sequential wavelet-ANN with embedded ANN-PSO hybrid electricity price forecasting model for Indian energy exchange. *Neural Computing and Applications* 28: 2277–2292.
Poma, Y., P. Melin, C. I. González, and G.E. Martinez. 2020a. Optimal recognition model based on convolutional neural networks and fuzzy gravitational search algorithm method. In *Hybrid intelligent systems in control, pattern recognition and medicine*, vol. 827, 71–81.
Poma, Y., P. Melin, C.I. González, and G.E. Martinez. 2020b. Filter size optimization on a convolutional neural network using FGSA. In *Intuitionistic and type-2 fuzzy logic enhancements in neural and optimization algorithms*, 391–403.
Sánchez, D., and P. Melin. 2014. Optimization of modular granular neural networks using hierarchical genetic algorithms for human recognition using the ear biometric measure. *Engineering Applications of Artificial Intelligence* 27: 41–56.
Sanchez, M.A., O. Castillo, J.R. Castro, and P. Melin. 2014. Fuzzy granular gravitational clustering algorithm for multivariate data. *Information Sciences* 279: 498–511.
Sánchez, D., P. Melin, J. Carpio, and H. Puga. 2017a. Comparison of optimization techniques for modular neural networks applied to human recognition. In *Nature-Inspired Design of Hybrid Intelligent Systems*, 225–241. Cham: Springer.

Sánchez, D., P. Melin, and O. Castillo. 2017b. Optimization of modular granular neural networks using a firefly algorithm for human recognition. *Engineering Applications of Artificial Intelligence* 64: 172–186.

Sánchez, D., P. Melin, and O. Castillo. 2020. Comparison of particle swarm optimization variants with fuzzy dynamic parameter adaptation for modular granular neural networks for human recognition. *Journal of Intelligent & Fuzzy Systems* 38 (3): 3229–3252.

Schmidhuber, J. 2015. Deep learning in neural networks: An overview. *Neural Networks* 61: 85–117.

Varela-Santos, S., and P. Melin. 2020. Classification of x-ray images for pneumonia detection using texture features and neural networks. In *Intuitionistic and type-2 fuzzy logic enhancements in neural and optimization algorithms: Theory and applications*, vol. 862, 237–253.

Xianwei, J., and L. Mingzhou. 2019. *An eight-layer convolutional neural network with stochastic pooling, batch normalization and dropout for fingerspelling recognition of Chinese sign language*, 19. Springer Link.

Xianwei, J., M. Lu, and S.-H. Wang. 2019. *An eight-layer convolutional neural network with stochastic pooling, batch normalization and dropout for fingerspelling recognition of Chinese sign language*, 1–19. Springer Multimedia Tools and Applications.

Xiaojing, Y., J. Qingju, and L. Xinke. 2019. Center particle swarm optimization algorithm. In *2019 IEEE 3rd Information Technology, Networking, Electronic and Automation Control Conference (ITNEC)*, 2084–2087.

Yannakakis, G.N., and J. Togelius. 2015. A panorama of artificial and computational intelligence in games. *IEEE Transactions on Computational Intelligence and AI in Games* 7 (4): 317–335.

Zou, Z., B. Shuai, and G. Wang. 2016. Learning contextual dependence with convolutional hierarchical recurrent neural networks. *IEEE Transactions on Image Processing* 25 (7): 2983–2996.

One-Dimensional Bin Packing Problem: An Experimental Study of Instances Difficulty and Algorithms Performance

Guadalupe Carmona-Arroyo, Jenny Betsabé Vázquez-Aguirre, and Marcela Quiroz-Castellanos

Abstract The one-dimensional Bin Packing Problem (BPP) is one of the best-known optimization problems, and it has a significant number of applications. For this reason, several strategies have been proposed to solve it, but only few works have focused on the study of the characteristics that distinguish the BPP instances and that could affect the performance of the algorithms that solve it. In this work, we present a comprehensive study of the performance of four well-known algorithms on a set of new BPP instances. First, the features of the instances and the performance of the algorithms are quantified by indices; next, an exploratory analysis of the indices is carried out in order to identify the characteristics that define the difficulty of a BPP instance and to understand the performance of the algorithms. The algorithmic behavior explanations obtained by the study suggest that the difficulty of the BPP instances is related to: (1) the bin capacity; (2) the dispersion of the items weights; and (3) the difference between the largest and the smallest items weight.

Keywords Bin packing · Instance difficulty · Algorithmic performance · Characterization

1 Introduction

Throughout the search for the best possible solutions for NP-hard problems, a wide variety of solution procedures have been proposed, however, there is no efficient algorithm capable of finding the best solution for all possible instances of a problem. For the development of better solution methods, a profound comprehension of the

G. Carmona-Arroyo (✉) · J. B. Vázquez-Aguirre · M. Quiroz-Castellanos
Universidad Veracruzana Artificial Intelligence Research Center, Xalapa, Mexico
e-mail: gcarmonaarroyo@gmail.com

J. B. Vázquez-Aguirre
e-mail: jennybey13@gmail.com

M. Quiroz-Castellanos
e-mail: maquiroz@uv.mx

O. Castillo and P. Melin (eds.), *Fuzzy Logic Hybrid Extensions of Neural and Optimization Algorithms: Theory and Applications*, Studies in Computational Intelligence 940,
https://doi.org/10.1007/978-3-030-68776-2_10

properties of problem instances and the performance of the algorithms that solve them is needed.

This work is focused on a classical combinatorial optimization problem: the one-dimensional Bin Packing Problem (BPP). The BPP consists of packing a set of items with a certain size within a set of bins that have the same capacity, the goal is to minimize the number of bins used to store all the items. BPP has been one of the most studied combinatorial optimization problems, and a significant number of strategies have been proposed to solve it, such as exact algorithms, approximate algorithms, heuristics, as well as metaheuristic algorithms.

The main idea of this work arises from the knowledge that, despite the existence of multiple solution techniques for this problem, few investigations have centered on the analysis of the behavior of the algorithms. In other words, good solutions have been obtained with the proposed strategies, but there are not enough studies about why some instances of the problem can be solved more easily than others. To address this issue, this chapter focuses on the characterization of the difficulty of the BPP instances and the relationship of these characteristics with the final result of the algorithms through exploratory analysis processes.

As part of this experimental analysis, three classic heuristics and a state-of-the-art grouping genetic algorithm are implemented to solve this problem on a set of 2800 new instances. This implementation aims to analyze the results obtained, also studying the characteristics of the test instances. For this, the properties of the instances and the performance of the algorithms are quantified using indices. The three classic heuristics are First Fit Decreasing (FFD), Best Fit Decreasing (BFD), and Minimum Bin Slack (MBS). The first two are simple heuristics and are often part of more complex algorithms. Minimum Bin Slack is a sophisticated heuristic that has given good results in experiments, it has inspired new heuristics and local searches. The last algorithm is the Grouping Genetic Algorithm with Controlled Gene Transmission (GGA-CGT), which is one of the best in the literature for BPP. This analysis has allowed us to identify some characteristics that define the difficulty of BPP instances and to understand the performance of the algorithms. Some of the more revealing features are the capacity of the bins and the ranges of the weights of the items. We could observe that the four studied algorithms decrease their effectiveness as the bin capacity increases, which confirms the relationships found in previous studies. Furthermore, we were able to confirm that instances with average weights around 0.3 of the bin capacity seem to be the hardest for simple heuristics. On the other hand, MBS and GGA-CGT were the algorithms with the best results, however, their execution times were affected by the increase in the bin capacity and by the range of the weights vector.

The document continues as follows: Sect. 2 presents the Bin Packing Problem, the indices, and the instances that are used for the proposed analysis. Section 3 describes the algorithms implemented to study their performance. Section 4 presents the results obtained by these algorithms according to their four performance measures. Then, in Sect. 5 the performance analysis of the algorithms is presented. And finally, Sect. 6 presents conclusions and future research directions.

2 The Bin Packing Problem

The one-dimensional Bin Packing Problem (BPP) is a well-known grouping combinatorial optimization problem defined as follows: given an unlimited number of bins with a fixed capacity c and a set of n items, each with a specific weight w_i, BPP aims to find the minimum number of bins needed to pack all the items without violating the capacity of any bin.

Formally, given a set $N = 1, \ldots, n$ of items, a set of bins with the same capacity $c > 0$, and let w_i be the respective weight of each item $i \in N$ with $0 < w_i \leq c$. BPP consists of grouping all the elements in bins in such a way that the sum of their weights does not exceed c, and the number of bins m used is minimal (Martello and Toth 1990). In other words, we want to create a partition of N into the minimum number of subsets B_j possible:

$$\bigcup_{j=1}^{m} B_j = N \tag{1}$$

such that:

$$\sum_{i \in B_j} w_i \leq c, \qquad 1 \leq j \leq m, \qquad i \in N$$

BPP belongs to the NP-hard class (Garey and Johnson 1979; Basse 1998) because it requires a significant amount of computational resources to be solved, and the problem complexity grows exponentially with the problem size since the number of possible partitions is higher than $(n/2)^{n/2}$. This implies there is no efficient algorithm to find an optimal solution for every instance of BPP.

This problem has a significant number of applications in the industry, so it has been highly studied, and multiple algorithms have been developed to solve it, from approximate and exact algorithms to metaheuristics. The performance of the algorithms has been evaluated with different types of published problems. However, for understanding which features of the instances match the strengths and weaknesses of different algorithms, there is much more work to be done.

Table 1 shows the description of the notation used in this work.

2.1 *Instances*

For the BPP study, most algorithms proposed in the literature have been evaluated using a well-studied trial benchmark (Delorme et al. 2018); it includes 1615 instances in which the number of items n varies within [50, 1000], the bin capacity c is within [100, 100000] and the ranges of the weights are within $(0, c]$.

Some researchers have investigated the impact of the weight vector, the bin capacity c, and the number of items n in the hardness of the instances (Schwerin and Wascher 1997; Quiroz 2014). Studies about the effect of the range of the weight

Table 1 Notation

Notation	Description
n	Number of items
c	Bin capacity
opt	Number of bins in the optimal solution
w_i	Weight of the item i ($i = 1, \ldots, n$)
$l(j)$	Accumulated weight in bin j
Z	List of items sorted in decreasing order according w_i
Sol	Solution, i.e., complete assignment of items to bins
m	Number of bins in the solution
$s(j)$	Free space in the bin j, i.e., $c - l(j)$
N	Set of n items
W	Set of n weights

w_i of the items have shown that a smaller range makes the instances more difficult for fast approximation heuristics, which have low performance on instances with weights distributed between 0.15 and 0.5 of the bin capacity c. Other ranges also create difficult instances when the average weight is between 0.3 and 0.5 of the bin capacity c. On the other hand, instances with average weights >0.5 of the bin capacity seem to be the easiest for most of BPP state-of-the-art algorithms. Instances with average weights <0.3 of the bin capacity have also been shown to be easy for algorithms that include random strategies. Another important parameter to consider when studying BPP instances is the bin capacity c, the analysis of the performance of the algorithms have shown that instances characterized by very large capacities are generally difficult to solve. The studies have also pointed out that one of the features that has a greater impact on the BPP instances difficulty is associated with the optimal solutions structure, instances whose optimal solution presents very little free space in bins are very difficult.

To contribute with the study of the features that define the difficulty of a BPP instance, we have performed some experimental comparisons over new instances with different range of the weights w_i and different bin capacities c. All the instances were generated with known optima: in any optimal solution, all the bins must be filled without leaving unused space (perfect packing). We generate 2800 instances, which are referred to as BPPv_u_c, and they are available at https://github.com/gcarmonaar/CO. The weights were generated in the interval $(0, v_u c]$ so that v_u correspond to the upper bound on the weights, relative to the bin capacity c. The instances can be grouped in four classes with $v_u = 0.25$, $v_u = 0.5$, $v_u = 0.75$ and $v_u = 1$, respectively. There are seven sets in each class, each consists of 100 instances with bin capacities $c = 10^2$, $c = 10^3$, $c = 10^4$, $c = 10^5$, $c = 10^6$, $c = 10^7$, and $c = 10^8$, respectively. The first class (BPP.25 class) includes instances where the number of items n varies within [110, 154] and the number of bins in the optimal solution is 15; the second class (BPP.5 class) comprises instances where the number of items n varies within [124, 167] and the number of bins in the optimal solution is 30; the third class (BPP.75 class) contains instances where the number of items n varies within [132, 165] and

Table 2 Description of the instances

Class	#	w_i	c	n	opt
BPP.25	700	$(0, 0.25c]$	$10^2, 10^3, 10^4, 10^5, 10^6, 10^7, 10^8$	[110,154]	15
BPP.5	700	$(0, 0.5c]$		[124,167]	30
BPP.75	700	$(0, 0.75c]$		[132, 165]	45
BPP1	700	$(0, c]$		[148, 188]	60

the number of bins in the optimal solution is 45; and the fourth class (BPP1 class) involves instances where the number of items n varies within [148, 188] and the number of bins in the optimal solution is 60. The difficulty of the proposed instances is caused by the lack of any slack in the optimal solution and by the large capacities of the bins. We summarize this information in Table 2 in which the number of instances per class (#), the distribution of the items weights (w_i), the bin capacities (c), the number of items (n), and the number of bins in the optimal solution (opt) are shown.

2.2 Index Description

Most of the works focused on analyzing the performance of heuristic algorithms are concentrated on comparative analysis of experimental results. However, to understand and explain the behavior of an algorithm, it is necessary to identify and study the factors that affect it, as mentioned in some works that highlight the importance of explaining the reason for the observed results (Chiarandini et al. 2007; Snodgrass and Ergalics 2010; Quiroz 2014).

The difficulty of the BPP instances has been previously studied. Different authors have proposed characterization indices; a study of the BPP structure was presented by Cruz (2004) who formulated a set of indices to measure the characteristics of BPP. On the other hand, Quiroz (2014) presents in her work a set of indices to measure the difficulty of BPP, which allow to identify trends on weights as well as important information on their distribution. Cruz and Quiroz make an extensive study on the selection of the indices that provide the most relevant information of the instances and that present a greater relationship with the performance of the algorithms. We have selected some of them after preliminary tests, and the selected ones are shown in Table 3.

2.3 Performance Measures

In order to measure the algorithms performance, we have evaluated three characteristics for the algorithms.

First, we have calculated the relative difference between the optimal (opt) and the number of bins (m) obtained by an algorithm. This measurement tells us how close to the optimal is a solution found by an algorithm. The above is defined by the Eq. 2.

Table 3 Indices presented by Cruz (2004) and Quiroz (2014)

Equations	Descriptions
$t = \frac{\sum_{i=1}^{n} \frac{w_i}{n}}{c}$	Capacity occupied by an average item
$d = \sqrt{\frac{\sum_{i=1}^{n}[t-(\frac{w_i}{c})]^2}{n}}$	Degree of dispersion of the quotient of the weight of the items and the bin capacity
$f = \frac{\sum_{i=1}^{n} factor(c,w_i)}{n}$	Proportion of items whose weight w_i is a factor of the bin capacity; where $factor(c, w_i)$ is 1 when $(c \bmod w_i) = 0$ and is 0 otherwise
$b = \begin{cases} 1 & \text{if } c \geq \sum_{i=1}^{n} w_i \\ \frac{c}{\sum_{i=1}^{n} w_i} & \text{otherwise} \end{cases}$	Proportion of the total weight than can be assigned in one bin with capacity c
$Range = \max(w_i) - \min(w_i)$	Difference between the largest and the smallest weight
$RC = \frac{Range}{c}$	Range of the weights relative to the bin capacity c
$Major = \frac{\max(w_i)}{c}$	The largest and smallest weight of the instance, respectively, regarding the bin capacity
$Minor = \frac{\min(w_i)}{c}$	
$Multiplicity = \frac{\sum_{i=1}^{n'} m_i}{n'}$	The average number of repetitions for each weight; where m_i is the number of items each with different weight
$Major\_Multiplicity = \max(M)$	Represent the maximum and minimum frequency of weight recurrence for the n items, respectively. Where M saves the frequency for each different weight
$Minor\_Multiplicity = \min(M)$	
$Pct\_\tilde{n} = \frac{\tilde{n}}{n}$	Proportion of large items; where $\tilde{n}$ is the number of items with weight larger than fifty percent of the bin capacity, i.e., $w_i > c/2$
$L2$	Lower bound for BPP solutions (Martello 1990; Martello and Toth 1990)

$$Dif = \frac{m - opt}{opt} \tag{2}$$

Another measurement is made through the cost function introduced by Falkenauer and Delchambre (1992). This function evaluates the average of the squaring of the bin efficiency given by $l(j)/c$, which measures the exploitation of bins capacity. Equation (3) shows this calculation.

$$F_{BPP} = \frac{\sum_{j=1}^{m}(l(j)/c)^2}{m} \tag{3}$$

In addition, we have measured the execution time of the algorithms (*time*). This measurement is given in seconds and the experiments were performed on a computer with an Intel (R) Core (TM) i5 CPU with 2.50 GHz and Microsoft Windows 10.

It is important to note that F_{BPP} and *Dif* are measures of effectiveness since they give us information about the solutions quality and *time* is a measure of efficiency, by providing us with information on the speed of the algorithms.

3 Algorithms

For the solution of BPP, elaborate procedures that incorporate various techniques have been designed. Algorithms proposed in the literature range from simple heuristics to hybrid strategies, including branch and bound techniques (Delorme and Martello 2016), evolutionary algorithms (Quiroz-Castellanos et al. 2015) and special neighborhood searches (Buljubašic and Vasquez 2016).

In the present work, four well-known methods to solve BPP have been analyzed with the new instances. Three of the methods are simple heuristics that have been incorporated into high performance algorithms and the other is one of the best state-of-the-art algorithms. These algorithms are mentioned below.

3.1 First Fit Decreasing (FFD)

First Fit Decreasing (FFD) is a classical heuristic to solve BPP based on the FF strategy (Johnson 1974), it is a simple heuristic that has been used in more complex strategies, such as creating initial solutions in evolutionary algorithms. This algorithm works as follows: before starting to place the items, they are sorted in non-increasing order of their sizes. Then, this heuristic packs each item in the first bin with enough capacity. If no bin is found, FFD creates a new bin to pack the item. This algorithm can be implemented to have a running time of $O(n \log n)$. Algorithm 1 shows FFD process.

Algorithm 1: First Fit Decreasing.

Input: Z,c
Output: m
for *each item* $i \in Z$ **do**
  for *each bin* $j \in Sol$ **do**
    if $w_i + l(j) \leq c$ **then**
      Pack item i in bin j
      Update $l(j)$
      Break for
  if *item i was not packed in any available bin* **then**
    Create a new bin for item i

3.2 Best Fit Decreasing (BFD)

Best Fit Decreasing (BFD) is another of the most popular heuristics to solve BPP (Johnson 1974) and like FFD, it has been used in several more complex algorithms as an aid to skew solutions. It consists of assigning each item to the bin that has the maximum load, where the item fits. If several bins have the maximum load, then the first one found is chosen. In the same way as the previous heuristic, it considers the items in decreasing order of the weights. BFD can be implemented in $O(n \log n)$ time. The above procedure is described in the Algorithm 2.

Algorithm 2: Best Fit Decreasing.

```
Input: Z,c
Output: m
for each item i ∈ l do
    best_bin = ∅
    for each bin j ∈ Z do
        if w_i + l(j) ≤ c and l(j) > l(best_bin) then
            best_bin = j
    if no bin has enough capacity to pack i then
        Create a new bin to pack i
    else
        Pack item i in the fullest bin (best_bin)
        Update l(j)
```

3.3 Minimum Bin Slack (MBS)

Minimum Bin Slack (MBS) was proposed by Gupta and Ho (1999). It is a bin-focused heuristic; at each execution of MBS, an attempt is made to find a set of items that fits the bin capacity as much as possible (Fleszar and Charalambous 2011). Fleszar and Hini (2002) have presented a recursive implementation of MBS. At each step, the items that have not been packed are considered in non-increasing order of their weights, then, this procedure tests all possible subsets of items and the subset whit the least free space is chosen. If the algorithm finds a subset of items whose sum of weights equals the bin capacity, then the search stops. Algorithm 3 shows the procedure to fill a bin, it begins whit $A^* = A = \emptyset$ and $q = 1$, where A^* represents the best subset (according to $l(A)$), A is the auxiliary subset to evaluate. The procedure is repeated removing the chosen items for a bin until the list Z is empty. The complexity of this procedure is $O(2^n)$.

Algorithm 3: Minimum Bin Slack.

Input: $q = 1$
Output: A^*
for $r = q$ *to* n **do**
  Let i be the $r-$th element $\in Z$
  if $w_i \leq s(A)$ **then**
    $A = A \cup \{i\}$
    Apply MBS($r + 1$)
    $A = A \setminus \{i\}$
    if $s(A^*) = 0$ **then**
      End
if $s(A) \leq s(A*)$ **then**
  $A* = A$

3.4 GGA-CGT

The Grouping Genetic Algorithm with Controlled Gene Transmission (GGA-CGT) is a method proposed by Quiroz-Castellanos et al. (2015) to solve BPP. It is one of the best algorithms found in the state-of-the-art to solve this problem; it focuses on the transmission of the best genes on the chromosomes, keeping a balance between selective pressure and diversity in the population, to favor the generation and evolution of high-quality solutions. This algorithm was executed once with the parameters that the author have established. See the procedure in Algorithm 4.

Algorithm 4: GGA-CGT.

Generate an initial population P with FF$-\tilde{n}$
while $generation < max\_gen$ *and Size*($best\_solution > L_2$) **do**
  Select n individuals to cross by means of Controlled_Selection
  Apply Gene_Level_Crossover + FFD to the n_c selected individuals
  Apply Controlled_Replacement to introduce progeny
  Select n_m individuals and clone elite solutions by means of Controlled_Selection
  Apply Adaptive_Mutation + RP to the best n_m individuals
  Apply Controlled_Replacement to introduce clones
  Update the $global\_best\_solution$

4 Results

As we mentioned before, in this study we want to identify the relationships between the characteristics of the BPP instances and the behavior of the algorithms. For this, in Sect. 2.3, we have described the performance measures of the algorithms, which are F_{BPP}, *Dif*, and *time*. Therefore in this section we present the results of the executions of the four algorithms described in the previous section. In addition, the number of optimal solutions obtained by each algorithm is presented here.

In Table 4, we have the number of instances that each algorithm solved optimally. It contains the total number of instances resolved by class and also a general total, that is, how many of the 2800 instances were solved. As we can see, FFD solves a total of 229 instances of the 2800, and BFD solves 234. On the other hand, MBS solves a total of 1337 and finally GGA-CGT solves 1549.

From the table, it can also be observed that between FFD and BFD there is a difference of 5 instances in which the optimal solution was found. In the first class of BPP.25 instances, there are a total of 100 instances (99 from the first set), and for the other classes, the optimal solutions number decreases for both algorithms, with the slight difference that BFD solves five more. It is important to mention that all the instances that FFD solves, BFD does too. For MBS and GGA-CGT a similar behavior is seen for the sets in which the solution was found. It should be noted that in the BPP.25 class, MBS solves more instances, but in the remaining three classes GGA-CGT solves more.

Furthermore, Fig. 1 shows the four graphs of the optimal ones found by each algorithm. On the horizontal axis are the sets, and the vertical axis has the number of instances that were optimally solved. The colors correspond to the 4 main classes of instances. We see that FFD (Fig. 1a) solves instances only for the capacity $c = 100$ in all four classes, in the same way as BFD (Fig. 1b). However, MBS (Fig. 1c) solves multiple instances where the bin capacity is lower, as well as several with the larger values of c. GGA-CGT (Fig. 1d) has similar behavior than MBS, with the difference that in 3 of the 4 main classes this algorithm does not solve several instances with the largest bin capacities.

As we have mentioned before, the relative difference between the number of bins of the solutions found by the four algorithms and the optimal was calculated through Eq. (2). Table 5 shows the total relative difference *Dif* per set, each row has the sum of the relative differences for the 100 instances of the set. Note than a *Dif* value equal to zero, since we are presenting the sum, means that the 100 instances of the set were optimally solved. Thus, at the end of the columns for each algorithm, we present the sum of the relative differences of the 2800 instances.

An important aspect to mention is that the relative difference measures how greater is the number of bins found by an algorithm compared to the optimal solution. So when the value of *Dif* grows, we can affirm that the solutions of the algorithms are further away from the optimal solution. We can see that the largest value of the total

Table 4 Number of optimal solutions found by each algorithm

Class	Set	FFD	BFD	MBS	GGA-CGT
BPP.25	100	99	99	100	100
	1000	1	1	100	100
	10000	0	0	100	100
	100000	0	0	100	100
	1000000	0	0	69	48
	10000000	0	0	15	0
	100000000	0	0	1	0
Total BPP.25		100	100	485	448
BPP.5	100	80	82	100	100
	1000	0	0	97	100
	10000	0	0	23	100
	100000	0	0	0	21
	1000000	0	0	0	0
	10000000	0	0	0	0
	100000000	0	0	0	0
Total BPP.5		80	82	220	321
BPP.75	100	25	25	100	100
	1000	0	0	81	100
	10000	0	0	1	42
	100000	0	0	0	1
	1000000	0	0	2	12
	10000000	0	0	18	17
	100000000	0	0	63	4
Total BPP.75		25	25	265	276
BPP1	100	24	27	100	100
	1000	0	0	40	100
	10000	0	0	0	21
	100000	0	0	10	59
	1000000	0	0	39	80
	10000000	0	0	82	83
	100000000	0	0	96	61
Total BPP1		24	27	367	504
Total		229	234	1337	1549

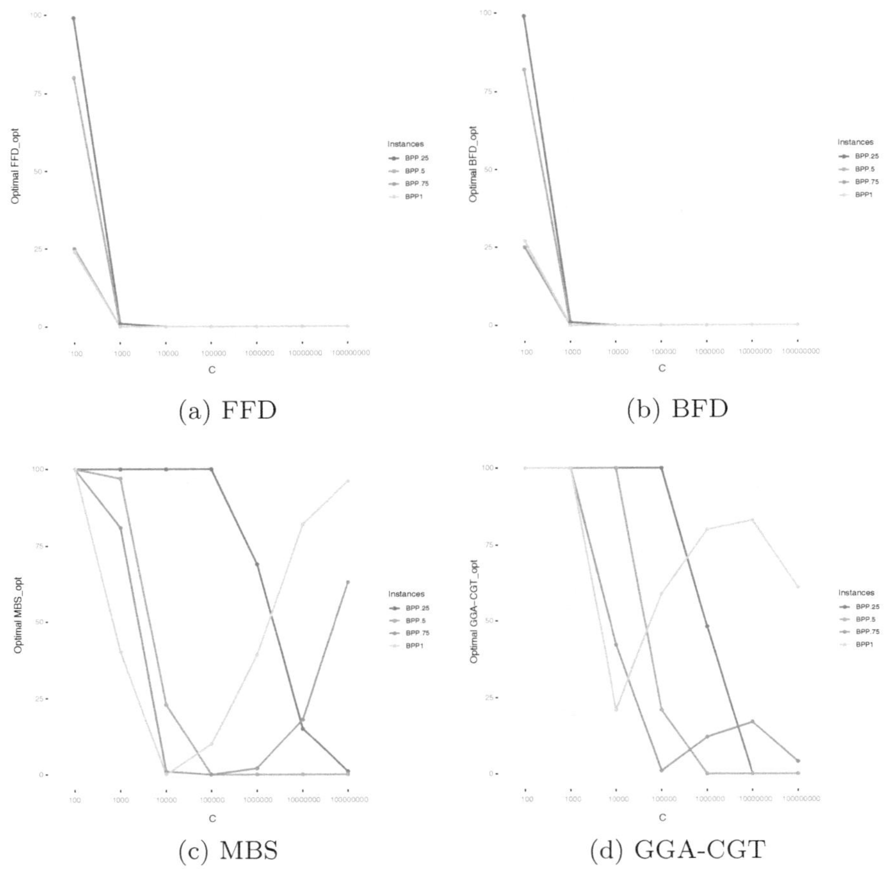

Fig. 1 Number of optimal solutions found by each algorithm

sum of the relative difference is presented by FFD, although the difference with BFD is very small. Furthermore, the smallest value is given by GGA-CGT, where most zeros are also found.

On the other hand, we have Table 6 with the average F_{BPP} (Eq. 3) for the solutions found. Note that if F_{BPP} equals 1, it means that the optimum was found for the 100 instances, in addition to the fact that all these instances have an optimal solution with a 100% of fullness rate. And as this number decreases, we can argue that the average fill of bins does as well. This can be seen in the results for MBS and GGA-CGT, where the value of 1 is found in the rows where the algorithm solved all instances. That is not the case for FFD and BFD, where for none of the sets all the optimal ones were found. At the end of the table the average of the total instances is shown.

Table 5 Relative difference sum between each algorithm and the best solution

Class	Set	FFD	BFD	MBS	GGA-CGT
BPP.25	100	0.067	0.067	0	0
	1000	6.6	6.6	0	0
	10000	6.667	6.667	0	0
	100000	6.667	6.667	0	0
	1000000	6.667	6.667	2.067	3.467
	10000000	6.667	6.667	5.667	6.667
	100000000	6.667	6.667	6.6	6.667
BPP.5	100	0.667	0.6	0	0
	1000	3.333	3.333	0.1	0
	10000	3.333	3.333	2.567	0
	100000	3.333	3.333	3.333	2.633
	1000000	3.333	3.333	3.333	3.333
	10000000	3.333	3.333	3.367	3.333
	100000000	3.333	3.333	3.333	3.333
BPP.75	100	1.667	1.667	0	0
	1000	2.222	2.222	0.422	0
	10000	2.222	2.222	2.244	1.289
	100000	2.222	2.222	2.311	2.2
	1000000	2.222	2.222	2.267	1.956
	10000000	2.222	2.222	1.933	1.844
	100000000	2.222	2.222	0.933	2.133
BPP1	100	1.267	1.217	0	0
	1000	1.667	1.667	1	0
	10000	1.667	1.667	1.683	1.317
	100000	1.667	1.667	1.567	0.683
	1000000	1.667	1.667	1.017	0.333
	10000000	1.667	1.667	0.317	0.283
	100000000	1.667	1.667	0.067	0.65
Total		85.28	85.17	46.13	42.12

Figure 2 presents the average value of F_{BPP} for each type of different bin capacities (sets). The different colors represent the four classes of instances. The horizontal axis contains the bin capacities, and the vertical axis shows the average value of the F_{BPP} function. Each graph corresponds to a different algorithm.

As before, we observe similar behavior in FFD and BFD (Fig. 2a and b), which is a constant value from capacity 10^4 in the four classes of instances. The value of F_{BPP} for MBS solutions decreases in three of the classes and then grows again. Contrary

Table 6 F_{BPP} average for the four algorithms

Class	Set	FFD	BFD	MBS	GGA-CGT
BPP.25	100	0.999	0.999	1	1
	1000	0.936	0.936	1	1
	10000	0.935	0.935	1	1
	100000	0.935	0.935	1	1
	1000000	0.935	0.935	0.977	0.967
	10000000	0.935	0.935	0.94	0.937
	100000000	0.935	0.935	0.929	0.937
BPP.5	100	0.993	0.993	1	1
	1000	0.963	0.963	0.998	1
	10000	0.962	0.962	0.969	1
	100000	0.962	0.962	0.958	0.994
	1000000	0.962	0.962	0.956	0.967
	10000000	0.962	0.962	0.954	0.967
	100000000	0.962	0.962	0.954	0.967
BPP.75	100	0.982	0.982	1	1
	1000	0.973	0.973	0.995	1
	10000	0.972	0.973	0.968	0.987
	100000	0.972	0.973	0.966	0.991
	1000000	0.972	0.972	0.967	0.981
	10000000	0.972	0.972	0.973	0.982
	100000000	0.972	0.972	0.987	0.979
BPP1	100	0.986	0.987	1	1
	1000	0.979	0.98	0.987	1
	10000	0.979	0.979	0.975	0.987
	100000	0.979	0.979	0.977	0.995
	1000000	0.979	0.979	0.986	0.997
	10000000	0.979	0.979	0.996	0.997
	100000000	0.979	0.979	0.999	0.994
Average		0.966	0.966	0.979	0.987

to GGA-CGT, where the performance decrease in classes BPP.25 and BPP.5 without recovering, and both have a variable behavior. Note that the BPP.25 class (with the blue color) has the most similar behavior in the last two algorithms.

Table 7 presents the execution times in seconds of each algorithm (average). At the end of the table, the average time spent by each algorithm is shown. Note that in the case of MBS and GGA-CGT, there are high execution times compared to simple heuristics. Also, this behavior occurs only in sets with larger bin capacities. GGA-CGT shows the highest average CPU time. As we mentioned earlier, $time$ is a measure of efficiency, and although the MBS and GGA-CGT algorithms have

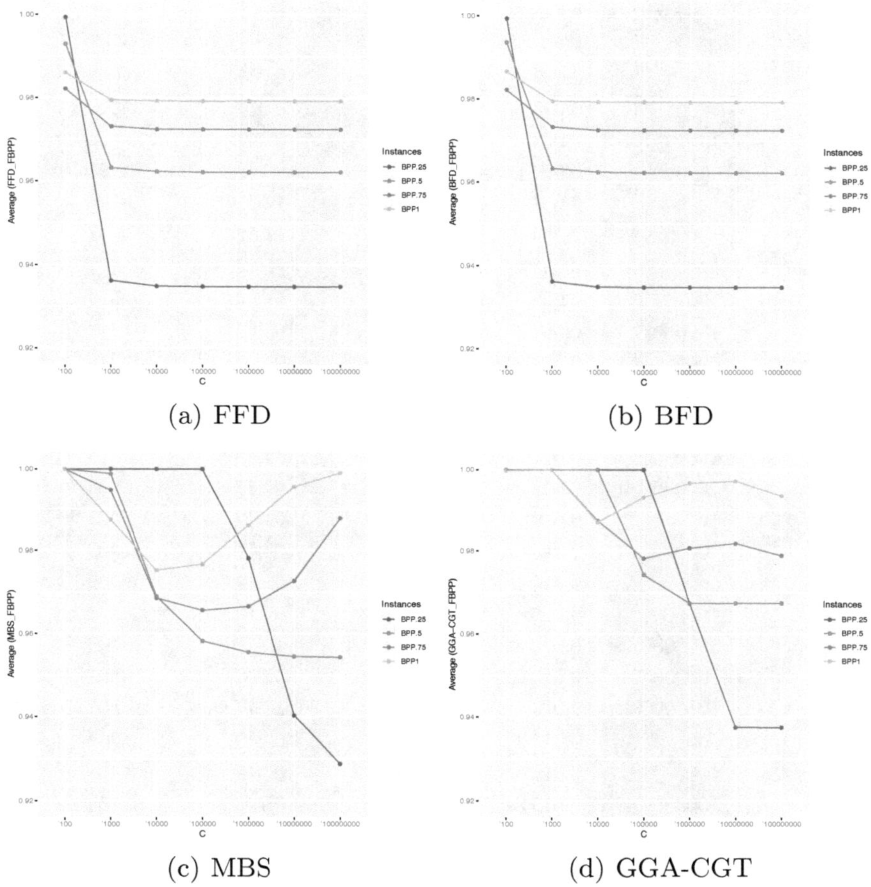

Fig. 2 Average of the F_{BPP} value for the solutions of the four algorithms

presented the best values in terms of effectiveness, the execution time is too high compared to the simple FFD and BFD heuristics. Note that the highest value occurs in the set of $c = 10^8$ in the BPP.25 class with the MBS heuristic.

With the tables and figures above, we were able to observe certain trends in the algorithms and their performance evaluation in terms of effectiveness and efficiency. Now, in the next section, these results will be studied to determine if they could be an effect of the characteristics of the BPP instances.

Table 7 Average CPU time in seconds for the four algorithms

Class	Set	FFD	BFD	MBS	GGA-CGT
BPP.25	100	0.003	0.003	0.037	0.002
	1000	0.003	0.003	0.037	0.026
	10000	0.003	0.003	0.047	0.436
	100000	0.003	0.003	0.148	10.92
	1000000	0.003	0.003	1.631	36.96
	10000000	0.003	0.004	12.52	27.48
	100000000	0.003	0.004	75.97	22.71
BPP.5	100	0.008	0.007	0.044	0.014
	1000	0.008	0.007	0.062	0.218
	10000	0.007	0.007	0.123	3.718
	100000	0.007	0.007	0.696	17.32
	1000000	0.007	0.007	4.631	15.28
	10000000	0.007	0.007	34.25	13.78
	100000000	0.007	0.007	67.32	12.34
BPP.75	100	0.010	0.023	0.042	0.023
	1000	0.010	0.023	0.071	0.309
	10000	0.010	0.023	0.101	6.163
	100000	0.010	0.023	0.359	9.232
	1000000	0.010	0.023	1.37	10.39
	10000000	0.010	0.023	4.702	10.02
	100000000	0.010	0.024	13.74	10.07
BPP1	100	0.012	0.030	0.043	0.044
	1000	0.012	0.030	0.049	0.407
	10000	0.013	0.030	0.108	6.137
	100000	0.013	0.030	0.282	5.191
	1000000	0.013	0.030	0.757	4.756
	10000000	0.013	0.030	0.664	4.245
	100000000	0.013	0.030	0.689	10.32
Average		0.007	0.016	7.875	8.518

5 Experimental Analysis

In this section, using an exploratory approach, we seek insights into the relationships between the BPP instances structure and the performance of the BPP algorithms. The relationships are studied by applying correlation and graphical analysis to have an explanation of the results obtained by the above algorithms. We have used the indices presented in Sect. 2 to analyze the existing correlation between the BPP features, the relative difference, the cost function, and the execution time for each algorithm. Figure 3 presents the correlation matrix, which includes the degree of linear rela-

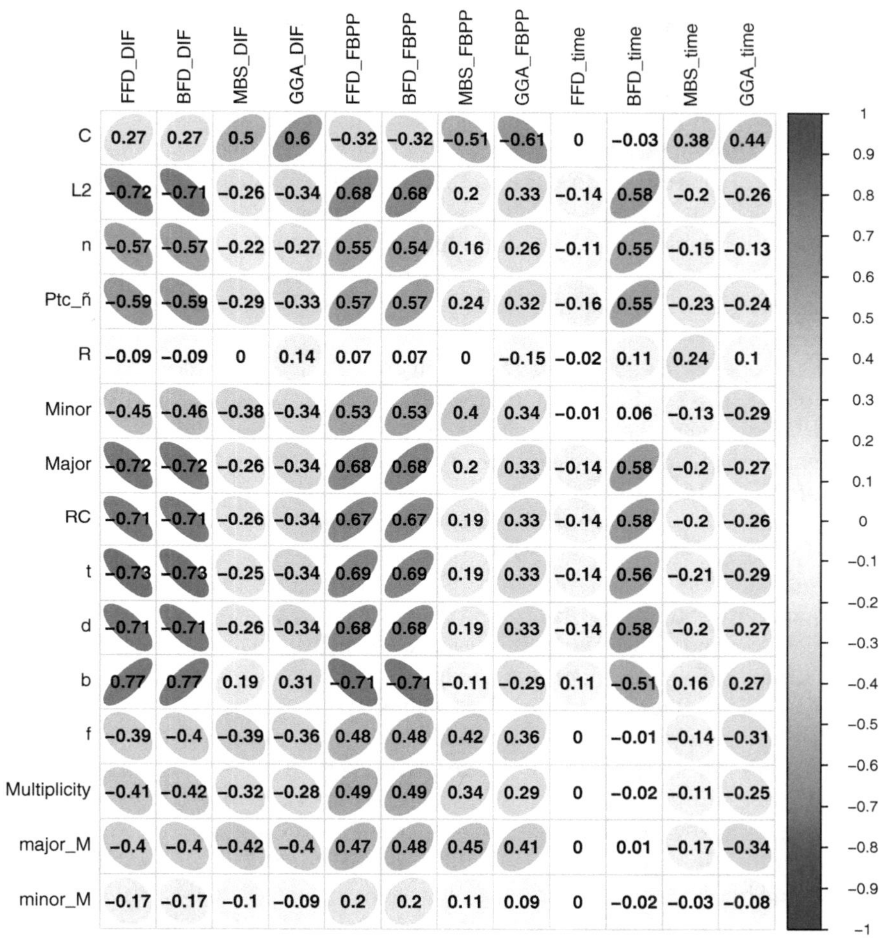

Fig. 3 Correlation matrix between the BPP indices and the three performance measures for the 2800 instances

tionship between the features of the instances and the algorithms performance. In the rows, we have the BPP indices (note that we have also included the number of items n, as well as the bin capacity c). In the columns, we have the relative difference *Dif* for the four algorithms, then the cost function F_{BPP} and finally the *time*. It is important to remember that a good value for the relative difference *Dif* is equal to zero and it grows as the error of the algorithm increases. However, the F_{BPP} value has a different behavior considering as optimal, the values equal to 1.

From Fig. 3 we can see that the highest correlations are associated with the relative difference *Dif* and the cost function F_{BPP} obtained by FFD and BFD. Remember that these are the algorithms that obtained the lowest results in the number of optimal solutions found. From the results presented in Table 4, it can be observed that as the average weights of the items and the bin capacity increase, the probability of these algorithms to find the optimal solutions decrease. We can also appreciate that the time for FFD presents a null correlation against some characteristics of the instances.

The values of the correlations in Fig. 3 suggest that, regardless of the algorithm, the relationships between the characteristics of the instances with *Dif* and F_{BPP}, always have the same meaning. These relationships are stronger in the simple FFD and BFD heuristics, which seems to be the consequence of the fact that these heuristics do not include strategies focused on maximizing the filling of the bins, as opposed to MBS and GGA-CGT.

The correlation matrix also shows which characteristics are more related to the values of F_{BPP} and *Dif*. For example, the correlations associated with $L2$, *Major*, RC, t, and d, show that as these are higher, the values of the relative difference decrease. This behavior is seen in the result of the four algorithms, but it is only strong for the result of FFD and BFD. In the same way, when the values of these metrics increase, the F_{BPP} value does so, maintaining a consistent value of association for FFD and BFD. To have a clearer notion of this description, we have plotted some of these indices to show the behavior mentioned. First, we decided to plot 3D graphs with the bin capacity c, the *Major* index, and the value of F_{BPP}.

Figure 4 contains four 3D scatter plots, one for each algorithm; in each plot, the x-axis indicates the bin capacity c, the y-axis depicts the value of the index *Major*, and the z-axis indicates the cost function F_{BPP} of the solutions generated for each instance. A different color is used to plot each instance according to the class it belongs, as can be seen in this figure, the instances of each set are together, according to the bin capacity c. From the figure, we can see that as the value of *Major* increases, the value of F_{BPP} also increases; that is, as the value of the highest weight in the instance relative to the bin capacity is higher, FFD and BFD have a higher value of F_{BPP} (Fig. 4a and b). Furthermore, these algorithms are the ones with the least dispersion in the values obtained per set. Also, we can note that MBS (Fig. 4(c)) presents the greatest variability in F_{BPP} values, both by class and by set.

Similarly, Fig. 5 shows the impact of the index b in the values of the cost function F_{BPP} obtained for each instance. This figure has the same structure of Fig. 4. Here, the $y-$axis indicates the b values of each instance. As seen in this figure, when the proportion of the total weight than can be assigned to a bin b grows, the F_{BPP} value decreases, especially for FFD and BFD (Fig. 5a and b), which are the algorithms with the least number of optimal solutions found. In the case of MBS (Fig. 5c), the dispersion associated with the value of the metric is very wide, both in the classes and in the sets. Finally, GGA-CGT (Fig. 5d) shows a similar behavior regarding the relation between b and F_{BPP}, however, for this algorithm most of the results are closer to the optimal solution, which indicates that characteristic b is not a problem for the algorithm to find the optimal solution.

The previous results allowed us to graphically appreciate that the result of the algorithms is different by class and set, therefore, we have done the correlation analysis for each class of instances. The following subsections present the particular

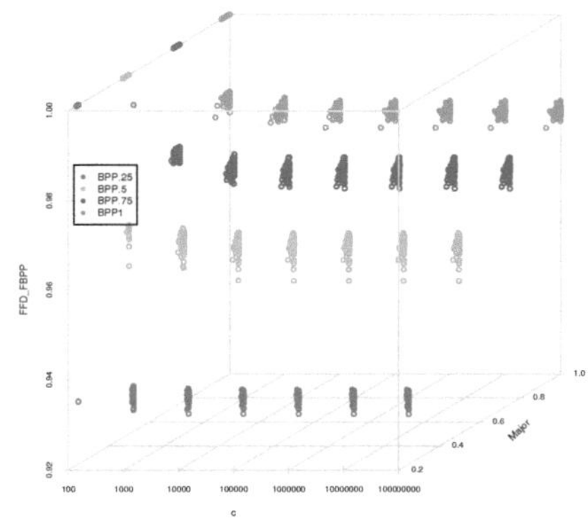

(a) FFD

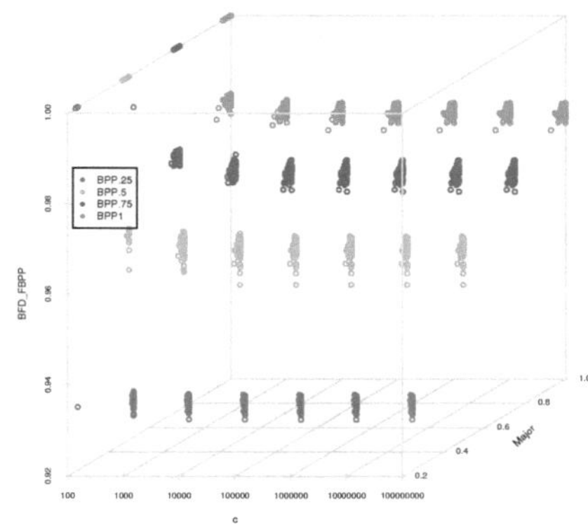

(b) BFD

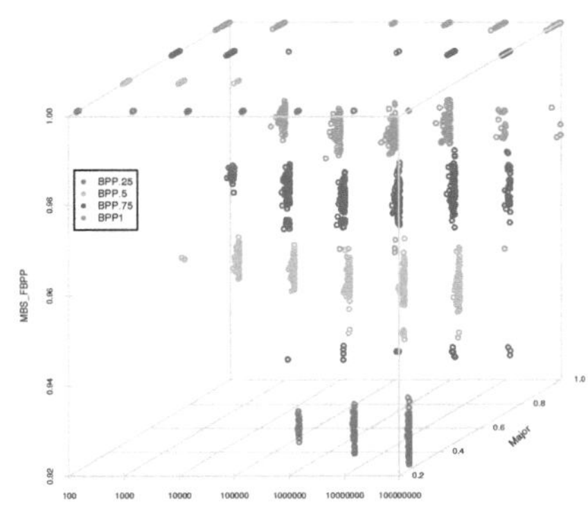

(c) MBS

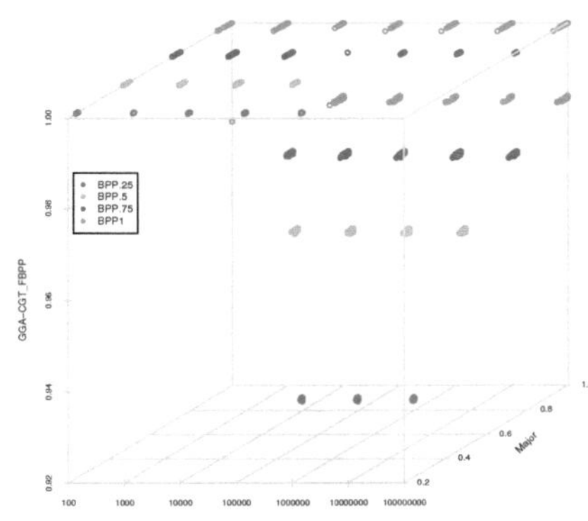

(d) GGA-CGT

Fig. 4 Relations between the index *Major*, the bin capacity c, and F_{BPP} for the 2800 instances

analysis of the instances of each of the classes. In each of the cases, we present the correlation analysis between the indices and the performance measures, with the difference that these correlations are only measured for the 700 instances of the respective class. It is important to clarify that some correlations could not be extracted because the value of some of the characteristics is the same for all the instances in the same class. Question marks (?) in the correlation matrix denote that situation. Also, we have chosen the f index (given the correlations presented) to have a graphical analysis of its behavior in each of the classes.

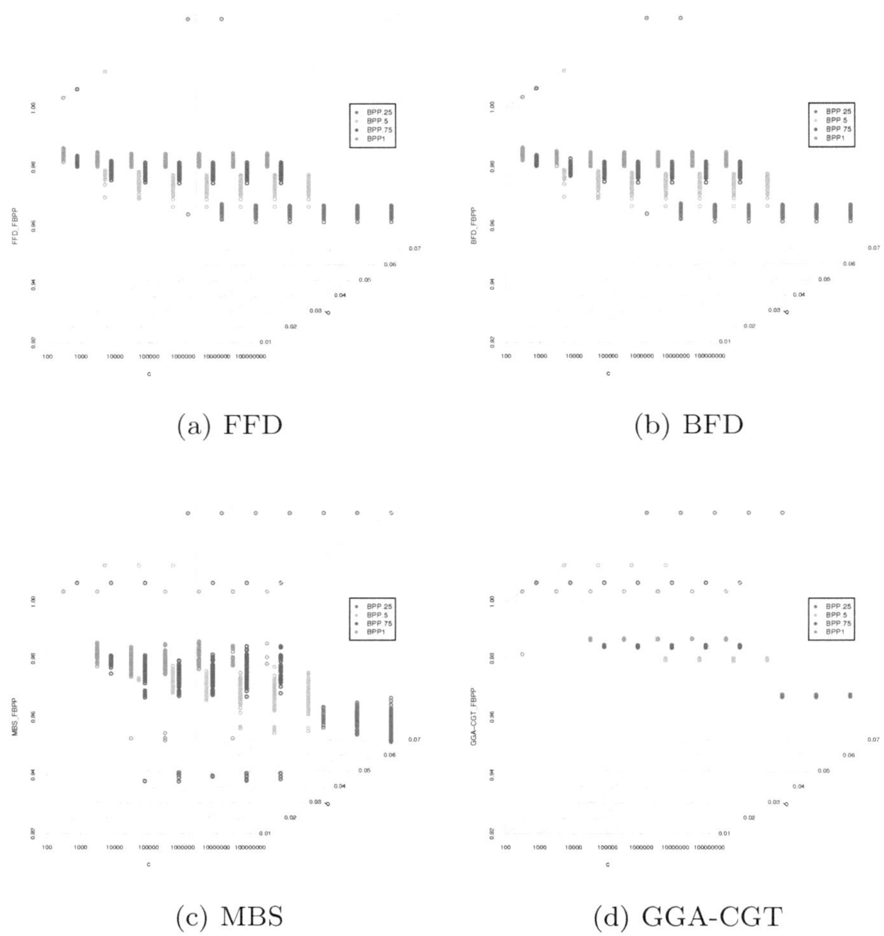

(a) FFD (b) BFD

(c) MBS (d) GGA-CGT

Fig. 5 Relations between the index b, the bin capacity c and, F_{BPP} for the 2800 instances

5.1 *Class BPP.25*

The class BPP.25 includes 7 sets of instances, each with 100 instances for every bin capacity c ($10^2, \ldots, 10^8$), where the number of items n varies within [110, 154], the weights are distributed between $(0, 0.25c]$ and the number of bins in the optimal solution is 15. For this class, the BPP indices that are proportional to the bin capacity ($Minor$, $Major$, RC, t and d) present similar values, the average values are: $Minor = 0.001$, $Major = 0.246$, $RC = 0.246$, $t = 0.115$ and $d = 0.072$. This class was the least difficult for the three heuristics (FFD, BFD and MBS), the algorithms found a higher number of optimal solutions than in the other classes.

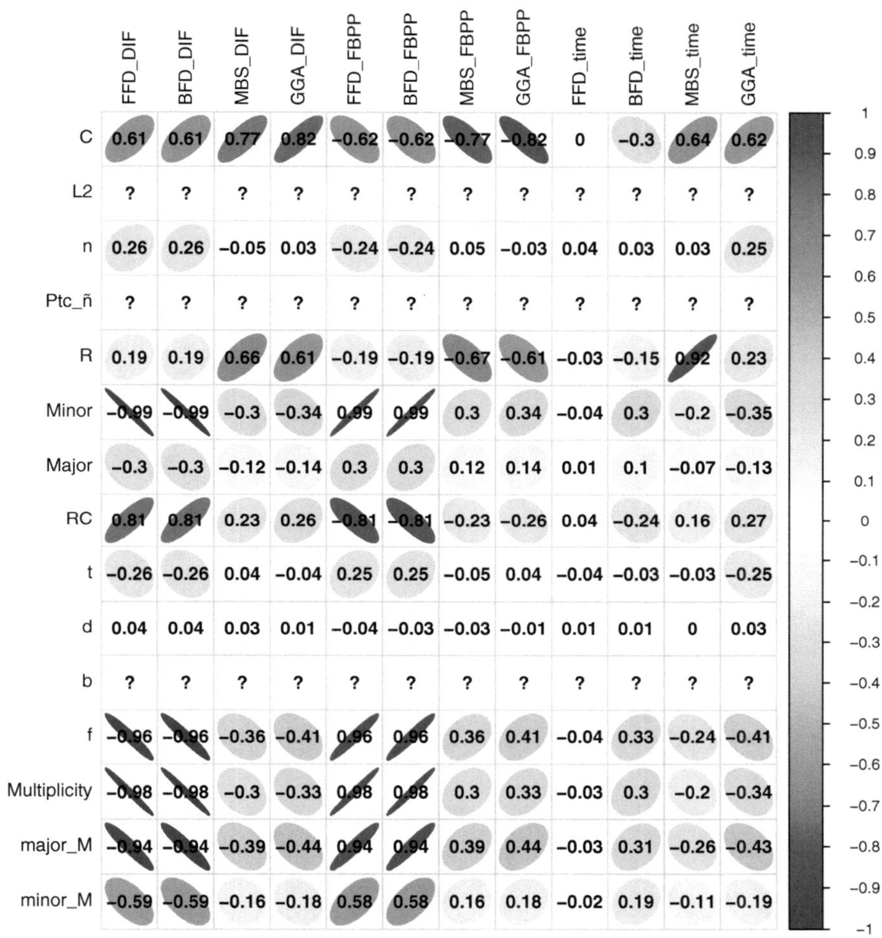

Fig. 6 Correlation matrix between the BPP indices and the three performance measures for the class BPP.25

The correlation analysis for this class shows interesting results (Fig. 6), for example, we can see that as *Minor*, f, *Multiplicity*, *Major_Multiplicity*, and *Minor_Multiplicity* increase, the FFD, BFD and MBS algorithms obtain better results. However, when the bin capacity c increases, these algorithms perform less well. In the same way, the algorithms MBS and GGA-CGT are affected by the bin capacity. An increasing time for MBS is also notable when the difference between the largest and the smallest weight R grows. This could be attributed to the fact that it is more difficult for the algorithm to find the appropriate grouping of items for each bin.

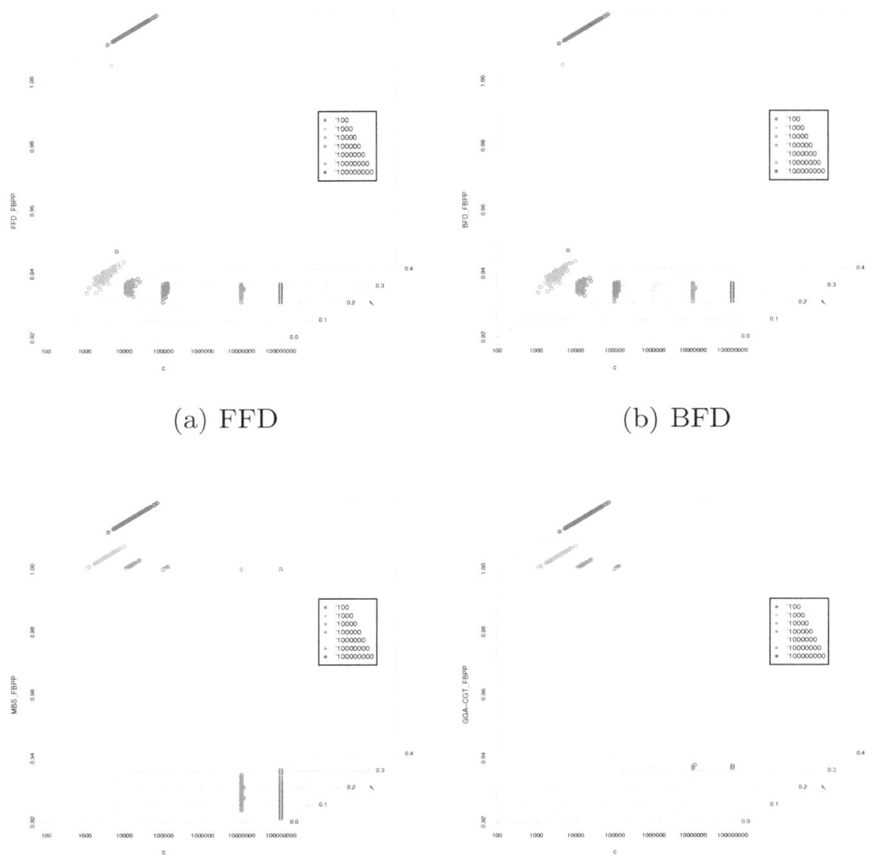

(a) FFD (b) BFD

(c) MBS (d) GGA-CGT

Fig. 7 Relations between the index f, the bin capacity c, and F_{BPP} for the instances in the BPP.25 class

To have a visual analysis of these relations, Fig. 7 present the scatters plots for the proportion of items whose weight is a factor of the bin capacity f, the bin capacity c and the cost function F_{BPP} achieved by each algorithm. In this figure, a different color is used for each of the seven sets.

From Fig. 7 we can observe how for the smallest bin capacity $c = 10^2$ the four algorithms found most of the optimal solutions, without being affected by the value of f. However, as the bin capacities increase, the effectiveness of FFD and BFD (Fig. 7a and b) drops dramatically; unlike MBS (Fig. 7c) which even solved some instances of all sets. On the other hand, GGA-CGT (Fig. 7d), despite being the algorithm that solves more instances on the other classes, did not manage to find an optimal solution in none of the two sets with the highest bin capacities ($c = 10^7$ and $c = 10^8$).

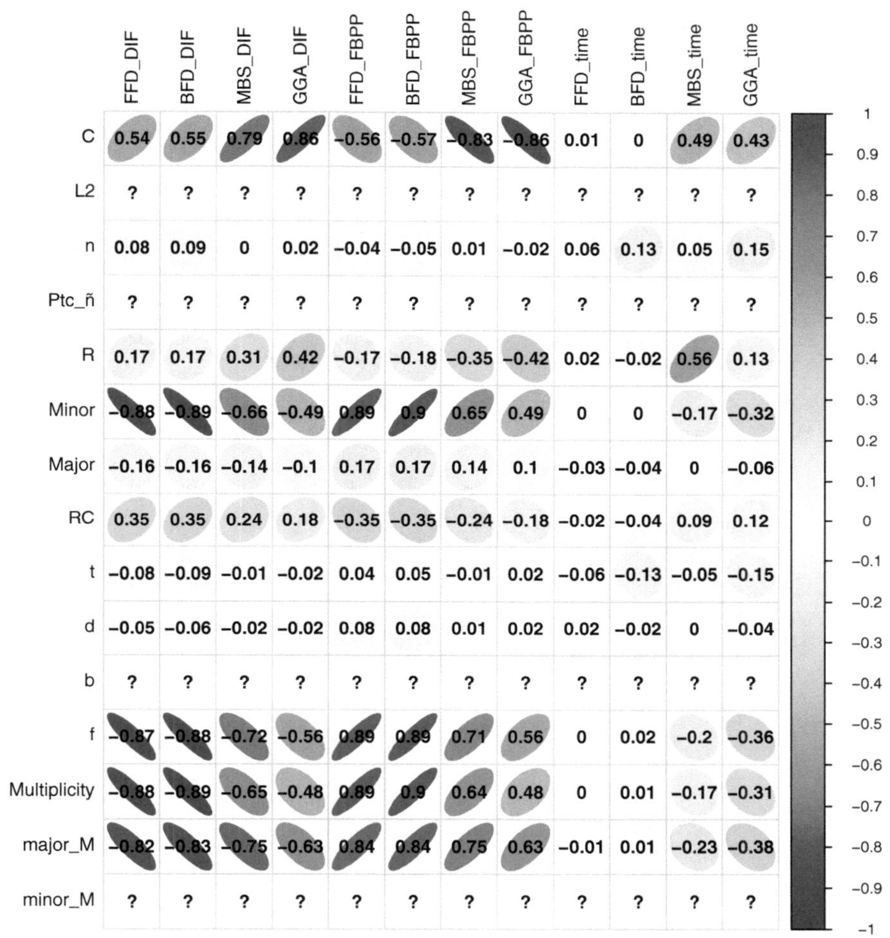

Fig. 8 Correlation matrix between the BPP indices and the three performance measures for the class BPP.5

5.2 *Class BPP.5*

For the class BPP.5, the number of items n varies within [124, 167], the weights are distributed between $(0, 0.5c]$ and the number of bins in the optimal solution is 30. For this class, the BPP indices that are proportional to the bin capacity ($Minor$, *Major*, RC, t and d) present similar values, the average values are: $Minor = 0.001$, $Major = 0.495$, $RC = 0.493$, $t = 0.213$ and $d = 0.140$. This class was the hardest for MBS heuristic, the algorithm only found 220 optimal solutions, the lowest number in the four classes. For the instances of this class, the correlation matrix (Fig. 8) shows that the relationships between the BPP indices and the performance of the

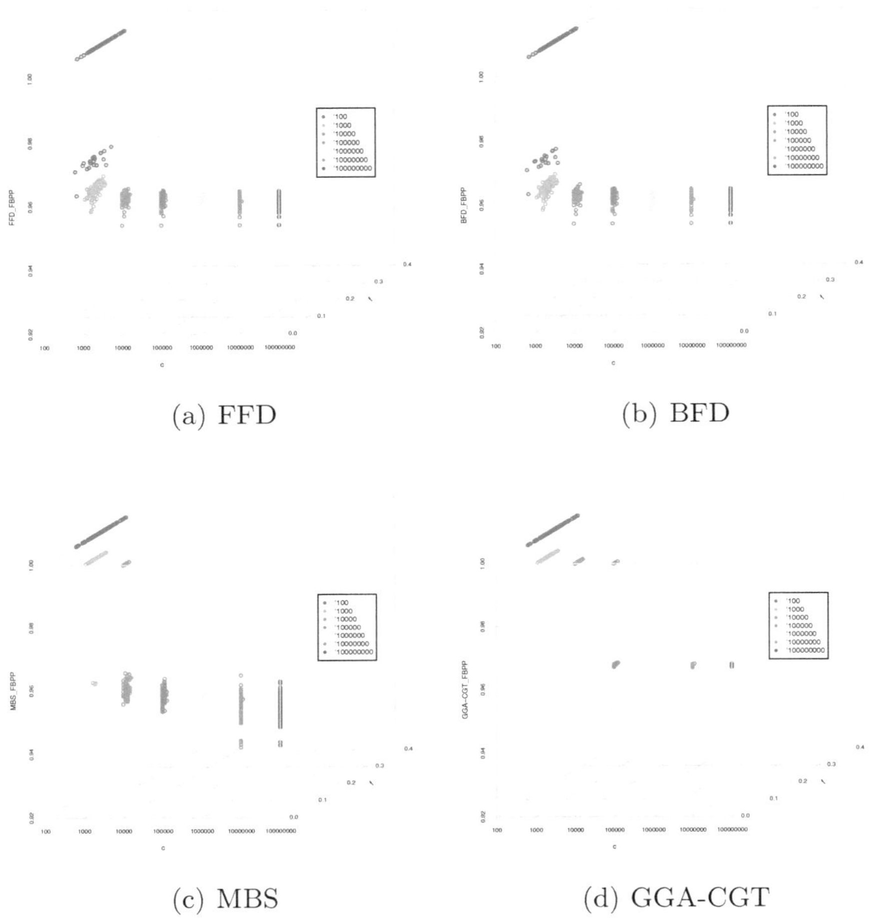

(a) FFD (b) BFD

(c) MBS (d) GGA-CGT

Fig. 9 Relations between the index f, the bin capacity c, and F_{BPP} for the instances in the BPP.5 class

algorithms are like those observed for the class BPP.25. However, it is evident that, for some indices, the associative level begins to be lower, although it does not decrease dramatically.

Similar to Figs. 7 and 9 present the scatters plots for f, c and F_{BPP}, for the class BPP.5 and we can see that for the characteristic f and the F_{BPP} value, there is a greater variability, especially for the FFD and BFD algorithms (Fig. 9a and b). On the other hand, it can be seen that the F_{BPP} values are more consistent by class and set since only for FFD and BFD there are two groups of F_{BPP} values obtained, one optimal and the other very low, this being similar for all sets. For its part, GGA-

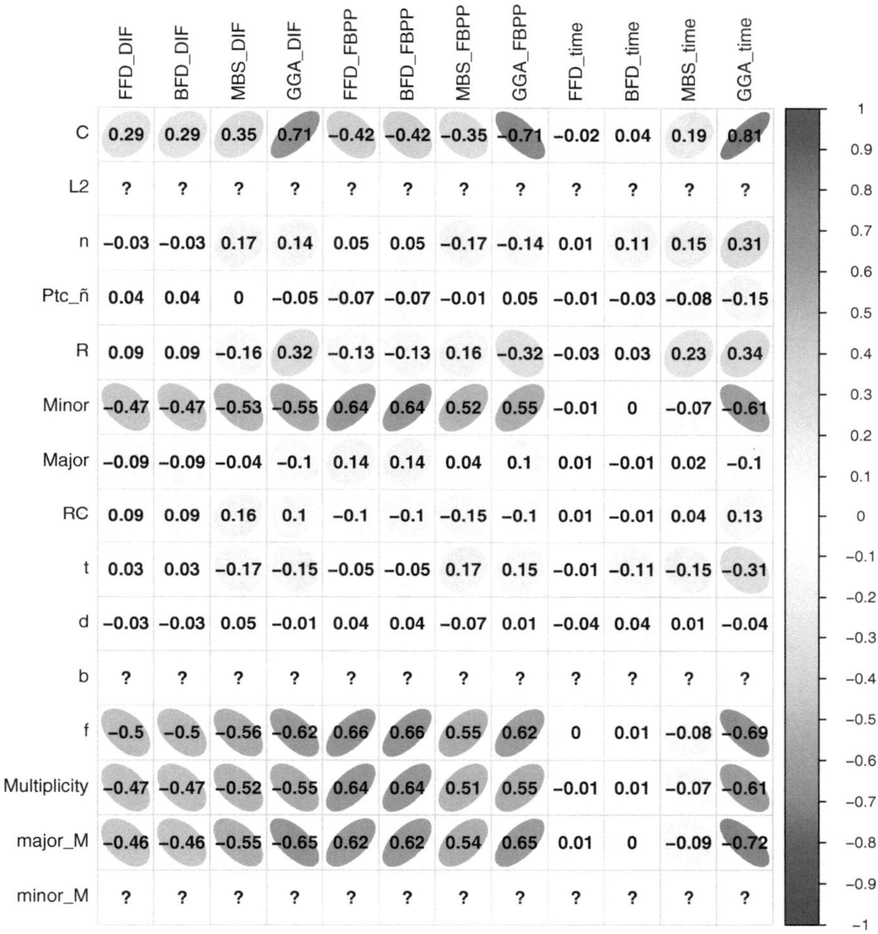

Fig. 10 Correlation matrix between the BPP indices and the three performance measures for the class BPP.75

CGT (Fig. 9d) presents almost zero variability within the sets, which indicates that the algorithm allows a better exploitation of the bin's capacity, which is what F_{BPP} represents.

5.3 *Class BPP.75*

For the class BPP.75, the number of items n varies within $[132, 165]$, the weights are distributed between $(0, 0.75c]$ and the number of bins in the optimal solution is

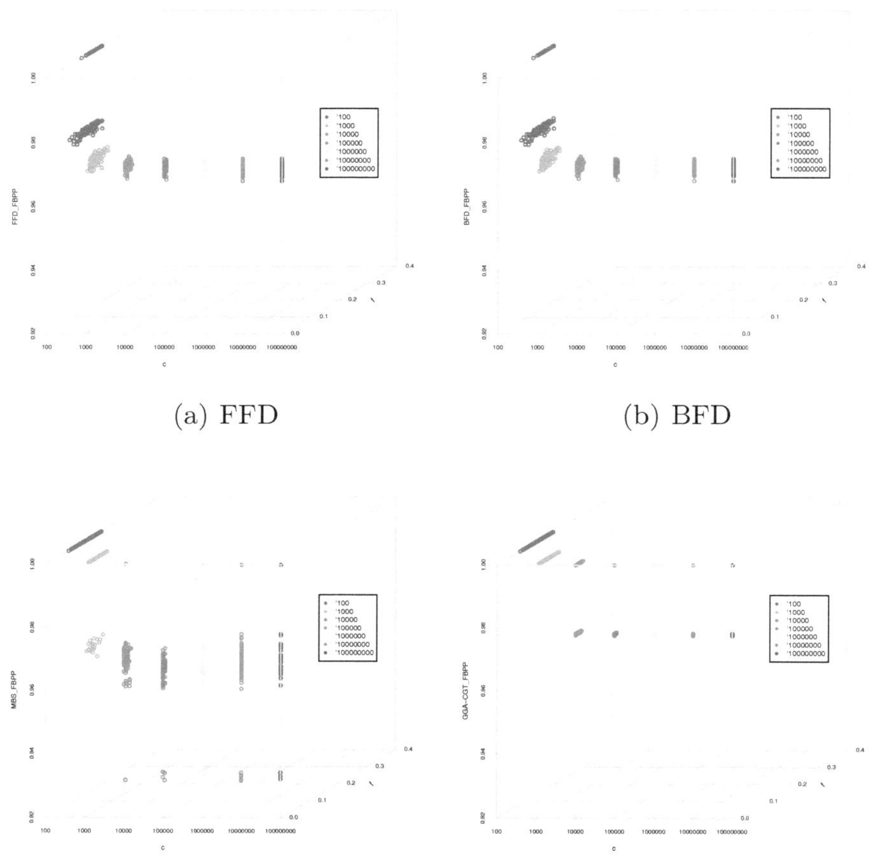

(a) FFD (b) BFD

(c) MBS (d) GGA-CGT

Fig. 11 Relations between the index f, the bin capacity c, and F_{BPP} for the instances in the BPP.75 class

45. For this class, the BPP indices that are proportional to the bin capacity ($Minor$, *Major*, RC, t and d) present similar values, the average values are: *Minor* = 0.001, *Major* = 0.741, RC = 0.739, t = 0.301 and d = 0.202. This class was the hardest for GGA-CGT, the algorithm only found 276 optimal solutions, the lowest number in the four classes. Figure 10 presents the correlation matrix for this class, the efficiency of the algorithms, measured by the index *time*, seems to be influenced by the features of the BPP instances. The value of *time* decreases when *Minor*, f, *Multiplicity*, *Major_Multiplicity*, and *Minor_Multiplicity* increase for the GGA-CGT algorithm. However, it increases when the bin capacity increases. It is remarkable how the correlations for the rest of the algorithms and characteristics have decreased against

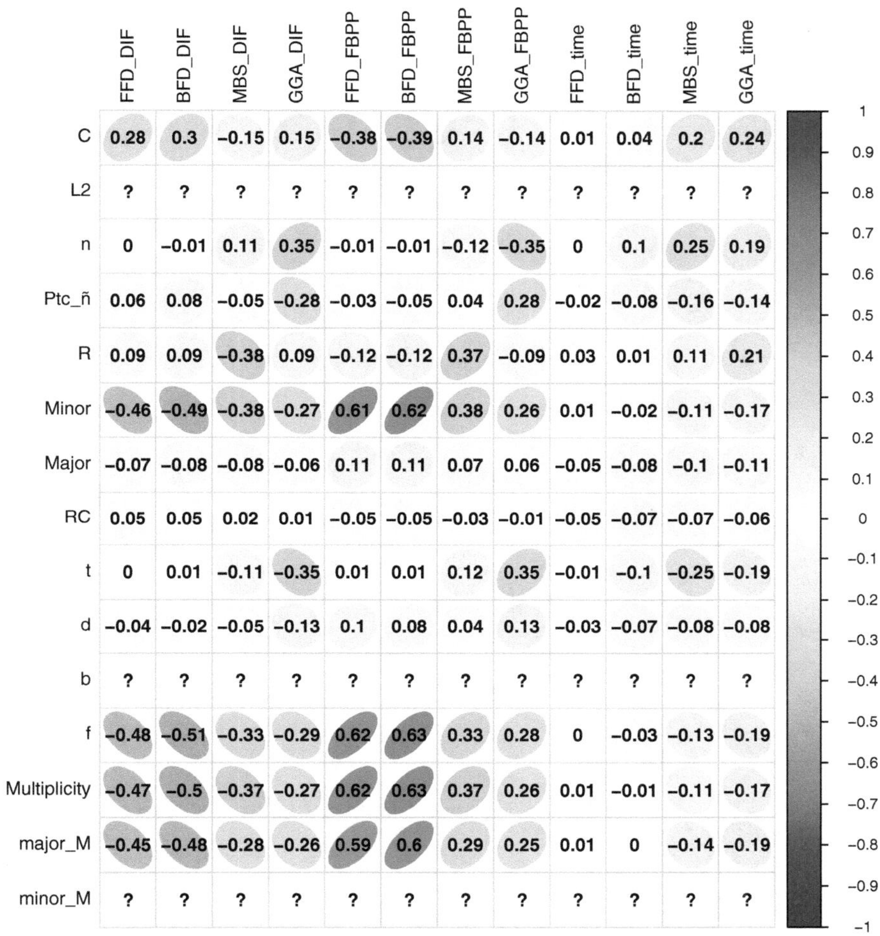

Fig. 12 Correlation matrix between the BPP indices and the three performance measures for the class BPP1

the first two classes. It is important to observe than even in this set of instances the F_{BPP} value is compromised when the bin capacity increases, especially for the GGA-CGT algorithm.

The graphs for class BPP.75 in the Fig. 11 present a different behavior from the two previous classes. In these, it is possible to appreciate how as f increases, the value of F_{BPP} increases, finding a much greater variability in the values obtained by this measure. It is evident that the highest values for f are found in small bin capacities, for this class the values of f are within $[0, 0.193]$, and this range does not seem to have a significant influence in the behavior of the algorithms, since the performance of these has been mainly affected by the capacity of the bins.

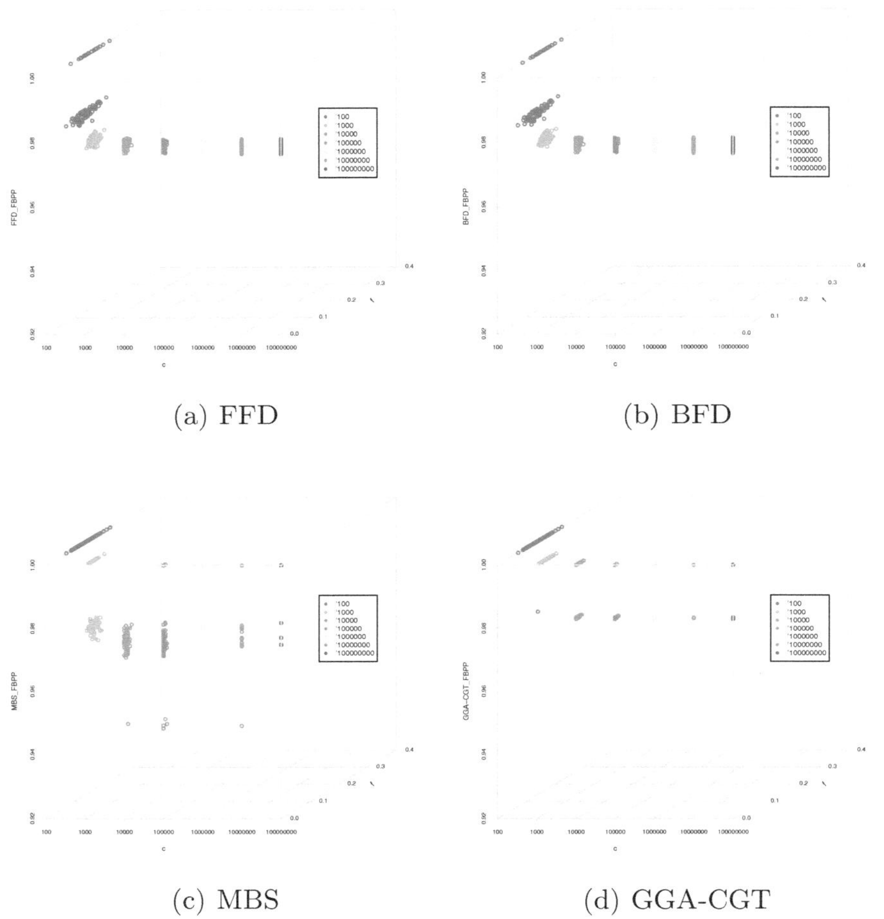

Fig. 13 Relations between the index f, the bin capacity c, and F_{BPP} for the instances in the BPP1 class

5.4 Class BPP1

For the class BPP1, the number of items n varies within [148, 188], the weights are distributed between $(0, c]$ and the number of bins in the optimal solution is 60. For this class, the BPP indices that are proportional to the bin capacity (*Minor*, *Major*, RC, t and d) present similar values, the average values are: *Minor* = 0.001, *Major* = 0.989, RC = 0.987, t = 0.368 and d = 0.269. This class was the least difficult for GGA-CGT, the algorithm found 504 optimal solutions, the highest number in the four classes. Figure 12 shows the correlation matrix for the class BPP1, as can be seen, this class presents the weakest association values between the BPP features and the performance of the algorithms, but the same characteristics are highlighted, which are: *Minor*, f, *Multiplicity*, *Major_Multiplicity*, and *Minor_Multiplicity* only for FFD and BFD heuristics.

On the other hand, we have the graphs for the class in Fig. 13, in this we can see that, as well as in the BPP.75 class, the bin capacity is an important characteristic that impact the effectiveness of the algorithms, since it can be observed that as the value of c increases the F_{BPP} values decrease, however, in this set we can highlight that even when the optimal solutions are not found, the F_{BPP} value is better than the other three classes, especially for MBS and GGA-CGT (Fig. 13c and d).

In the studies separated by classes, we could observe that the behavior is very different in each of them. In other words, while in one class some indices present a high correlation, in another class this may differ. We can highlight that the index *time* is the only measure that behaves similarly in all groups, without having a great correlation with the indices. As well as in each of the subsections, we were able to identify some characteristics that show high correlations with each of the classes of instances.

6 Conclusions and Future Work

This work addressed the characterization of new hard instances for the one-dimensional Bin Packing Problem (BPP) and the analysis of the algorithm behavior of four well-known algorithms using data analysis techniques. This study aimed to gain a deeper understanding of the difficulty of the BPP instances through the investigation of the relationships between the features of the instances and the algorithms performance. The results that were obtained by the four algorithms confirm that the new instances have a high degree of difficulty; for most of these instances, the included strategies in sophisticated algorithms like MBS and GGA-CGT do not appear to lead to better solutions. The study revealed interesting relationships between the characteristics of the new instances and the four algorithms implemented here. One of the first observations is the fact that the performance of the algorithms is affected by the bin capacity and this happens regardless of the algorithm. We have also observed that simple heuristics such as FFD and MBS did not perform well and they present the highest correlations with the characteristics of the BPP instances, in comparison with MBS and the GGA-CGT.

As we could observe, FFD and BFD presented a similar behavior in all classes of instances, both heuristics solved the least number of instances, which represents in general, the difficulty of the new instances for simple heuristics. The bin capacity and the range in which the weights are distributed were the characteristics that make these algorithms difficult. Furthermore, f, *Multiplicity*, *Major_Multiplicity*, *Minor_Multiplicity* and *Minor* indices were the most correlated with the value of the cost function and the relative difference in the results obtained by said heuristics. Another interesting result is the similar behavior presented by the MBS and GGA-CGT algorithms, which presented correlations with the bin capacity in some of the sets and classes. Despite solving a greater number of instances compared to simple heuristics, the bin capacity also affected the performance of the algorithms. We could also see that the average execution times for these algorithms is the highest, which is associated with some factors such as the range R and the bin capacity c.

Something important to mention is that these evaluated instances were created with very particular characteristics, however, the same relationships found in previous studies are observed. For example, the identification of the general difficulty of the instances when the bin capacity is large, as well as when the average weights of the items is around 0.3 of the bin capacity.

The knowledge obtained in this work and the introduction of the new sets of instances open up an interesting range of possibilities for future research. First, it is expected that the new instances presented in this work can be used for studying the performance of the state-of-the-art BPP algorithms. On the other hand, new strategies can be developed incorporating knowledge of the problem domain. Future directions also include predictions of performance and additional analysis of hard BPP instances, combining different characterization techniques to obtain better explanations regarding the internal behavior and the performance of the algorithms and the difficulty of the instances solved.

References

Basse, S. 1998. *Computer Algorithms*. Introduction to Design and Analysis: Editorial Addison-Wesley Publishing Company.

Buljubašić, M., and M. Vasquez. 2016. Consistent neighborhood search for one-dimensional bin packing and two-dimensional vector packing. *Computers & Operations Research* 76: 12–21.

Chiarandini, M., L. Paquete, M. Preuss, and E. Ridge. 2007. Experiments on metaheuristics: Methodological overview and open issues. Technical Report DMF-2007-03-003, The Danish Mathematical Society.

Cruz, L. 2004. Caracterización de Algoritmos Heuísticos Aplicados al Diseño de Bases de Datos Distribuidas. PhD Thesis, Centro Nacional de Investigación y Desarrollo Tecnológico, Cuernavaca, Morelos, México.

Delorme, M., M. Iori, and S. Martello. 2016. Bin packing and cutting stock problems: Mathematical models and exact algorithms. *European Journal of Operational Research* 255 (1): 1–20.

Delorme, M., M. Iori, and S. Martello. 2018. BPPLIB: A library for bin packing and cutting stock problems. *Optimization Letters* 12 (2): 235–250.

Falkenauer, E. 1992. The grouping genetic algorithm-widening the scope of the GAs. *Belgian Journal of Operations Research, Statistics and Computer Science* 33: 79–102.

Falkenauer, E., and A. Delchambre. 1992, May. A genetic algorithm for bin packing and line balancing. In *ICRA*, 1186–1192.

Fleszar, K., and C. Charalambous. 2011. Average-weight-controlled bin-oriented heuristics for the one-dimensional bin-packing problem. *European Journal of Operational Research* 210 (2): 176–184.

Fleszar, K., and K.S. Hindi. 2002. New heuristics for one-dimensional bin-packing. *Computers & operations research* 29 (7): 821–839.

Garey, M.R., and D.S. Johnson. 1979. *Computers and Intractability*, vol. 174. San Francisco: Freeman.

Gupta, J.N., and J.C. Ho. 1999. A new heuristic algorithm for the one-dimensional bin-packing problem. *Production planning & control* 10 (6): 598–603.

Johnson, D.S. 1974. Fast algorithms for bin packing. *Journal of Computer and System Sciences* 8 (3): 272–314.

Quiroz, M. 2014. Caracterización del proceso de optimización de algoritmos heurísticos aplicados al problema de empacado de objetos en contenedores. Instituto Tecnológico de Tijuana.

Quiroz-Castellanos, M., L. Cruz-Reyes, J. Torres-Jimenez, C. Gómez, H.J.F. Huacuja, and A.C. Alvim. 2015. A grouping genetic algorithm with controlled gene transmission for the bin packing problem. *Computers & Operations Research* 55: 52–64.

Martello, S. 1990. *Knapsack problems: algorithms and computer implementations*. Wiley-Interscience series in discrete mathematics and optimization.

Martello, S., and Toth, P. 1990. Lower bounds and reduction procedures for the bin packing problem. *Discrete Applied Mathematics* 28 (1): 59–70.

Schwerin, P., and G. Wascher. 1997. The bin-packing problem: A problem generator and some numerical experiments with FFD packing and MTP. *International Transactions in Operational Research* 4 (56): 377–89.

Snodgrass R. 2010. Ergalics: A Natural Science of Computation.

Looking for Emotions in Evolutionary Art

Francisco Fernández de Vega, Cayetano Cruz, Patricia Hernández, and Mario García-Valdez

Abstract Although evolutionary art has demonstrated its potential over the past two decades, based primarily on interactive versions of the evolutionary algorithm (IEAs), new implementations of the algorithm continue to be developed and applied both to produce works of art and to study computer-added creative processes. Recently, human artists have been asked to participate more directly in every step included in the algorithm so that we learn from them. Nevertheless, the behavior shown is so complex that retrieving useful information that allows to understand better the creative process they develop is not an easy task. Moreover, emotions have not been analysed previously in this context. Although useful information has already been obtained and conclusions reached regarding the future improvement of EAs devoted to art, the role of emotions as conveyed by artists and understood by the audience is still unexplored. This paper paves the way towards that specific goal, that we summarize as follows: (i) applying harsh constraints on the creative processes so that the number of formal elements to be analyzed is notably reduced, thus allowing an easier analysis and at the same time to draw useful conclusions when evolutionary algorithms are applied to art and creativity; (ii) being able to analyze both, the human emotions shown by artists and those felt by the audience; together with their connections with formal elements present in the artwork. The main goal is to derive new measures that may allow in the future to introduce changes to the fitness functions

F. F. de Vega · C. Cruz
University of Extremadura, Mérida, Spain
e-mail: fcofdez@unex.es

C. Cruz
e-mail: ccruz@unex.es

P. Hernández
University of Seville, Seville, Spain
e-mail: patriciapahr@gmail.com

M. García-Valdez (✉)
Tecnológico Nacional de México, Tijuana, Mexico
e-mail: mario@tectijuana.edu.mx

O. Castillo and P. Melin (eds.), *Fuzzy Logic Hybrid Extensions of Neural and Optimization Algorithms: Theory and Applications*, Studies in Computational Intelligence 940,
https://doi.org/10.1007/978-3-030-68776-2_11

that allow the artificial evolution of artwork conveying emotions. Finally, the quality of the artwork produced is formally tested by submitting it to an international art competition.

Keywords Evolutionary art · Interactive evolutionary algorithm · Creativity

1 Introduction

Evolutionary art and design has become a mature area within the Evolutionary Algorithms community (Romero and Machado 2008). Several conferences exclusively devoted to the subject, such as EvoMusart[1] the european conference on evolutionary art and music; the special session at IEEE CEC organized by the Computational Intelligence Task Force on Creative Intelligence[2]; or tracks at international conferences, such as ACM Gecco DETA track,[3] publish every year the latest results.

Being computer creativity, one of the Holy Grails of computer science today, pursued but not reached yet, evolutionary art relies on the application of interactive versions of EAs. It is helping in the search and development of creative computers (Takagi 2001). Nevertheless, difficulties for accurately distinguishing when a result is worthwhile has led researchers to not simply trust in human beings when assessing the aesthetic quality of results within the interactive process, but also include artists in every operation of the evolutionary loop (de Vega et al. 2014).

Thus, the unplugged version of the EA (UEA) was proposed with two main goals: (i) to include artists in the evolutionary loop, thus producing collective evolutionary art (so that all of the genetic operations, such as crossover and mutation, were used by artists as an inspiration when producing new artworks); (ii) to allow a detailed analysis of creativity displayed by artists, so that the lessons learned may help to design better software tools for computer-based assisted creativity. While human creativity and machine creativity certainly do not have to be the same, a better understanding of human creativity and its underlying processes from the point of view of evolutionary algorithms may benefit both in the future.

The first goal (to produce quality collective artworks by means of evolutionary approaches) has already been attained with interesting results-such as the ACM GECCO Art, Design and Creativity 2011 award to the project entitled *XY* (de Vega et al. 2013); yet, the second goal is still far, although some proposals were described in de Vega et al. (2014).

Researchers have certainly obtained useful information from the experiments. However, the possibilities for artists, who can choose among materials employed, concepts, and ideas to be developed, apply their experience, among other things are so large that it is not easy to extract clear conclusions from the experiments. It should

[1]http://www.evostar.org/2019/cfp_evomusart.php.

[2]http://cilab.cs.ccu.edu.tw/ci-tf/.

[3]https://gecco-2019.sigevo.org/index.html/Program+Tracks.

be noted that when artists apply genetic operations in the evolutionary context, they are free to understand and apply what a mutation or crossover is, and therefore the possibilities for the new work to be produced from two previous ones acting as parents is so large that drawing clear conclusions is a hard task.

On the other hand, given the importance of public opinion on the results obtained, surveys are frequently used to analyze them, and this produces fatigue (Frade et al. 2010): the larger the number of formal elements to be analyzed in a given artwork, the greater the number of questions to be included in surveys we must collect from the audience, this is a well known problem in IEAs. Moreover, if we want to include in the analysis emotions that artworks convey, this problem greatly increases. Few times emotions displayed by the artists and perceived by the audience have been analyzed in the context of evolutionary algorithms.

This chapter thus presents to the best of our knowledge, the first of such emotion analysis in the context of evolutionary art, when a collective artwork is produced using UEAs with a series of hard constraints. Although emotional models have been described previously in the literature (Posner et al. 2005; Takagi 1998) we decided to simplify them as much as possible. Results allows us to obtain clear ideas on how to improve EAs devoted to art. These conclusions have been reached confronting information extracted through surveys and forms completed by artists when producing the artwork. Moreover, this methodology could be extended to include well-known emotional models in the future, and adding elements to be analyzed, as described below.

The rest of the chapter is organized as follows: Section 2 presents the literature review. Section 3 proposes the methodology aimed at analyzing public reaction and the experiment performed, while Sect. 4 presents results. Finally, we draw our conclusions in Sect. 5.

2 In Search of Lost Emotions

Computer creativity is an active area since the sixties, when the Serendipity exhibition in London showed a series of interesting results in a wide set of artistic areas: poetry, music, design, art (Reichardt 1969). All these areas have matured and nowadays we can find books written by computers, such as *Mexica* (Pérez and Sharples 2001), poetry (Gervás 2001), and music (Diaz-Jerez 2011; de Vega 2017).

Yet, regarding the wide set of technologies available when looking for computers creativity (McCormack and d'Inverno 2012), which includes different flavors of Artificial Intelligence (AI), such as evolutionary approaches, few works can be found in the literature that consider the importance of emotions in this context.

Already raised as one important question in the future of AI by Minsky (2007), we could safely state that the pursuing of emotions is an underdeveloped area in the AI context, and particularly in computer-generated art. Although some models for computer emotions have already been described Wehrle (2001), few authors have tried to make use of them, the work by Pérez is one example in the narrative context (y Pérez 2007).

In pictorial art, the reflections offered by Kandinsky are well-known examples of the importance of emotions, and how different components, such as colors or lines, can be used to produce emotions in the audience (Kandinsky 2012). As referred above, some authors have proposed a categorization for emotion (Takagi 1998). Although we could take Kandinsky's reflections on the line and its effects as a starting point on both a formal and a psychological level, some of the many categorizations carried out by specialists in the field, have been based on a proposal that, on a formal level, usually obeys intuitions. In contrast, on a psychological level, they have not clung to a single theory or classification of emotions.

Both on a formal and psychological level, this proposal obeys more to intuition than to a rational theory.

Colton presented an interesting approach to analyze audience feelings and use this analysis to produce real-time portraits in Colton et al. (2008). This is one of the few works that have considered the interest of analyzing audience emotions in the context of computational creativity.

Our main goal in the experiment we describe below is to research emotions and find ways to convey them to the audience when producing artworks by evolutionary means. Of course, given the difficulty of the approach, we understand that a series of constraints must be established.

2.1 Humans in the EA Loop

As described above, evolutionary art is based on Interactive EAs: users are allowed to interact with the EA being in charge of fitness evaluations. The reason behind is the difficulty for properly encoding aesthetic evaluations. Moreover, new means to allow users to play different roles within the EA has being developed where human beings are in charge of all of the steps in the algorithm (de Vega et al. 2013).

The basic idea behind the Unplugged Evolutionary Algorithm is to allow artists to apply the EA as a creative methodology along the art creation process. Therefore, all operations, including evaluation, selection, crossover and mutation, are performed by human beings -artists- that simultaneously complete a form describing the reasons behind every decision they take and the operation they apply. Thus, once the experiment has been completed, we have a bunch of information to be analyzed and contextualized under the EA perspective, which could help understand artists' creativity better. Moreover, the work can be then shown to the audience so that additional information is collected, and finally may be confronted with the *critic*, as it is typically the case in the traditional art world.

We could summarize the main steps of the methodology provided by UEAs as follows: (i) Unplugging the EA to be executed by artists; (ii) artists produce a collective work while applying the algorithm; (iii) operations are then analyzed, and the creative process evaluated; (iv) Audience perception is analyzed and compared with data collected from the previous analysis; (v) finally, art exhibits or competitions are useful to test the quality of the work.

However, difficulties arise in the third step when we try to analyze results: given the complex behaviors, motivations, and interests artists display, obtaining clear conclusions from the artworks produced is not easy.

Nonetheless, the methodology has been satisfactorily applied in 2012 when producing *XY* (de Vega et al. 2013) and then in 2014 (de Vega et al. 2014). However, no specific analysis of emotions displayed by artists nor the audience were performed. Moreover, we never submit the work to any international art competition, so the work produced never competed on an equal footing with other types of art. The new goal is to test the quality of the work produced.

Therefore, the main goal is to analyze and confront emotions that artist display, the connection with formal elements in the drawings, and the connection with emotions the audience shows. And then, to check the artistic value of the work produce by submitting it to an international art competition.

3 Methodology: Analysis of Emotions in the Era of Evolutionary Art

The idea in this experiment that makes use of UEAs, is to be able to connect emotions with formal elements in an artistic work. If we could achieve such a goal, in a constrained framework, some basic operations could be derived for evolutionary algorithms devoted to art so that particular emotions could be embodied within the evolutionary artworks.

The basic UEA works as follows: A team of artists is in charge of applying every genetic operation in the evolutionary context to produce a collective artwork. Every artist is asked to produce an initial work, thus giving rise to the first generation; then every week artists select two works from the previous generation -parents- and produce a child, which is part of the next generation. Given that works are not signed, no artist knows who of his colleagues has produced the selected work. After a number of weeks, the collective work is finished, which includes as many works per generation as artists are part of the team, and as many generations, as weeks the experiment lasted.

As described above, after an experiment is completed, a quite complex analysis follows. In order to make the posterior analysis of the results easier, we decided to establish several constraints for this new experiment: the size (A4 vertical format), just 40 lines as the only graphic component to be employed, and the expressive operations allowed on the lines: they can be broken in three connected segments, and two lines can never cross. The idea is to focus more on the analysis of the components of interest in the experiment: emotions conveyed by artists and their relationship with selected formal elements. On the other hand, a software tool was employed by the team, as described below, for a better coordination of the experiment.

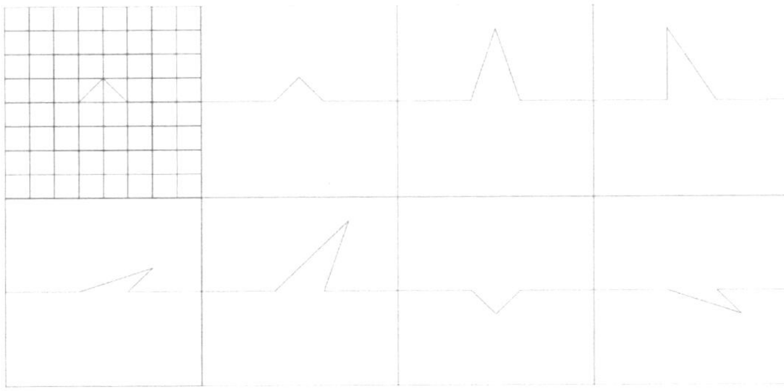

Fig. 1 Some samples of basic lines available for building drawings

Artists were asked to express emotions in each of the drawings, given that a relationship between emotions, graphics, and formal components, and the final output is the primary goal in the experiment. Artist freedom is narrowed when compared with previous experiments, but the idea is that the restricted framework established will help in the final analysis. Therefore, all the images generated will share expressive coherence: a vertical area that includes 40 broken lines designed for every generation as the result of applying genetic operators and human creativity over parents selected in the previous generation.

3.1 The Line

Figure 1 shows some of the basic lines employed as the building blocks for the drawings, and the grid employed to understand how a line can be traced.

The line -only tool available in this work- is made up of four sections of straight lines joined together, which break at two points, i.e., the line is made up of two horizontal straight lines and two broken lines.

The zigzag line can be made up of two connected diagonal lines or a vertical line connected to a diagonal line. From left to right, the line is formed by a first horizontal segment that starts from the right edge of the drawing, and its width is variable. A second segment is anchored, a diagonal line, whose inclination will be up or down, and the length will be variable. The third section of the line is also a diagonal line, with an inclination, opposite to the previous one. Finally, the fourth section is a horizontal straight line whose width depends on the opening of the angle formed by the two broken straight lines.

3.2 *Simplifying the Problem*

Artists were also asked to complete a form stating the emotion translated to the drawing, the reasons for selecting specific parents, and the emotion expressed in every work produced. A number of very limited emotional categories were allowed. The idea was to make the analysis as simple as possible. Yet, artists described a high number of emotions that we have categorized as follows:

- Positive (includes joy, passion, efficacy, love...)
- Negative (includes fear, anxiety, inferiority, remorse, sadness, obsession, resentment, anger...)
- Neutral (includes doubt, ambiguity, concentration...)

This simplification and categorization allow us to trace a genealogical tree of emotions as well as an analysis correlating graphical elements with emotions displayed. Moreover, the audience can also be asked about the category of emotion perceived and thus we can test whether artists message is properly understood. Finally, the main goal is to find easy genetic operations that could in the future be included within an evolutionary algorithm.

3.3 *Evospace-Interactive Module*

In order to run an unplugged EA experiment, the project director needs to manage and coordinate the activities of artists, the delivery of artworks, and the collection of surveys applied to participants of each step of the algorithm. Although there are not computers involved in the generation and selection of solutions, we need a software system for the management of the experiment. The system must aid the deployment of the UEA and later on the display of the generated artwork, in the form of a digital gallery. In this section, we describe in detail the unplugged module of Evospace-Interactive (García-Valdez et al. 2013) used in this experiment. For a detailed description of the technical and implementation aspects of the web application, please refer to de Vega et al. (2014). Once configured, the system has two stages, as mentioned before the unplugged evolutionary loop and then the presentation stage where the result is shown to the audience. These steps are described below.

3.3.1 UEA Stage

In this stage there are two user roles: firstly, the artist, responsible for the creation of the artworks according to the agreed rules. Secondly, the administrator is responsible for configuring the web application, creating a profile for each artist, setting up the beginning and end date for each generation, creating the surveys and coordinating the participants' efforts. As a first step, the administrator must configure the first generation (see Fig. 2).

Fig. 2 Web form for the administration of a new generation

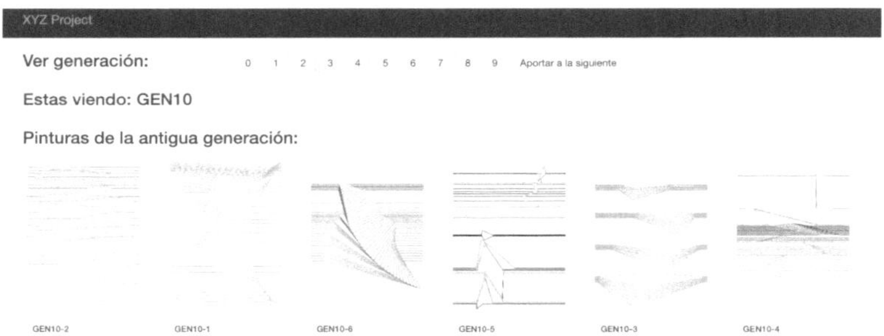

Fig. 3 Selection screen for the artist

The first task artists perform in every generation, and only once the previous one has finished, is to select which individuals will become parents for their new creation. Artists must select two images from the ones produced in the previous generation. The selected artworks will be considered as the parents or inspiration for the new creation. The selection is made in the web application, and the form is shown in Fig. 3.

Artists must make their selection and then have the option to download the image or see it in detail, as seen in Fig. 4.

After a previously accorded time, artists can upload their new creation, stating the emotion they transmit. In this experiment, an external component was used. After all the artists uploaded their newest file, they answered a questionnaire where they reflected about the emotions caused by the artworks from the previous generation. The questionnaire was implemented as a Google Form.

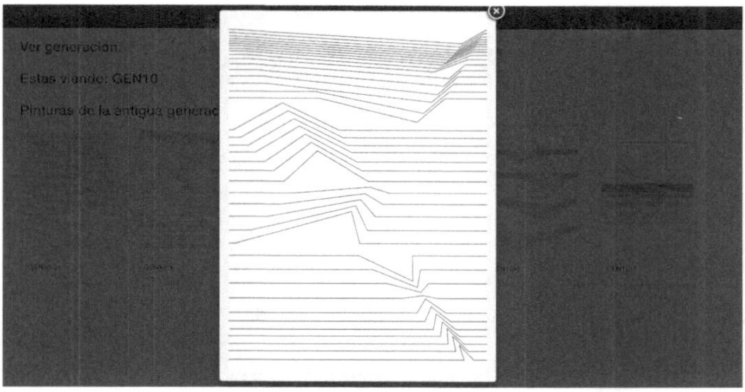

Fig. 4 Artwork is shown in detail

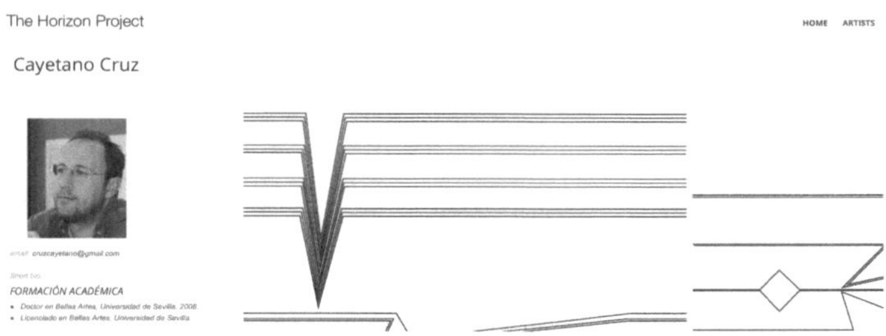

Fig. 5 Profile for an artist, showing her artworks

3.3.2 Gallery Stage

Once the evolutionary stage has ended, the collaborative artwork can be presented to the general public. In this stage, there is also an admin, who is in charge of updating the artist's profile. Each participant has an individual page where a short bio and contact information is presented along with the artworks of their authorship. An example is shown in Fig. 5.

Visitors of the virtual gallery can explore each generation. For instance in Fig. 6, a generation is presented, and the visitor selected an artwork to see it in detail.

Visitors can also follow the genealogical tree for an artwork, exploring their parents and descendants.

3.3.3 Audience Analysis

After visiting the collection, visitors can fill a survey where again certain artworks are presented, and they must express their emotions.

An emotional analysis can thus be applied, using information collected from the audience through these surveys.

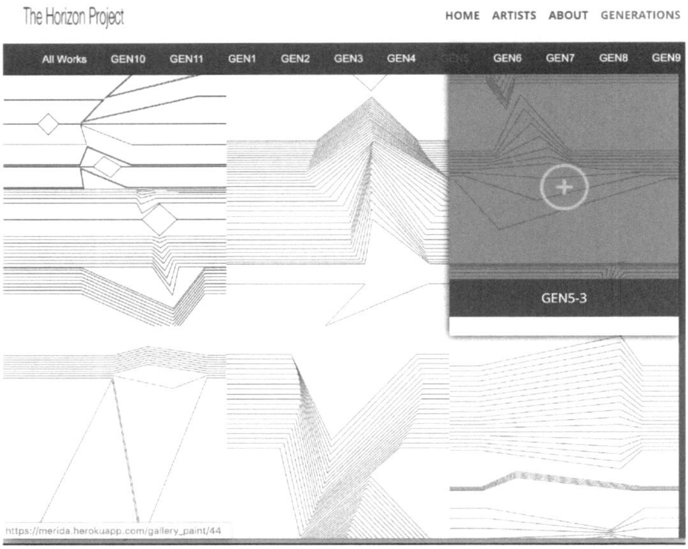

Fig. 6 Gallery view for a generation, in this case a visitor is selecting an artwork to see it in detail

To do so, 10 out of 60 works were selected for the survey, trying thus to avoid fatigue that may cause the inclusion of the 60 works. Figure 7 shows the works selected, with some comments that describe the reason for the selection. When two image share some features, they are grouped in the same column and displayed with a background color.

We have collected information from 22 surveys, and the analysis is also included in the next section.

4 Results

Figure 8 shows some of the works produced in the first generation. The collective work includes 60 drawings: six artists working for 10 weeks.[4]

Figure 9 shows the emotional-category that artists expressed when producing the work and completing the corresponding form. A "+" sign means a "positive" emotion; the sign "–" a negative one; and "#" means "neutral".

[4](Accessible at http://merida.herokuapp.com/masonry).

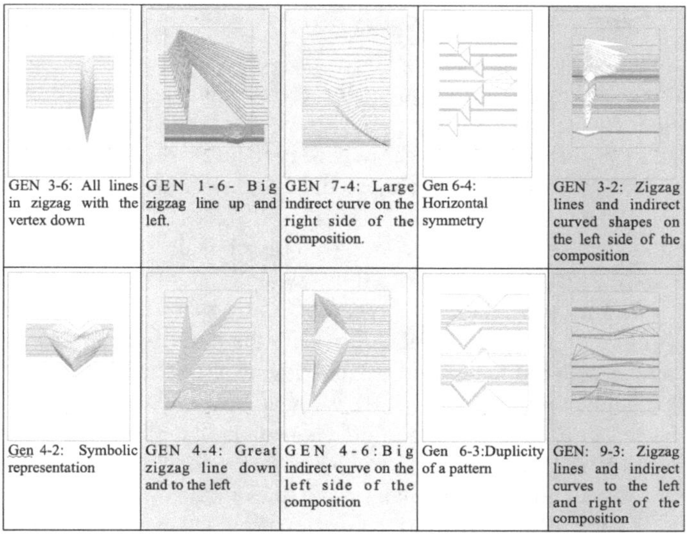

Fig. 7 Ten images selected for the survey

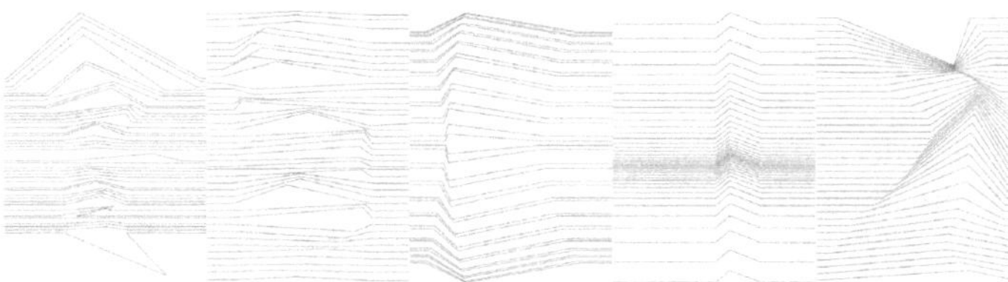

Fig. 8 Some of the works produced in the first generation

	GEN 1	GEN 2	GEN 3	GEN 4	GEN 5	GEN 6	GEN 7	GEN 8	GEN 9	GEN 10
1	#	-	+	-	#	+	+	+	+	-
2	#	-	-	+	#	+	#	+	-	-
3	-	+	+	+	+	+	+	-	+	+
4	+	+	+	-	+	#	+	-	-	+
5	+	-	+	-	+	-	+	#	-	-
6	-	#	-	-	#	+	+	+	#	#

Fig. 9 Emotions selected by artists when producing images (+ positive, – negative, # neutral.)

4.1 Analyzing Formal Elements

The main thing we have tried to study is the correlation among emotions expressed by artists and formal elements that can be seen in every drawing. Given the difficulty of such an analysis, in this paper, we have tried to focus on formal elements as simple as possible, and we hope the data collected will allow a more in-depth analysis in the future.

Therefore, to make things easier, we have only considered at this stage the general direction of zigzag lines: we have thus counted the number of zigzag lines up and down first. This is probably the most straightforward measure we can apply. Moreover, programming genetic operations with the aim of generating lines within a digital drawing with particular upward and downward patterns is an easy task.

In our first study, displayed in Fig. 10, we have begun with an initial hypothesis: correspondence of "downwards direction" as negative emotion (–), and "'upwards direction" as positive (+), (# corresponds to neutral) each of the cells includes any of these three codes attending to the majority of lines in the drawing. Then colors are used to express coincidence with the information provided by the authors (see Fig. 9): dark grey when we have positive-positive, neutral-neutral or negative-negative (complete coincidence); white when positive-negative or negative-positive is found, the remaining when any neutral value is found, which could somehow be paired with both positive or negative. This kind of "thermal" photograph allows us to study the percentage of correlation visually. As we notice, the initial hypothesis led to a low number of coincidences (a high number of white cells in the table).

We have thus changed the hypothesis as follows: we consider "downwards direction" as positive (+), and "upwards direction" as negative (–), (# corresponds to neutral). Figure 11 allows us to confront the formal element selected, broken lines'

	GEN 1	GEN 2	GEN 3	GEN 4	GEN 5	GEN 6	GEN 7	GEN 8	GEN 9	GEN 10
1	-	+	-	+	-	+	+	#	-	#
2	+	+	+	-	+	#	+	-	#	+
3	#	-	-	-	-	-	#	#	+	-
4	+	+	#	-	+	#	#	-	#	+
5	+	+	-	-	-	#	#	-	#	+
6	+	-	-	#	#	#	+	#	-	-

Fig. 10 Thermal map of emotion coincidences: comparison among broken-lines general direction (up-positive, down-negative) and authors emotional category selected; from white = no-coincidence, dark gray = whole coincidence

	GEN 1	GEN 2	GEN 3	GEN 4	GEN 5	GEN 6	GEN 7	GEN 8	GEN 9	GEN 10
1	+	-	+	-	+	-	-	#	+	#
2	-	-	-	+	-	#	-	+	#	#
3	#	+	+	+	+	+	#	#	-	+
4	-	-	#	+	-	#	#	+	#	-
5	-	-	+	+	+	#	#	+	#	-
6	-	+	+	#	#	#	-	#	+	+

Fig. 11 Thermal map of emotion coincidences: comparison among broken-lines (up-negative, down-positive) general direction and authors emotional category selected; from white = no-coincidence, dark gray = whole coincidence

general directions, and emotional category selected by authors when producing each of the works. And now things are different: only 24% of cells include the white color assigned (no-coincidence), and this means that in 76% of the cases, we have a total or partial coincidence among emotions and broken-lines direction.

Certainly, this study could be fine-tuned in the future by addressing more particular issues according to the typology or classification of the resulting design; and although coincidences or correlation among components does not necessarily imply a causal relationship, we think the findings showed above open doors to interesting future research on emotions and formal elements that can be properly addressed in evolutionary art. In any case, this starting point is employed in the next sections to analyze emotional perception.

4.2 Are Emotions Properly Understood?

Once emotions displayed by artists have been studied in relation to some formal elements, we have also tried to analyze how their peers perceive emotions. This can be accomplished by comparing what artists perceive in a given work produced by one of their peers, and the sentiment the author tried to convey. We must bear in mind that this information is provided by artists when they produce new artwork, and the emotional response induced by a work is only conveyed by the formal elements included in that same work, given that the information provided by artists, i.e. the emotional category, is not shown.

Figure 12 shows results obtained after comparing the emotion an artist expressed in a work with the sentiment category perceived by his colleagues when observing the work. As described before, neutral emotions assigned by an artist to the artwork are

	G1	G2	G3	G4	G5	G6	G7	G8	G9	G10
1	# 100%	- 100%	+ 20%	- 80%	# 40%	+ 60%	+ 20%	+ 60%	+ 20%	- 80%
2	# 80%	- 80%	- 80%	+ 60%	# 80%	+ 60%	# 40%	+ 20%	- 20%	- 60%
3	- 20%	+ 40%	+ 40%	+ 40%	+ 60%	+ 60%	+ 40%	- 40%	+ 80%	+ 40%
4	- 80%	+ 100%	+ 60%	- 100%	+ 60%	# 20%	+ 80%	- 0 %	- 40%	- 20%
5	+ 100%	- 60%	+ 0%	- 20%	+ 100%	- 20%	+ 40%	# 100%	- 80%	- 100%
6	- 100%	# 100%	- 80%	- 80%	# 40%	+ 60%	+ 60%	+ 80%	# 60%	# 60%

Fig. 12 Thermal map of emotion coincidences: comparison among what artists try to express and what their peers understand

always considered coincident (with any positive, negative or neutral). Percentages of coincidence and colors are used to see that large correspondences have been easily found. As we can see, there is a positive correlation between what artists try to express and what their peers understand. The average computed over the values displayed returns 60.66%. Although positive, this value is not too high.

Of particular interest are some drawings that receive a 0%. We do not have yet any cut, and clear conclusion, and more analysis will be devoted in the future to these specific cases.

4.3 Audience Analysis

The same analysis can be applied using information collected from the audience by means of surveys. We have collected information from 22 surveys, and a similar percentage analysis as with artists has been performed, so that we study coincidence among emotions artist display and the ones perceived by the audience. 10 out of 60 works were selected for the survey, as described before, trying thus to avoid fatigue that may cause the inclusion of the 60 works, and hit percentages computed.

If we compute the success rate considering all the images shown to the audience and information collected from the surveys, we obtained a success rate that is slightly higher for the artists (84%) (see Fig. 13) than for the audience (81.02%), and much higher than the value shown in Fig. 12.

Considering all the results described above, the main conclusion here is that both artists and the audience can recognize emotions. The element we analyzed, the line, and the layout created was useful for allowing artists to express their emotions.

If we may perform a more detailed analysis of results. For instance, if we focus in two pairs of designs with very pronounced zigzag lines: Pair #1) -GEN 1-6-GEN 4-4-, selected for having a large zigzag line up and down to the left, and Pair #2): -GEN 4-6-GEN 7-4-, chosen because of the sizeable indirect curve created on the

Fig. 13 Comparing audience and artists' perception. Blue and mustard colors mean a good understanding (more than 80%) by only one group, artists or audience; Mustard color means good understanding and coincidence by both groups, artists and audience. Neutral perception has been computed as a (partial-)hit to any emotion. On average, 81,04%

	ARTWORK	EMOTION	AUDIENCE	ARTISTS
GEN 1-6		-	50%	100%
GEN 3-2		-	72,70%	80%
GEN 3-6		-	81,80%	80%
GEN 4-2		+	81,40%	60%
GEN 4-4		-	68,20%	100%
GEN 4-6		-	90,90%	80%
GEN 6-3		+	91,50%	60%
GEN 6-4		#	100%	20%
GEN 7-4		+	86,40%	80%
GEN 9-3	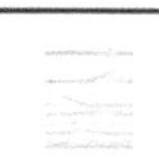	+	87,50	80%

right and left side of the composition; we see that the pair with a large indirect curve with zigzag lines has been better understood by both the public and the artists (Pair #2). On the other hand, if we focus on those works that combine zigzag lines with indirect curves (such as GEN 3-2 and GEN 9-3), the one with the highest number of curves (GEN 9-3) was better perceived (100% hit rate for the artists and 87.50% for the audiance). Although we have not explicitly studied the symbolic representation in this experiment, we may see that in some cases, such as GEN 4-2, it helps for better understanding, particularly for the audience which we found striking. The same is true of the design containing a duplicate pattern (GEN 6-3). Finally, in the case of the design with horizontal symmetry (GEN 6-4), we were surprised by the high percentage of success of the public (100%) and the artists (100%). The fact that this is an emotion that has been classified as neutral may have influenced the result.

Summarising, we can say first, that a correlation has been found among formal elements in the artworks (direction of lines) and emotions artists try to express; secondly, the emotion perceived by both artists and audience when analyzing works, correctly correlates with authors' intentions. This means that such a simple genetic operation consisting on deviating lines in a drawing up or down may somehow influence the emotions perceived by the team of artists, and thus an EA devoted to art -thus becoming an artist expressing emotions- that employs this simple element -the line- could express an emotion that is well understood by the audience.

On the other hand, we have seen that applying strong constraints in the way artwork is produced allowed us to perform an analysis that may easily improve EAs experiments. The constraint of using lines in a given direction, and how they can be easily changed with a mutation operator, has proven to by enough to convey emotional messages. Although only positive, negative or neutral emotions have been considered in this experiment, in future work a more specific set of emotions could be analyzed in the same way.

4.4 International Art Competitions

Although more in-depth analysis will provide useful information and will allow us to reach broader conclusions, we decided to submit the work to some international art competition: beyond the possible formal analysis, conclusions on algorithmic operations, etc., we also need to prove that artistic work produced by evolutionary means has value.

We submitted *the horizon project* to the international *Show Your World 2017* art competition.[5] The collective artwork was selected as one of the finalists and took part in the art exhibit at the MC Gallery, Manhattan, New York, in November 2017.

5 Conclusion

This paper presents an evolutionary art experiment developed using the unplugged evolutionary algorithm. We have been mainly interested in emotion analysis under this perspective, which is not frequently addressed in computational creativity.

The experience obtained in previous experiments based on UEAs, where artists display a broad set of behaviors, materials, ways of expressions, etc., led us to apply strong constraints on this new experiment. Forty zigzag lines that cannot cross are the only formal elements that artists employ to generate drawings inspired by a given emotion.

We have first analyzed the emotions translated and have categorized them into three groups: Positive, Neutral or Negative. Then, a detailed analysis on information

[5]http://www.reartiste.com/juried-exhibition-show-your-world-2017/.

provided by artists as well as several surveys filled by the audience have allowed us to reach some interesting conclusions: (i) there is a correlation about categories analyzed and general directions of the zigzag lines in every work; (ii) when artists analyze the work produced by a colleague, they generally understand the emotion portrayed; however (iii) the audience have shown an even better understanding of the works from the emotional point of view.

This analysis leads us to conclude firstly that the audience perception may be of interest to analyze the output produced by an evolutionary art experiment, particularly when no human artists are behind to explain what is being produced. Secondly, the connections found among formal elements, and emotional categories can be exploited to design specific operations -such as a general direction of broken lines in our case, that may be changed by means of mutation- that can easily be embodied within evolutionary algorithms devoted to art. Yet, we understand that this simplistic approach is just a beginning, and a more extended and detailed study should be made.

Finally, to asses the quality of the collective artwork produced, that we consider crucial in any art project, we sent it to an international artwork competition, no specifically devoted to evolutionary art or any other technique or movement, and the work was selected as a finalist. This is an additional interesting result that few times have been considered: evolutionary art may become a competitive art movement.

Acknowledgements We acknowledge support from Spanish Ministry of Economy and Competitiveness under project TIN2017-85727-C4-{2,4}-P, Regional Government of Extremadura, Department of Commerce and Economy, the European Regional Development Fund, a way to build Europe, under the project IB16035 and Junta de Extremadura, project GR15068, and CONACYT-PEI Project No. 220590.

References

Colton, S., M.F. Valstar, and M. Pantic. 2008. Emotionally aware automated portrait painting. In *Proceedings of the 3rd international conference on digital interactive media in entertainment and arts*, 304–311. ACM

Diaz-Jerez, G. 2011. Composing with melomics: Delving into the computational world for musical inspiration. *Leonardo Music Journal* 13–14.

Frade, M., Fernández de Vega, F., and C. Cotta. 2010. Evolution of artificial terrains for video games based on accessibility. In *European conference on the applications of evolutionary computation*, 90–99. Springer.

García-Valdez, M., L. Trujillo, F.F. de Vega, J.J. Merelo Guervós, and G. Olague. 2013. Evospace-interactive: A framework to develop distributed collaborative-interactive evolutionary algorithms for artistic design. In *Evolutionary and Biologically Inspired Music, Sound, Art and Design*, ed. P. Machado, J. McDermott, and A. Carballal, 121–132. Berlin: Springer.

Gervás, P. 2001. An expert system for the composition of formal Spanish poetry. In *Applications and innovations in intelligent systems VIII*, 19–32. Springer.

Kandinsky, W. 2012. *Concerning the spiritual in art*. Courier Corporation.

McCormack, J., and M. d'Inverno. 2012. Computers and creativity: The road ahead. In *Computers and creativity*, 421–424. Springer.

Minsky, M. 2007. *The emotion machine: Commonsense thinking, artificial intelligence, and the future of the human mind*. Simon and Schuster.

y Pérez, R.P. 2007. Employing emotions to drive plot generation in a computer-based storyteller. *Cognitive Systems Research* 8 (2): 89–109.

Pérez, R.P.Ý., and M. Sharples. 2001. Mexica: A computer model of a cognitive account of creative writing. *Journal of Experimental & Theoretical Artificial Intelligence* 13 (2): 119–139.

Posner, J., J.A. Russell, and B.S. Peterson. 2005. The circumplex model of affect: An integrative approach to affective neuroscience, cognitive development, and psychopathology. *Development and Psychopathology* 17 (3): 715–734.

Reichardt, J. 1969. *Cybernetic serendipity: The computer and the arts*. Praeger.

Romero, J.J., and P. Machado. 2008. *The art of artificial evolution: A handbook on evolutionary art and music*. Springer Science & Business Media.

Takagi, H. 1998. Development and validation of brief measures of positive and negative affects: The pannas scales. *Journal of Personality and Social Psychology* 54 (6): 1063–1070.

Takagi, H. 2001. Interactive evolutionary computation: Fusion of the capabilities of EC optimization and human evaluation. *Proceedings of the IEEE* 89 (9): 1275–1296.

de Vega, F.F., C. Cruz, L. Navarro, P. Hernández, T. Gallego, and L. Espada. 2014. Unplugging evolutionary algorithms: An experiment on human-algorithmic creativity. *Genetic Programming and Evolvable Machines* 15 (4): 379–402.

de Vega, F.F., L. Navarro, C. Cruz, F. Chavez, L. Espada, P. Hernandez, and T. Gallego 2013. Unplugging evolutionary algorithms: On the sources of novelty and creativity. In *2013 IEEE congress on evolutionary computation (CEC)*, 2856–2863. IEEE.

de Vega, F.F., T. Gallego, M. García-Valdez, L. Espada, L. Navarro, and V. Albarrán. 2014. When artists met evospace-i. In *2014 IEEE congress on evolutionary computation (CEC)*, 2282–2289. IEEE.

de Vega, F.F. 2017. Revisiting the 4-part harmonization problem with gas: A critical review and proposals for improving. In *2017 IEEE congress on evolutionary computation, CEC 2017*, Donostia, San Sebastián, Spain, June 5–8, 2017, 1271–1278. IEEE. https://doi.org/10.1109/CEC.2017.7969451

Wehrle, T. 2001. The grounding problem of modeling emotions in adaptive artifacts. *Cybernetics & Systems* 32 (5): 561–580.

Review of Hybrid Combinations of Metaheuristics for Problem Solving Optimization

Marylu L. Lagunes, Oscar Castillo, Fevrier Valdez, and Jose Soria

Abstract This article describes a review of the state of the art of some of the different metaheuristic combinations that exist for solving problems, using two or more methods in combination or hybrid form. There are different nature-inspired or metaheuristic algorithms, which have been classified to solve certain types of problems, although currently, trying to solve problems with a single method has been somewhat neglected, and modifications, hybridizations, among others, have been carried out. Combinations to solve a common problem, provides an opportunity to exploit the methods in different scenarios to be solved, focusing not only on the solution but also on the use of algorithms.

Keywords Metaheuristics · Hybrid- combined metaheuristics · Optimization

1 Introduction

Combinations and hybridizations have become very popular in the area of computer science, as they have proven to be effective in real-world implementations. Sometimes combinations are made to strengthen the weaknesses of metaheuristics. This combination can be between two or more methods, either data combination, combination of some equation with another one, among other hybridization techniques, in order to generate a hybrid algorithm that is better than the separate methods.

As an example we can discuss the following works. In Chaimatanan (2014), a hybrid methodology for strategic trajectory planning is developed that aims to minimize interaction between aircraft on a European continent scale. In addition, in Mortazavi et al. (2018) it modifies and combines the characteristics of two developed metaheuristic methods, called Integrated Particle Swarm Optimization (IPSO) and Teaching and Learning Based Optimization (TLBO). Among other interesting hybridizations, we find in Kaveh et al. (2014) a combination of Swallow Swarm Optimization (SSO) with Particle Swarm Optimization (PSO) to form the Hybrid Particle

M. L. Lagunes · O. Castillo (✉) · F. Valdez · J. Soria
Tijuana Institute of Technology, Tijuana, BC, Mexico
e-mail: ocastillo@tectijuana.mx

O. Castillo and P. Melin (eds.), *Fuzzy Logic Hybrid Extensions of Neural and Optimization Algorithms: Theory and Applications*, Studies in Computational Intelligence 940,
https://doi.org/10.1007/978-3-030-68776-2_12

Swarllow Swarm Optimization (HPSSO) algorithm. A hybrid algorithm formed in Niknam and Amiri (2010) using Particle Swarm Optimization (PSO) and Ant Colony Optimization (ACO) for data grouping making a better search for optimal grouping. There are different forms of hybridizations used for the improvement and optimization of methods or results as described in the following references (Mafarja and Mirjalili 2017; Lanza-Gutierrez and Gomez-Pulido 2015; Potluri and Singh 2013; López-Camacho et al. 2015; Yi et al. 2013). The selection of articles considered in this review was made using the online system called EBSCO Connect, where articles can be searched by subject name or author. A search for documents on hybridizations or metaheuristic combinations was performed.

The article is structured as follows: Sect. 2 outlines a review of hybrid or combined metaheuristics. Section 3 presents the discussion of the most relevant papers found in the current literature. Finally, Sect. 4 presents the conclusions.

2 Review of Hybrid or Combined Metaheuristics

This section outlines a group of papers where two or more metaheuristics are combined with the idea of producing powerful synergies that enable achieve better optimization results. Figure 1 describes the taxonomy of some Metaheuristics methods, in which it can be seen that they have two important classifications: specifies heuristics and metaheuristics; In those classified as specific, there is an algorithm widely used such as Tabu search, on the other hand, the methods that work in population to reach a global optimum since they are multidimensional; Each of the illustrated heuristics and metaheuristics have been part of important contributions and has been used to solve different problems, as it will be possible to read in this investigation the way that the meta heuristic algorithms have been adapted as hybridizations have

Fig. 1 Taxonomy of some heuristic methods

Metaheuristics

Solution Based	**Based on population**
Hill Climbing	Genetic Algorithms (GA)
Simulated annealing (SA)	Particle Swarm Optimization (PSO)
Tabu Search (TS)	Ant Colony Optimization (ACO)
	FireFly Algorithm (FA)
	Differential Evolution (DE)
	Others

been developed to achieve a more efficient result. Combining its best characteristics with others to achieve a better method.

It is important to know how metaheuristics are being used and in what types of problems are giving better results, based this idea this research work was developed taking 10 hybridization articles or combination of metaheuristics that have been developed between the years 2010 and 2020, taking the most relevant articles in combination of swarm intelligence.

Metaheuristics	
Solution based	Based on population
Hill climbing	Genetic Algorithms (GA)
Simulated Annealing (SA)	Particle Swarm Optimization (PSO)
Tabu Search (TS)	Ant Colony Optimization (ACO)
	FireFly Algorithm (FA)
	Differential Evolution (DE)
	Others

Ludo game-based metaheuristics for global and engineering optimization (Singh 2019). The focus of this article was to improve the ability of swarm algorithms, using Ludo game-based swarm intelligence (LGSI) (Parlett 2020), using several metaheuristics to simulate players; Moth Flame Optimization (MFO) (Mirjalili 2015), Grey Wolf Optimizer (GWO) (Mirjalili et al. 2014), Sine Cosine Algorithm (SCA) (Mirjalili 2016) and Grasshopper Optimization Algorithm (GOA) (Saremi et al. 2017), are used to solve optimization problems with the Exploration and exploitation to simulate the Ludo game rules. In this case, two or four players are used to perform the update process for different intelligent swarming behaviors. Each player is represented by methods that share the same platform with this strategy, so in this case, competitive behavior may not be underestimated. The proposed LGSI algorithm shares positions among all the algorithms used during the search for the optimal solution. The performance of the LGSI algorithm is tested on a set of CEC2005 benchmark problems and engineering problems and compared against the original versions of the algorithms used and a variety of other next-generation algorithms. The results show that the LGSI algorithm can provide promising and competitive results.

Metaheuristic algorithms for approximate solution to ordinary differential equations of longitudinal fins having various profiles (Sadollah et al. 2015). To solve ordinary differential equations (ODE) in engineering many methods have been applied for approximating analytical solutions, these approximation methods can be metaheuristic algorithms, which can be changed depending on the user's preference. In this case particle swarm optimization are used (PSO), genetic algorithm (GA) (Holland 1992) and search for harmony (HS) (Woo Geem et al. 2001), are used since these methods provide highly accurate approximate solutions. The values of the main parameters for PSO were modified for each method: population size, inertia weight (w) for speed and cognitive and social components. In the case of GA population size and crossing rates and mutation, and finally in the HS the size of the harmonies, all with the intention of minimizing the residual function of the ODE error.

A hybrid metaheuristic and kernel intuitionistic fuzzy c-means algorithm for cluster analysis (Kuo et al. 2018). This article uses three widely used metaheuristics, particle swarm optimization (PSO), genetic algorithm (GA) and artificial bee colony algorithms (ABC) (Karaboga and Basturk 2007) to propose an evolution-based clustering algorithm that combines metaheuristic algorithms with an intuitionistic fuzzy c-means algorithm (KIFCM) (Lin 2014). The proposed algorithms, hybridized to PSO-KIFCM, GA-KIFCM, and ABC-KIFCM algorithms, are evaluated using six reference data sets to demonstrate that the proposed algorithm shows better precision when using local and global search. The metaheuristic algorithms search for the optimal centroids, the population represents the centroid of each group. Here, a solution in the initial population is taken from the KIFCM result. The other solutions in the initial population are chosen randomly from the data points. This initial population is updated using the PSO, GA and ABC algorithms. The good of each solution is the sum of the distance squared within the group, the distance between two data points is calculated using the Euclidean distance.

An ACO hybrid metaheuristic for close–open vehicle routing problems with time windows and fuzzy constraints (Brito et al. 2015). Currently, tasks are being carried out where some companies provide services to others and the traveled distance is paid, as having to return to their place of origin causes a routing problem, the objective of this article is to formulate a model with windows of imprecise time and constraints and propose a fuzzy optimization approach and hybrid metaheuristics for your solutions. The vehicle routing problem with time windows (COVRPTW) model is proposed to minimize the costs of all operations, solving route planning problems. The proposal uses a hybrid ACO-GRASP-VNS metaheuristic to solve COVRPTW with fuzzy constraints. The application of Ant Colony Optimization (ACO) improves performance through the exploration of new solutions. The combination enables a constructive search for solutions that provide new random solutions by Greedy Randomized Adaptive Search Procedure (GRASP) (Feo and Resende 1995) and the historical memory process introduced by ACO to converge to the global optimum. Variable Neighborhood Search (VNS) (Hansen 2010) is presented as a systematic process of local search to improve solutions. ACO is combined with GRASP. The use of ACO is used for the ants in each iteration to build the solution paths in conjunction with GRASP, and VNS improves the solution obtained through movements from nested neighborhoods.

A hybrid metaheuristic for multiobjective unconstrained binary quadratic programming (Liefooghe et al. 2014). This article proposes the hybridization of an elitist multiobjective evolutionary algorithm and a local search algorithm, this to form a model that generates mUBQP instances (Neri et al. 2012) to obtain results with two or three objectives. unconstrained binary quadratic programming (UBQP) is a unified modeling and solution framework for many combinatorial optimization problems. For this reason, this conventional problem of single-objective UBQP is extended to the multiobjective case mUBQP, where multiple targets must be optimized simultaneously. Hybridization of elitist evolutionary multiobjective optimization algorithms is proposed as a state-of-the-art single-objective and tabu search (TS) based procedure by using an scalar function (Knowles et al. 2006; Glover et al. 2010).

A comparison of five hybrid metaheuristic algorithms for unrelated parallel-machine scheduling and inbound trucks sequencing in multi-door cross docking systems (Liao et al. 2014). This article describes the hybridization of five metaheuristics for a comparative study in the solution of two problems. The first is the parallel programming of unrelated machines and the second one is the sequencing of incoming trucks in a cross coupling of several doors system in consideration of the sequence dependent setting, and the zero and non-zero release time. Of the algorithms used for the study, three are hybrid ant colony optimization (ACO) algorithms (Behnamian et al. 2009; Arnaout et al. 2010; Keskinturk et al. 2012) and the remaining two are hybrid simulated annealing algorithms. Another one is a simulated annealing Differential Evolution (DE) (Angira and Babu 2006), while the other is a simulated annealing tabu search hybrid (TS).

PSOGSA-Explore: A new hybrid metaheuristic approach for beam pattern optimization in collaborative beam forming (Jayaprakasam et al. 2015). In a System where nodes are used, it is very common that there is a need to use meta heuristics for the reduction of randomness, since due to the positioning, the result may or may not be satisfactory. As this article describes the hybridization of Particle Swarm Optimization and Gravitational Search Algorithm-Explore (PSOGSA-E) to lower the maximum level of peak sidelobe level (PSL) in a conventional collaborative beamforming (CB) (Ochiai et al. 2005). The hybrid method seeks to find the best weight for each node. With the local search capabilities of the Gravitational Search Algorithm (GSA) (Rashedi et al. 2009) with the Social Thinking Skills of Particle Swarm Optimization (PSO) that enables exploration to avoid premature convergence. The hybrid approach will also help optimize the beam pattern of collaborative beamforming, which focuses on optimizing the beam current weight vector to achieve lower side lobes.

Hybridizations of genetic algorithms and neighborhood search metaheuristics for fuzzy bus terminal location problems (Babaie-Kafaki et al. 2016). This article proposes the hybridization of various methods based on neighborhood search with genetic algorithms (Han et al. 2005; Gao et al. 2008; Drezner 2008), to test their effectiveness, and the proposed algorithms are applied to problems of location of fuzzy bus terminals (Ghanbari and Mahdavi-Amiri 2011). The fuzzy model has a fuzzy number of passengers corresponding to the nodes, as well as fuzzy neighborhoods, along with preassigned upper and lower limits for the number of terminals required. The algorithms are tested on a variety of randomly generated large-scale fuzzy bus terminal location problems with fuzzy cost coefficients. The fuzzy lens is transformed into a sharp one by using a sort function. Hybridization was developed by some of the most popular neighborhood search metaheuristics, such as the genetic algorithm (GA), the simulated annealing (SA), the variable neighborhood search (VNS), the ant colony optimization (ACO) and tabu search (TS).

Two metaheuristic approaches for the multiple traveling salesperson problem (Venkatesh and Singh 2015). In this work, two metaheuristics are described that are used to solve the problem of the multiple street salesperson (MTSP) (Tang et al. 2000). In this problem there is more than one salesperson to visit the cities, although each city must be visited exactly once by a single salesperson. Two different objectives have been considered for this problem. The first is to minimize the total distance traveled by all salespersons, and the second is to minimize the maximum distance traveled by any vendor. The artificial bee colony algorithm (ABC), and the invasive weed optimization algorithm (IWO) (Mehrabian and Lucas 2006) swarm intelligence-based metaheuristics for the MTSP. In the first option, neighboring solutions are generated more or less the same distance from the original solution throughout the execution of the ABC algorithm, while the second option, the expected distance of the neighboring solution from the original solution is gradually reduced from a predefined start value to a predefined end value over the algorithm iterations.

Metaheuristics optimization applied to PI controllers tuning of a DTC-SVM drive for three-phase induction motors (Galvão Costa et al. 2018). The optimization proposed in this article is for direct torque control with space vector modulation (DTC-SVM) of a three-phase induction motor. Using the Ant Colony Optimization (ACO) and Differential Evolution (DE) algorithms to obtain an optimization of the tuning of proportional-integral (PI) controllers in the DTC-SVM control loops, such as rotor speed, electromagnetic torque, stator flux and link stator flux estimation (Restrepo et al. 2011; Hari Krishna, et al. 2012; Kim 1996). The metaheuristics were chosen because they have low computational complexity, few adjustment parameters and a good convergence rate.

3 Discussion

In this article an overview of hybrid or combined metaheuristics for optimization problem solving was presented. Metaheuristics are an extension of heuristics since they look beyond a local optimum and for this reason they are used more to provide more effective and concise results. As being inspired algorithms they make the search more efficient by having two main characteristics such as exploitation and exploration, and it is also important to mention that randomness and stochasticity provide a better convergence. The results obtained from this review show that most of the methods are focused on swarm intelligence metaheuristics because from the 10 reviewed articles, 9 are based on swarm intelligence, as can be noted in Fig. 2. Also in Fig. 3 the frequency of the methods used more than twice in the reviewed papers are illustrated, and Fig. 4 shows the frequency of the number of hybridized or combined metaheuristics in each article.

Figure 5. shows the frequency of readers that each article of this review has, as it is observed in Two metaheuristic approaches for the multiple traveling salesperson problem (Venkatesh and Singh 2015) has had over 50 readers, the one that has had

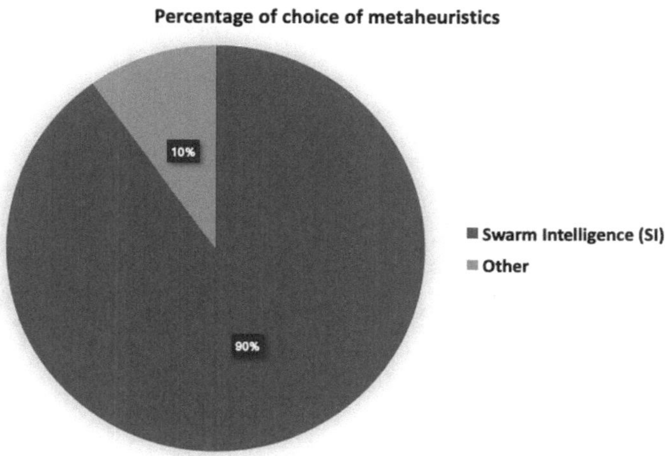

Fig. 2 Percentage of choice of metaheuristics

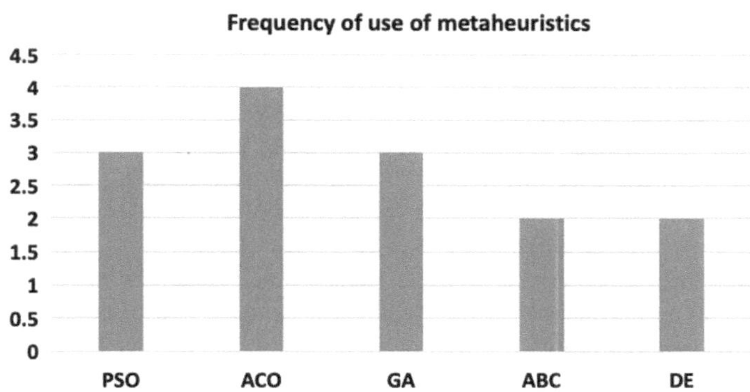

Fig. 3 Frequency of use of metaheuristics

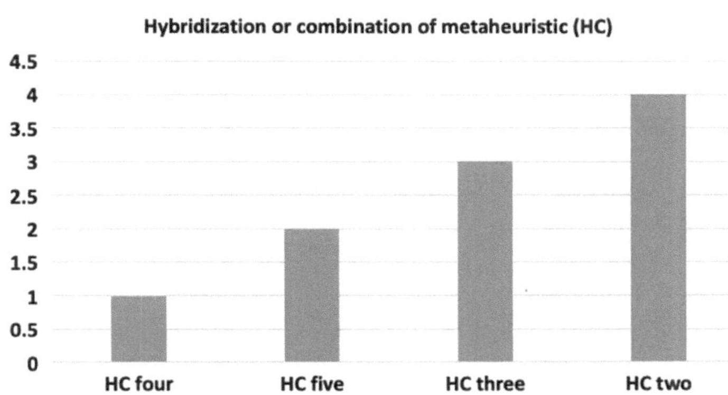

Fig. 4 Hybridization or combination of metaheuristic

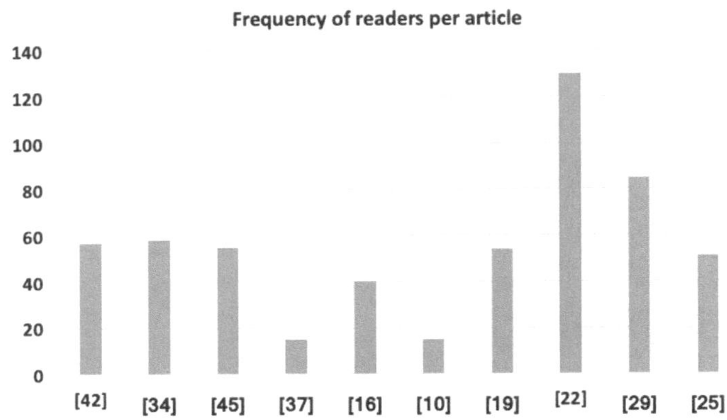

Fig. 5 Frequency of readers per article

more reading concurrence is the article An ACO hybrid metaheuristic for close–open vehicle (Brito et al. 2015) having over 120 readers, among the less read are Hybridizations of genetic algorithms and neighborhood search metaheuristics for fuzzy bus terminal location (Babaie-Kafaki et al. 2016) and Ludo game-based metaheuristics for global and engineering optimization (Singh 2019) with less than 20, continuing with the analysis we can note that of 10 articles reviewed, 6 are between 40 to 60 readers, 2 are with less than 20, and 2 are above 80 readers. It is important to highlight that hybridization or combination of metaheuristics is still a field that has a lot to offer, as new algorithms with better characteristics are being developed every time, which can be combined with other methods and make the most of each one, that is why the importance of knowing each new existing hybrid method (Fig. 6).

As for article export-saves, in this review we can see that of articles reviewed, only 3 of them were export-saves, therefore, we can deduce that researchers in this type of article prefer to read online.

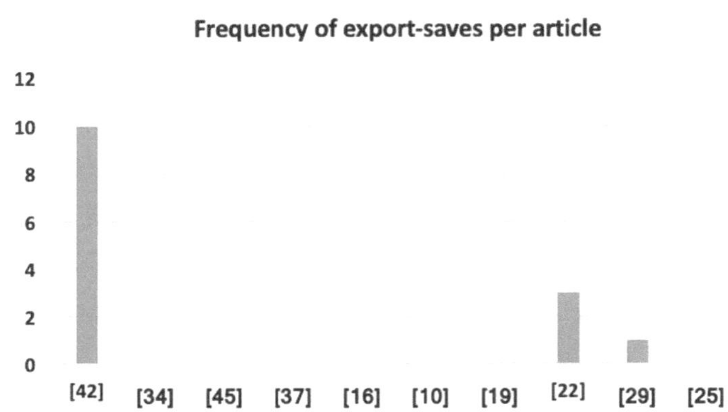

Fig. 6 Frequency of export-saves per article

Most researchers when looking for a specific topic, the first thing they focus on is the title, because it gives an idea of what it can deal with, but where they make sure that the article has what they are looking for, it is the abstract as it describes the focus of the article, also gives a broad vision of the objective to be achieved and future. As for example in Fig. 7 we can notice the number of views to the abstract of each article in this review, the one with the highest views is Ludo game-based metaheuristics for global and engineering optimization (Singh 2019) with more than 60.

Figure 8 shows that an ACO hybrid metaheuristic for close—open vehicle routing problems with time windows and fuzzy constraints (Brito et al. 2015), has above 12 link-out.

In Fig. 9 the frequency of citation of the articles is observed, the most cited hybridization methods have been the reference Two metaheuristic approaches for the multiple traveling salesperson problem (Venkatesh and Singh 2015) and PSOGSA-Explore: A new hybrid metaheuristic approach for beampattern optimization in collaborative (Jayaprakasam et al. 2015) where you can find hybrid combinations of ABC with IWO and GSA with PSO respectively.

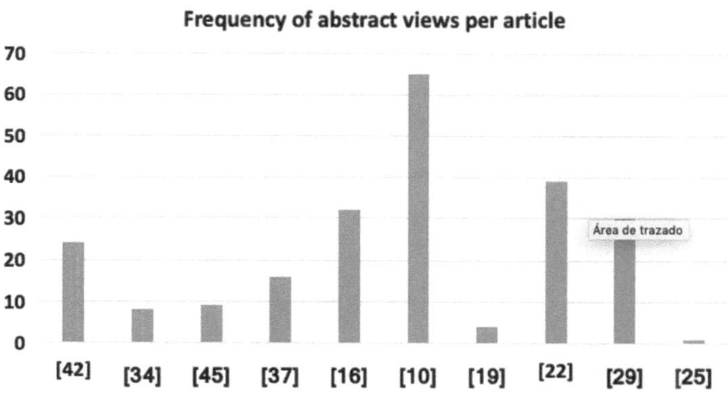

Fig. 7 Frequency of abstract views per article

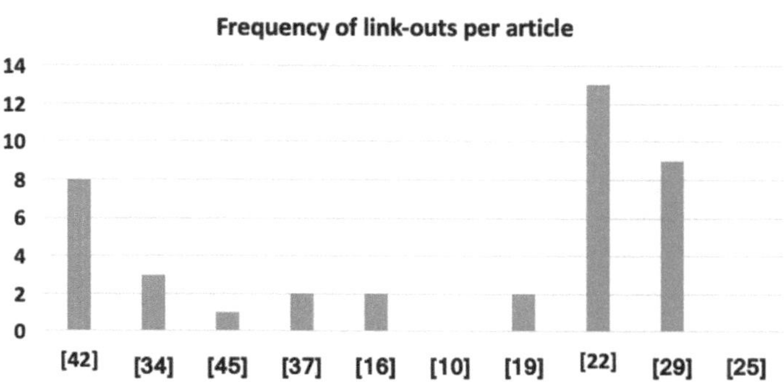

Fig. 8 Frequency of link-out per article

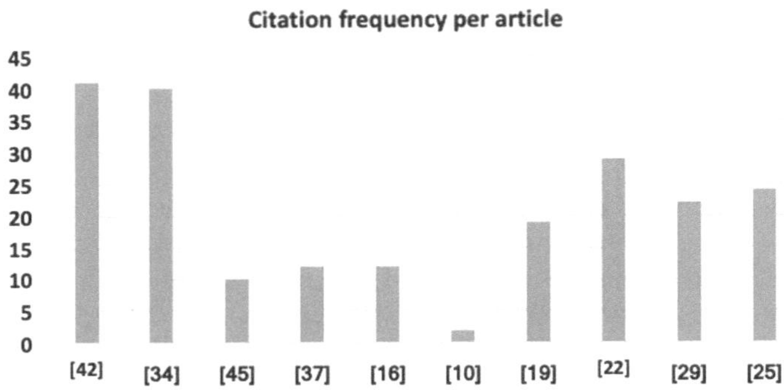

Fig. 9 Citation frequency per article

4 Conclusions

As can be noted in the graphs mentioned above, most of the metaheuristic combinatorial methods were including some swarm intelligence algorithm, because of the articles found in the literature from 2010 to 2020, their optimization approach was using SI, thus verifying that metaheuristic algorithms came to revolutionize the way to find solutions. This was not only in its search form using exploitation and exploration, but also in its multidimensional search space, giving this feature a great advantage in finding more possible solutions. In addition, being inspired by nature makes them easier to understand their behavior, for this reason when we think of hybridizing or combining some metaheuristics, analyzing each one is faster and when we put together the characteristics of each one in combination we can achieve more reliable and concise results. Future trends in this area are the use of new metaheuristics for the solution of benchmark problems developing hybridizations, to later use them in the different areas where it is believed that they can give good performance based on previous experimentation.

References

Angira, R., and B.V. Babu. 2006. Optimization of process synthesis and design problems: A modified differential evolution approach. *Chemical Engineering Science* 61 (14): 4707–4721.

Arnaout, J.P., G. Rabadi, and R. Musa. 2010. A two-stage Ant Colony optimization algorithm to minimize the makespan on unrelated parallel machines with sequence-dependent setup times. *Journal of Intelligent Manufacturing* 21 (6): 693–701.

Babaie-Kafaki, S., R. Ghanbari, and N. Mahdavi-Amiri. 2016. Hybridizations of genetic algorithms and neighborhood search metaheuristics for fuzzy bus terminal location problems. *Applied Soft Computing* 46: 220–229.

Behnamian, J., M. Zandieh, and S. M. T. Fatemi Ghomi. 2009. Parallel-machine scheduling problems with sequence-dependent setup times using an ACO, SA and VNS hybrid algorithm. Expert Systems with Applications 36 (6): 9637–9644.

Brito, J., F.J. Martínez, J.A. Moreno, and J.L. Verdegay. 2015. An ACO hybrid metaheuristic for close-open vehicle routing problems with time windows and fuzzy constraints. *Applied Soft Computing* 32: 154–163.

Chaimatanan, S., D. Delahaye, M. Mongeau. 2014. Hybrid metaheuristic optimization algorithm for strategic planning of 4D aircraft trajectories at the continent scale A hybrid metaheuristic optimization algorithm for strategic planning of 4D aircraft trajectories at the continental scale. 9 (4): 46–61. ieeexplore.ieee.org.

Drezner, Z. 2008. Extensive experiments with hybrid genetic algorithms for the solution of the quadratic assignment problem. *Computers & Operations Research* 35 (3): 717–736.

F. Neri, C. Cotta, and P. Moscato, Handbook of Memetic Algorithms. 2012.

Feo, T.A., and M.G.C. Resende. 1995. Greedy randomized adaptive search procedures. *Journal of Global Optimization* 6 (2): 109–133.

Galvão Costa, B. L., C. Luiz Graciola, B. A. Angélico, A. Goedtel, and M. F. Castoldi. 2018. Metaheuristics optimization applied to PI controllers tuning of a DTC-SVM drive for three-phase induction motors. *Applied Soft Computing* 62: 776–788.

Gao, J., L. Sun, and M. Gen. 2008. A hybrid genetic and variable neighborhood descent algorithm for flexible job shop scheduling problems. *Computers & Operations Research* 35 (9): 2892–2907.

Ghanbari, R., and N. Mahdavi-Amiri. 2011. Solving bus terminal location problems using evolutionary algorithms. *Applied Soft Computing Journal* 11 (1): 991–999.

Glover, F., Z. Lü, and J. K. Hao. 2010. Diversification-driven tabu search for unconstrained binary quadratic problems. *4OR*, 8 (3): 239–253.

Han, S., W. Pedrycz, and C. Han. 2005. Nonlinear channel blind equalization using hybrid genetic algorithm with simulated annealing. *Mathematical and Computer Modelling* 41 (6–7): 697–709.

Hansen, P., N. Mladenović, and J. A. Moreno Pérez. 2010. Variable neighbourhood search: Methods and applications. *Annals of Operations Research* 175 (1): 367–407.

Hari Krishna, C., J. Amarnath, and S. Kamakshaiah. 2012. Simplified SVPWM algorithm for neutral point clamped 3-level inverter fed DTC-IM drive. In *2012 International conference on advances in power conversion and energy technologies, APCET (2012).*

Holland, J.H. 1992. Genetic algorithms understand genetic algorithms. *Scientific American* 267 (1): 66–73.

Jayaprakasam, S., S. Rahim, and C. Yen Leow. 2015. PSOGSA-Explore: A new hybrid metaheuristic approach for beampattern optimization in collaborative beamforming. *Applied Soft Computing* 30, 229–237.

Karaboga, D., and B. Basturk. 2007. A powerful and efficient algorithm for numerical function optimization: Artificial bee colony (ABC) algorithm. *Journal of Global Optimization* 39 (3): 459–471.

Kaveh, A., T. Bakhshpoori, and E. Afshari. 2014. An efficient hybrid Particle Swarm and Swallow Swarm Optimization algorithm. *Computers & Structures* 143: 40–59.

Keskinturk, T., M.B. Yildirim, and M. Barut. 2012. An ant colony optimization algorithm for load balancing in parallel machines with sequence-dependent setup times. *Computers & Operations Research* 39 (6): 1225–1235.

Kim, J.-S., and S.-K. Sul. 1996. A novel voltage modulation technique of the space vector PWM. *IEEJ* Journal of Industry *Applications* 116 (8): 820–825.

Knowles, J., and D. Corne. 2006. Memetic algorithms for multiobjective optimization: Issues, methods and prospects. In *Recent advances in memetic algorithms*. Springer-Verlag. 313–352.

Kuo, R.J., T.C. Lin, F.E. Zulvia, and C.Y. Tsai. 2018. A hybrid metaheuristic and kernel intuitionistic fuzzy c-means algorithm for cluster analysis. *Applied Soft Computing* 67: 299–308.

Lanza-Gutierrez, J.M., and J.A. Gomez-Pulido. 2015. Assuming multiobjective metaheuristics to solve a three-objective optimisation problem for relay node deployment in wireless sensor networks. *Applied Soft Computing* 30: 675–687.

Liao, T.W., P.C. Chang, R.J. Kuo, and C.-J. Liao. 2014. A comparison of five hybrid metaheuristic algorithms for unrelated parallel-machine scheduling and inbound trucks sequencing in multi-door cross docking systems. *Applied Soft Computing* 21: 180–193.

Liefooghe, A., S. Verel, and J.-K. Hao. 2014. A hybrid metaheuristic for multiobjective unconstrained binary quadratic programming. *Applied Soft Computing* 16: 10–19.
Lin, K.P. 2014. A novel evolutionary kernel intuitionistic fuzzy C-means clustering algorithm. *IEEE Transactions on Fuzzy Systems* 22 (5): 1074–1087.
López-Camacho, E., M. Jesús, G. Godoy, J. García-Nieto, A.J. Nebro, and J.F. Aldana-Montes. 2015. Solving molecular flexible docking problems with metaheuristics: A comparative study. *Applied Soft Computing* 28: 379–393.
Mafarja, M.M., and S. Mirjalili. 2017. Hybrid Whale Optimization Algorithm with simulated annealing for feature selection. *Neurocomputing* 260: 302–312.
Mehrabian, A.R., and C. Lucas. 2006. A novel numerical optimization algorithm inspired from weed colonization. *Ecological Informatics* 1 (4): 355–366.
Mirjalili, S. 2015. Moth-flame optimization algorithm: A novel nature-inspired heuristic paradigm. *Knowledge-Based System* 89: 228–249.
Mirjalili, S. 2016. SCA: A sine cosine algorithm for solving optimization problems. *Knowledge-Based System* 96: 120–133.
Mirjalili, S., S.M. Mirjalili, and A. Lewis. 2014. Grey Wolf optimizer. *Advances in Engineering Software* 69: 46–61.
Mortazavi, A., V. Toğan, and A. Nuhoğlu. 2018. Interactive search algorithm: A new hybrid metaheuristic optimization algorithm. *Engineering Applications of Artificial Intelligence* 71: 275–292.
Niknam, T., and B. Amiri. 2010. An efficient hybrid approach based on PSO, ACO and k-means for cluster analysis. *Applied Soft Computing Journal* 10 (1): 183–197.
Ochiai, H., P. Mitran, H.V. Poor, and V. Tarokh. 2005. Collaborative beamforming for distributed wireless ad hoc sensor networks. *IEEE Transactions on Signal Processing* 53 (11): 4110–4124.
Parlett, S. David. The Oxford History of Board Games,... - Google Académico. https://scholar.google.com.mx/scholar?hl=es&as_sdt=0%2C5&q=Parlett%2C+S.+David%2C+The+Oxford+History+of+Board+Games%2C+Oxford+University+Press%2C+USA%2C+1999.&btnG. Accessed 29 May 2020.
Potluri, A., and A. Singh. 2013. Hybrid metaheuristic algorithms for minimum weight dominating set. *Applied Soft Computing* 13: 76–88.
Rashedi, E., H. Nezamabadi-pour, and S. Saryazdi. 2009. GSA: A gravitational search algorithm. *Information Sciences (Ny)* 179 (13): 2232–2248.
Restrepo, J., J.M. Aller, A. Bueno, V.M. Guzmán, and M.I. Giménez. 2011. Generalized algorithm for pulse width modulation using a two-vectors based technique. *EPE Journal* 21 (2): 30–39.
Sadollah, A., Y. Choi, G. Yoo, and J.H. Kim. 2015. Metaheuristic algorithms for approximate solution to ordinary differential equations of longitudinal fins having various profiles. *Applied Soft Computing* 33: 360–379.
Saremi, S., S. Mirjalili, and A. Lewis. 2017. Grasshopper optimisation algorithm: Theory and application. *Advances in Engineering Software* 105: 30–47.
Singh, P. R., M. Abd Elaziz, and S. Xiong. 2019. Ludo game-based metaheuristics for global and engineering optimization. *Applied Soft Computing*, 84, 105723.
Tang, L., J. Liu, A. Rong, and Z. Yang. 2000. A multiple traveling salesman problem model for hot rolling scheduling in Shanghai Baoshan Iron & Steel Complex. *European Journal of Operational Research* 124 (2): 267–282.
Venkatesh, P., and A. Singh. 2015. Two metaheuristic approaches for the multiple traveling salesperson problem. *Applied Soft Computing* 26: 74–89.
Woo Geem, Z., J. Hoon Kim, and G. V Loganathan. 2001. *A new heuristic optimization agorithm: Harmony search.*
Yi, H., Q. Duan, and T.W. Liao. 2013. Three improved hybrid metaheuristic algorithms for engineering design optimization. *Applied Soft Computing* 13: 2433–2444.

GPU Accelerated Membrane Evolutionary Artificial Potential Field for Mobile Robot Path Planning

Ulises Orozco-Rosas, Kenia Picos, Oscar Montiel, and Oscar Castillo

Abstract This work presents a graphics processing unit (GPU) accelerated membrane evolutionary artificial potential field (MemEAPF) algorithm implementation for mobile robot path planning. Three different implementations are compared to show the performance, effectiveness, and efficiency of the MemEAPF algorithm. Simulation results for the three different implementations of the MemEAPF algorithm, a sequential implementation on CPU, a parallel implementation on CPU using the open multi-processing (OpenMP) application programming interface, and the parallel implementation on GPU using the compute unified device architecture (CUDA) are provided to validate the comparative and analysis. Based on the obtained results, we can conclude that the GPU implementation is a powerful way to accelerate the MemEAPF algorithm because the path planning problem in this work has been stated as a data-parallel problem.

Keywords Membrane computing · Genetic algorithms · Artificial potential field · Path planning · Mobile robots · Graphics processing unit

---

U. Orozco-Rosas · K. Picos
CETYS Universidad, Centro de Innovación y Diseño (CEID), Av. CETYS Universidad. No. 4., Fracc, 22210 El Lago, Tijuana, Baja California, Mexico
e-mail: ulises.orozco@cetys.mx

K. Picos
e-mail: kenia.picos@cetys.mx

O. Montiel (✉)
Instituto Politécnico Nacional, CITEDI-IPN, Av. Instituto Politécnico Nacional No. 1310, 22435 Nueva Tijuana, Tijuana, Baja California, Mexico
e-mail: oross@ipn.mx

O. Castillo
Tecnológico Nacional de México, Calzada Del Tecnológico S/N, Tomas Aquino, 22414 Tijuana, Baja California, Mexico
e-mail: ocastillo@tectijuana.mx

O. Castillo and P. Melin (eds.), *Fuzzy Logic Hybrid Extensions of Neural and Optimization Algorithms: Theory and Applications*, Studies in Computational Intelligence 940,
https://doi.org/10.1007/978-3-030-68776-2_13

1 Introduction

At the present time, mobile robots have presented an increasing demand in different fields of application, such as service robots, material transport, monitoring, military applications, self-driving cars, among other critical and challenging applications. In most of the applications, the necessity of computations is demanding since the information of the environment must be processed in an accelerated form. In that sense, the path planning task is one of the most critical processes in terms of computation time for mobile robot navigation.

In this work, we present an extension of the work presented in Orozco-Rosas et al. (2019). In that work, the membrane evolutionary artificial potential field (MemEAPF) algorithm is presented for solving mobile robot path planning problems. The MemEAPF algorithm combines membrane computing with a genetic algorithm and the artificial potential field method to find the parameters to generate a safe and feasible path for mobile robot navigation.

In that sense, in this work, we present an extension that implements the MemEAPF algorithm for mobile robot path planning in a graphics processing unit (GPU) to accelerate the path planning computation. Therefore, the main contribution of this work is the implementation of the parallel evaluation on GPU to accelerate the MemEAPF algorithm.

The organization of this work is as follows; Sect. 2 presents a brief description of the main components of the MemEAPF algorithm, the membrane computing, the genetic algorithms, and the artificial potential field method. Section 3 describes the MemEAPF algorithm and the implementation of the parallel evaluation on GPU. Section 4 presents the experiments and results for the three different implementations of the MemEAPF algorithm. Finally, Sect. 5 summarizes the conclusions.

2 Fundamentals

In this section, we present the main components of the MemEAPF algorithm for mobile robot path planning. Firstly, the core component of the MemEAPF algorithm, membrane computing. Then, the metaheuristic employed in each elementary membrane, the genetic algorithms. Finally, the artificial potential field method, whose function in the MemEAPF algorithm is to act as an evaluation function, as well as a path planning method.

2.1 Membrane Computing

Membrane computing is part of natural computing, and it was initiated in 1998 by Păun (2000). Membrane computing is also called P systems or membrane systems, from the compartmentalized structure and interactions of living cells (Wang et al. 2015). The obtained membrane computing models are distributed parallel schemes that evolve through rules and process the multisets of objects into the compartments that are hierarchically defined (Păun and Rozenberg 2002).

In general, membrane computing or P systems are membrane structures with objects into their membranes, which have specific evolution rules like transformation and communication to merge and divide membranes (Păun 2000). P systems are classified into three main types, cell-like P systems that contain one-membrane cell, tissue-like P systems consisting of several one-membrane cells in a common environment, and neural-like P systems that consider neurons as their cell (Zhang et al. 2013). In this work, a cell-like P system is the core component of the MemEAPF algorithm since it is simple and practical to implement with parallel computing. Cell-like P systems present a hierarchical structure of membranes, type of rules, e.g., transformation and communication, and inherent parallelism (Orozco-Rosas et al. 2019). These characteristics and strengths are desirable from a computational point of view and attractive for modeling complex problems (Zhang et al. 2014).

Therefore, the MemEAPF algorithm for mobile robot path planning consists of a cell-like P system that evolves the set of parameters (k_r, k_a) required to generate the path with the artificial potential field method. In that sense, the MemEAPF algorithm employs a dynamic structure with active membranes, in specific a one-level membrane structure (Zhang et al. 2008) with rules, such as membrane merger and division (Zhang et al. 2014). The membrane merger is helpful to enhance the information communication among individuals, i.e., the set of parameters (k_r, k_a), and the membrane division is beneficial to improve the search capability (Liu 2010).

2.2 *Evolutionary Computation*

Evolutionary computation is a subfield of artificial intelligence that concentrates population-based algorithms with a metaheuristic or stochastic optimization nature for global optimization. These algorithms are inspired by biological evolution, where the genetic algorithms are the most representative example (Mitchell 2001). Genetic algorithms emulate the processes of natural selection and natural genetics. The advantages of genetic algorithms over other metaheuristics are flexibility in defining constraints and quality measures, effectiveness in large search spaces, the capability of working with discrete and continuous variable, the power of providing multiple optimal or suboptimal solutions, and a high potential for applying parallel computing techniques to speed up the computation (Dao et al. 2017).

A genetic algorithm is an adaptive heuristic search algorithm designed to simulate the evolution processes existing in natural systems (Orozco-Rosas et al. 2017). In its most basic form, a genetic algorithm contains the next three leading genetic operators, selection, crossover, and mutation (Orozco-Rosas et al. 2020).

The selection operator drives the genetic algorithm to enhance the population fitness over the successive generations (Orozco-Rosas et al. 2015). In the selection process, the best individuals are chosen according to their fitness value. Individuals with higher fitness values have a better chance to be selected than those individuals with lower fitness values. Consequently, the selection process tends to eliminate those individuals that present lower fitness values. The selection process determines which individuals will be retained as parents for being employed by the crossover operator, and their core genetic information will be passed to their offspring (Fogel 1998).

The crossover operator is applied over the population after the selection operator completes its process, and it is the primary search operator (Holland 1992). This operator randomly chooses a locus and exchanges the subsequences before and after the chosen locus between two individuals to create two offspring.

The mutation operator randomly modifies some percentage of the genetic material in the individuals (Orozco-Rosas 2020). The mutation process is the second way in which a genetic algorithm explores a cost surface. This mutation process can introduce characteristics that are not in the original population. Consequently, the mutation operator prevents the genetic algorithm from converging prematurely before sampling the entire cost surface.

2.3 Artificial Potential Field Method

The artificial potential field method establishes that for a configured space, the mobile robot is considered as a particle under the influence of an artificial potential field, whose local variation reflects the free space structure (Khatib 1985). The idea is to characterize a potential function based on the sum of an attractive potential that pulls the mobile robot toward the goal position, and a repulsive potential that pushes the mobile robot away from the obstacles (Masehian and Sedighizadeh 2007).

The artificial potential field method has many advantages for mobile robot path planning. The artificial potential field method is capable of finding feasible paths to the goal position allowing the mobile robot to reach it autonomously.

A disadvantage of the artificial potential field method is that it must be provided with the adequate proportional gain values; otherwise, feasible paths are not found. Therefore, the importance of feeding the artificial potential field method with adequate proportional gains (Orozco-Rosas 2018). It is well known that such a task in mobile robot path planning is hard work because the problem leads to several computational problems. At present, some of these problems are still being open for large spaces; there are no convincing solutions since the existing ones can last much time making the mobile robot controllability unfeasible. Hence, the importance of having a method, such as the MemEAPF, that can find the optimal parameters at high-computational speeds, making the mobile robot controllable in large environments.

3 GPU Accelerated MemEAPF

In this section, we present and explain the GPU accelerated MemEAPF algorithm for mobile robot path planning, see Fig. 1. In the beginning, an initial population $P(t)$ is randomly created. The population $P(t)$ is composed of m subpopulations $P_i(t)$, $1 \leq i \leq m$, where m is the number of elementary membranes S_i contained in the skin membrane S_0. The subpopulations forms multisets with unique individuals codified with the proportional gains k_r and k_a.

Next, the membrane structure is initialized by the number of elementary membranes, the skin membrane S_0 contains the merged membrane S_F that is divided into m elementary membranes S_i, we set the counter i to 1, and the subpopulation $P_i(t)$ is assigned to the elementary membrane S_i.

The genetic algorithm and the artificial potential field method are the main components of each elementary membrane. The artificial potential field method defined by the colored background blocks represents the evaluation method that is evaluated in parallel on the GPU. The backbone of the MemEAPF algorithm is the membrane computing for finding dynamically the optimal proportional gains k_r and k_a values required to generate the path for mobile robot navigation.

The artificial potential field method acts as an evaluation function. First, the potential field $U_{total}(q)$ is computed using the following function,

$$U_{total}(q) = \frac{1}{2}\left[k_a\left(q - q_f\right)^2 + k_r\left(\frac{1}{\rho} - \frac{1}{\rho_0}\right)^2\right] \tag{1}$$

where q indicates the mobile robot position vector in a two-dimensional workspace, $q = [x, y]^T$. The vector q_f indicates the goal position and k_a is a positive scalar-constant that indicates the proportional attraction gain of the function. The expression $q - q_f$ is associated with the linear distance between the mobile robot and the goal position. The repulsive potential function was given by Khatib (Khatib 1985), it has a limited range of influence that prevents the movement of the mobile robot from being affected by a remote obstacle, where ρ_0 indicates the limit distance of influence of the repulsive potential field and ρ is the shortest distance to the obstacle; the positive scalar-constant k_r denotes the repulsion proportional gain.

The total force $F_{total}(q)$ which is used to drive the mobile robot is found by the negative gradient of the total potential function (Orozco-Rosas et al. 2019), this force is expressed as follows

$$F_{total}(q) = -\nabla U_{total}(q) \tag{2}$$

After the potential force $F_{total}(q)$ is computed, the path length is evaluated by the function described in Eq. 3, where n indicates the number of mobile robot configurations to reach the goal position.

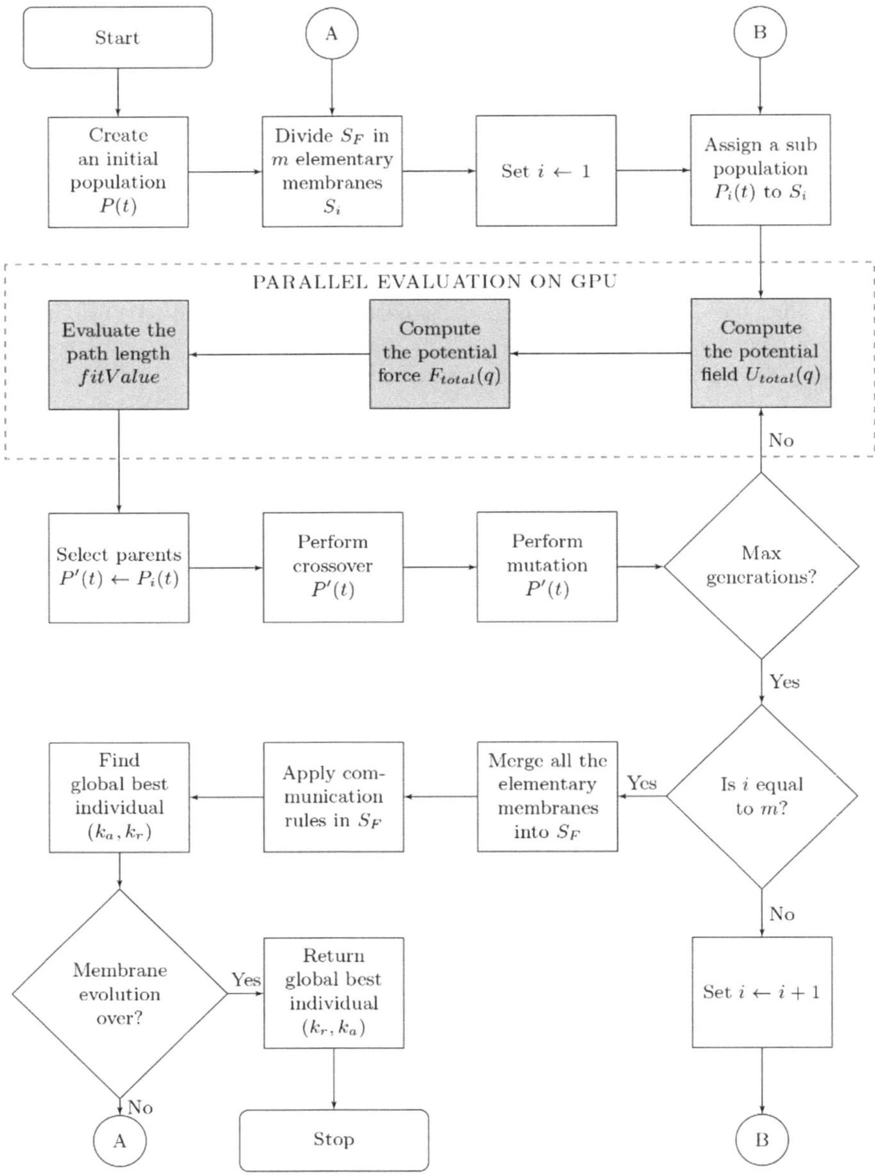

Fig. 1 Flowchart of the GPU accelerated MemEAPF algorithm

$$fitValue = \sum_{i=0}^{n} q_{i+1} + q_i \tag{3}$$

In Eq. 1, all the parameters are known except for the repulsion proportional gain k_r and the attraction proportional gain k_a. Therefore, the artificial potential field is combined with a genetic algorithm to obtain the best values for the proportional gains. In the most basic form, the genetic algorithm can be algorithmically modeled for computer simulation employing the following difference equation,

$$P'(t+1) = s\big(v\big(P'(t)\big)\big) \tag{4}$$

where t denotes the time, the next population $P'(t+1)$ is found from the existing population $P'(t)$ after it was worked by random variation (crossover and mutation) v, and selection s (Fogel 1998).

The evaluation through the artificial potential field and the evolution through the genetic algorithm iterates until the maximum number of generations is achieved. Consequently, the elementary membrane S_i evolves the individuals to obtain the best proportional gains k_r and k_a of the subpopulation $P_i(t)$. This process is continued in each elementary membrane until the counter i reaches the value of m.

After all the elementary membranes have been evolved, all the elementary membrane S_i are merged into a membrane S_F where the communication rules are applied. These rules include a copy of the best-selected individual from each elementary membrane S_i into the merged membrane S_F. In the MemEAPF algorithm, the primary purpose of the iterative process is to obtain the global best individual (k_r, k_a) from the best-selected group of individuals; this global best individual is copied and this copy is sent to the skin membrane to maintain the current global best solution. Through the communication rules, the merged membranes exchange information that will be evolved in the next generation. During the merge process, each subpopulation is maintained, and the worst individuals are replaced by a copy of the best-selected individuals to improve the subpopulation in each elementary membrane. Finally, when the evolution process has been completed, the global best individual is returned.

4 Results

In this section, we describe the experiments, and we present the results for mobile robot path planning in terms of path length for four different test environments. Moreover, we present a comparative analysis of three different implementations of the MemEAPF algorithm to assess the acceleration of the path planning computation. The sequential implementation on CPU (in C++), the parallel implementation on CPU (in C++ using the Open Multi-Processing (OpenMP) application programming interface), and the parallel implementation on GPU (in C++ using the Compute Unified Device Architecture, CUDA).

The experiments were performed on a quad-core Intel i7-4710HQ CPU @ 2.50 GHz with 8.00 GB of RAM, running Windows 10 with Microsoft Visual Studio Community 2019, and CUDA 10.1. The GPU employed is an NVIDIA GeForce GTX 860 M (Maxwell architecture) with 640 CUDA cores, and compute capability 5.0.

The next setup parameters yielded the best results for the three different implementations:

- The population size was set to 16 individuals and doubled in each different test up to 512 individuals to evaluate the MemEAPF algorithm performance.
- The individuals were codified using two genes, k_r and k_a.
- The proportional gains were constrained with values greater than zero and less than or equal to ten.
- A single point crossover was employed.
- The mutation rate was set to 0.20.
- The elitist selection was employed.
- The selection rate was set to 0.50.
- The stop condition is the maximum number of generations; it was set to 10 for the genetic algorithm.
- The membrane selection rate was set to 0.25.
- The stop condition was set to 10 for the skin membrane.

The different test environments are shown in Fig. 2, where the origin or the (0.00, 0.00) coordinates are at the bottom left corner, and the dimensions of each test environment are 10 × 10 meters. The test environments show the start and goal position for the mobile robot and the localization of the obstacles. Each test environment has been labeled for its identification as Map 1, Map 2, Map 3, and Map 4.

Figure 2a shows Map 1, which is constituted by three obstacles, two L-shaped obstacles, and one obstacle with rectangular form. Where the path planning problem is to go from the start position $q_0 = (5.50, 9.00)$ to the goal position $q_f = (4.50, 3.00)$. Figure 2b shows Map 2, which is composed of three obstacles. The obstacles in Map 2 are like Map 1 but in a different configuration. In this environment, the path planning problem is to go from the start position $q_0 = (2.00, 7.00)$ to the goal position $q_f = (8.00, 3.00)$. Figure 2c shows Map 3, which is formed by four obstacles, two L-shaped obstacles, one obstacle with rectangular form, and one obstacle with an isometric form. In this map, the path planning problem is to go from the start position $q_0 = (3.00, 6.50)$ to the goal position $q_f = (6.75, 3.75)$. Figure 2d shows Map 4, which is comprised of two obstacles with a rectangular form. The path planning problem in this map is to go from the start position $q_0 = (5.00, 9.00)$ to the goal position $q_f = (5.00, 1.00)$.

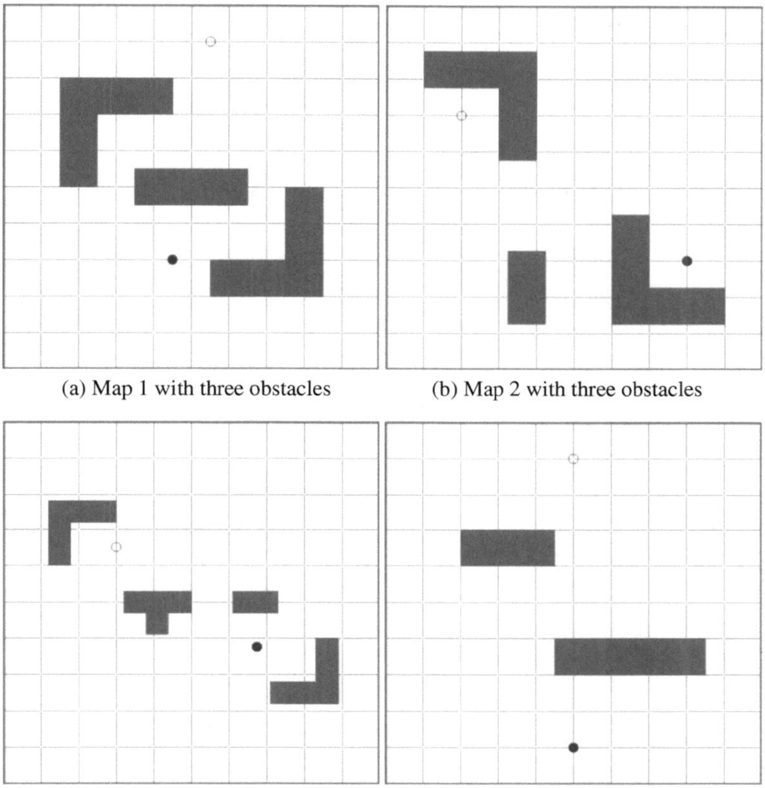

(a) Map 1 with three obstacles
(b) Map 2 with three obstacles
(c) Map 3 with four obstacles
(d) Map 4 with two obstacles

Fig. 2 Test environments

4.1 Path Planning Results

Figure 3 shows the best resultant path in terms of path length obtained with the MemEAPF algorithm in each test environment. The best resultant path for Map 1 is shown in Fig. 3a, it has a path length of 7.7230 meters, and it was generated with the proportional gains $k_r = 0.969$ and $k_a = 0.360$. For Map 2, the best resultant path is shown in Fig. 3b. The path has a length of 8.2830 meters, and it was generated with the proportional gains $k_r = 1.818$ and $k_a = 0.706$. Figure 3c shows the best resultant path for Map 3, the path has a length of 4.6920 meters, and it was generated with the proportional gains $k_r = 0.264$ and $k_a = 0.449$. The best resultant path for Map 4 is shown in Fig. 3d, it has a path length of 8.4310 meters, and it was generated with the proportional gains $k_r = 2.338$ and $k_a = 1.141$.

Table 1 shows the overall results obtained by the MemEAPF algorithm with different population sizes. The results show the proportional gains k_r and k_a for the best resultant path obtained in terms of path length. The results show the best, mean,

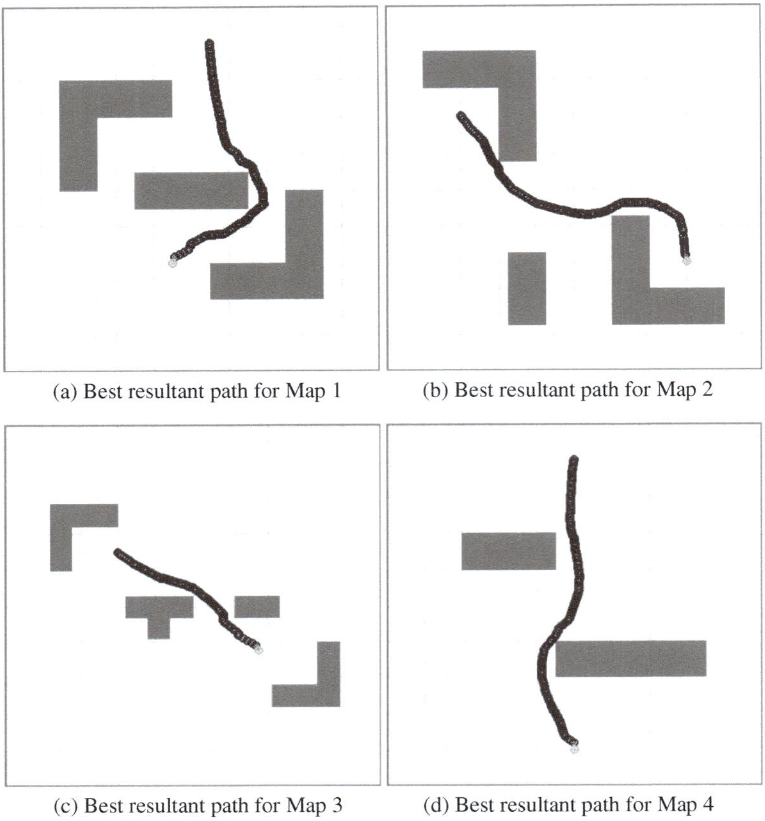

Fig. 3 The best resultant path for each test environment

worst, and standard deviation of thirty independent tests in each population size. The overall results show that the best path was found by the MemEAPF algorithm when the population size is 512 individuals. It can be observed that the quality of the solutions, i.e., the shorter path, it improves as the population size grew up.

Figure 4 shows the information on the variability or dispersion of the path planning results in terms of path length with the MemEAPF algorithm considering the different population sizes. We can observe a comparative in the range and distribution of the path length results for the different population sizes employed. In the overall results for the four test environments, we can observe that there is a more significant variability for small population sizes. Also, we can compare the interquartile ranges to examine how the data is dispersed between each sample.

We can observe a more extended box for small population sizes, which indicates more dispersed data. Opposite to large population sizes, where the box is less extended, that indicates less dispersed data. Additional information is presented by the overall spread, as shown by the extreme values at the end of two whiskers. In

Table 1 Overall results obtained by the MemEAPF algorithm with different population sizes

	Population size	Prop. gains (k_r,k_a)	Path length in meters			
			Best	Mean	Worst	Std. Dev.
Map 1	16	(1.575, 0.585)	7.7660	7.7884	7.8320	0.0183
	32	(1.565, 0.588)	7.7550	7.7711	7.7840	0.0102
	64	(0.931, 0.347)	7.7440	7.7653	7.7820	0.0096
	128	(0.975, 0.363)	7.7270	7.7542	7.7680	0.0122
	256	(0.458, 0.167)	7.7260	7.7487	7.7700	0.0106
	512	(0.969, 0.360)	7.7230	7.7479	7.7630	0.0106
Map 2	16	(1.034, 0.386)	8.3040	8.3311	8.3960	0.0269
	32	(0.900, 0.346)	8.2940	8.3145	8.3330	0.0119
	64	(1.813, 0.707)	8.2890	8.3009	8.3170	0.0091
	128	(1.832, 0.711)	8.2850	8.2957	8.3250	0.0104
	256	(1.815, 0.707)	8.2840	8.2894	8.2940	0.0031
	512	(1.818, 0.706)	8.2830	8.2885	8.2990	0.0044
Map 3	16	(2.272, 3.647)	4.7320	4.7679	4.7910	0.0173
	32	(0.697, 1.129)	4.7050	4.7397	4.7600	0.0173
	64	(0.706, 1.148)	4.7020	4.7305	4.7490	0.0138
	128	(0.816, 1.313)	4.6960	4.7191	4.7540	0.0167
	256	(0.948, 1.534)	4.6930	4.7104	4.7250	0.0082
	512	(0.264, 0.449)	4.6920	4.7093	4.7180	0.0072
Map 4	16	(1.692, 0.801)	8.4460	8.4611	8.4780	0.0087
	32	(2.658, 1.330)	8.4430	8.4544	8.4730	0.0082
	64	(3.479, 1.772)	8.4400	8.4505	8.4600	0.0055
	128	(1.854, 0.876)	8.4390	8.4488	8.4540	0.0041
	256	(2.124, 1.040)	8.4370	8.4463	8.4530	0.0041
	512	(2.338, 1.141)	8.4310	8.4432	8.4470	0.0047

some cases of the overall results, we can observe larger ranges, that indicate wider distribution, that is, more scattered data. Furthermore, some outliers are defined as data points and are located outside the whiskers of the boxplot.

4.2 Performance Results

Figure 5 shows the performance results of thirty independent run tests to obtain the average time of execution for the MemEAPF algorithm in each implementation, the sequential CPU, the parallel CPU, and the parallel GPU. The total of evaluated individuals in each test is based on the multiplication of the population size, the

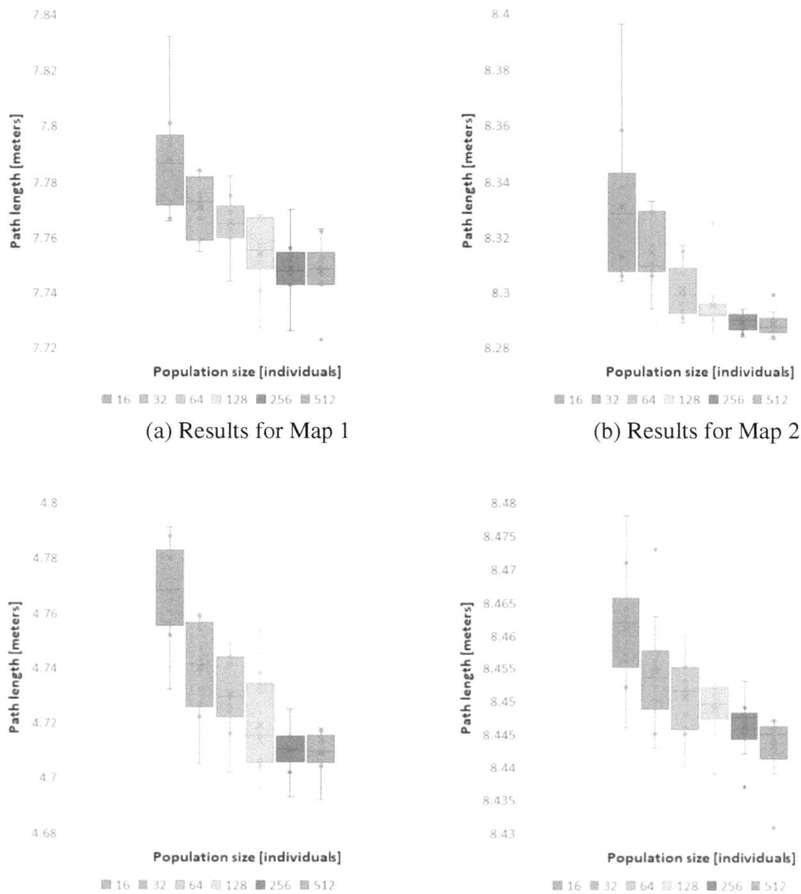

(a) Results for Map 1

(b) Results for Map 2

(c) Results for Map 3

(d) Results for Map 4

Fig. 4 Variation in path planning results considering the population size

maximum number of generations in the genetic algorithm, the number of elementary membranes, and the maximum number of generations in the skin membrane, e.g., for a population size of 16, a maximum number of generations in the genetic algorithm of 10, a number of elementary membranes of 2, and a maximum number of generations in the skin membrane of 10, we have $16 \times 10 \times 2 \times 10 = 3,200$ evaluated individuals.

The objective of the performance evaluation is to provide the speedup among the two different parallel implementations, parallel CPU and parallel GPU. Figure 5 shows the overall performance results. We can observe that the speedup is increased in both cases, for the parallel CPU implementation, and the parallel GPU implementation when the number of evaluations grows. These results are due to the increment of the total evaluations. We can clearly observe the advantages of the parallel CPU

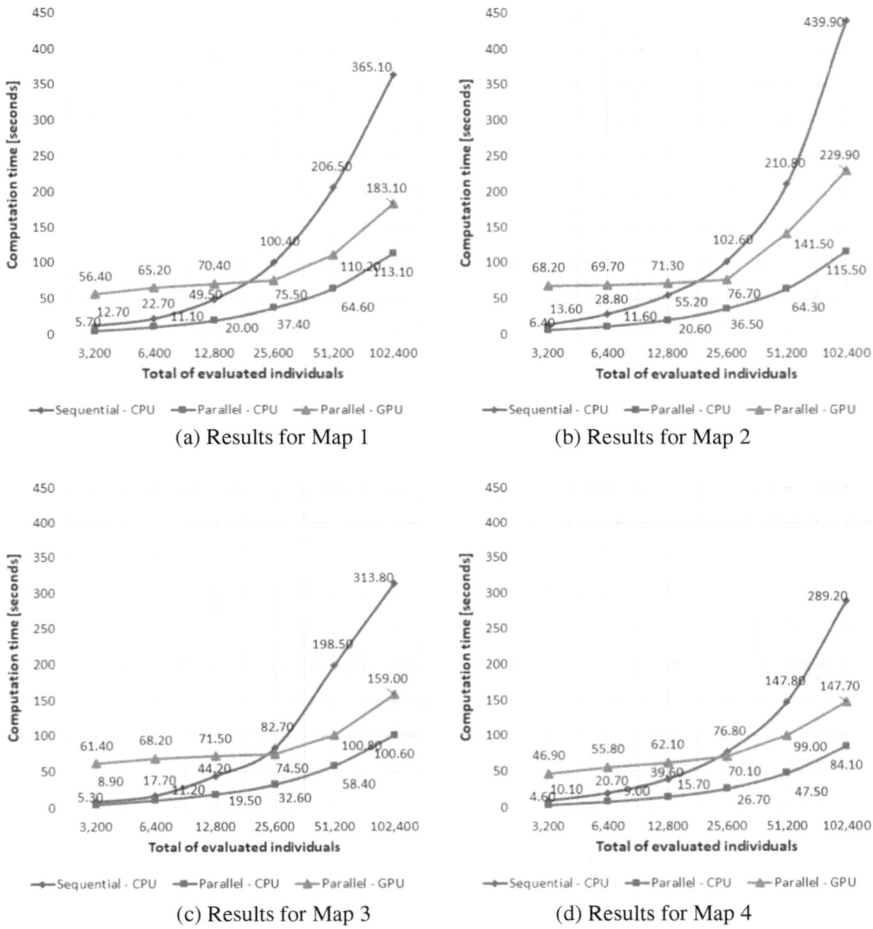

Fig. 5 Performance results

implementation for a small number of evaluations, and the advantage is more significant for the GPU implementation with many individuals to evaluate. The speedup is computed by the relation of the execution time in a processor and the runtime on a certain number of processors. Therefore, the computation of the speedup is obtained considering the execution time taken by the sequential implementation of the MemEAPF algorithm and the execution time taken by the parallel implementation in CPU and GPU, respectively.

In Fig. 5a, we can observe that the best speedup achieved for the parallel implementation in CPU is $365.10/113.10 = 3.23$, and the parallel implementation in GPU is $365.10/183.10 = 1.99$ in Map 1 for 102,400 evaluations. In Fig. 5b, the results show that the best speedup achieved for the parallel implementation in CPU is $439.90/115.50 = 3.81$, and the parallel implementation in GPU is $439.90/229.90 =$

1.91 in Map 2 for 102,400 evaluations. Figure 5c shows the results where the best speedup achieved for the parallel implementation in CPU is 198.50/58.40 = 3.40, and the parallel implementation in GPU is 198.50/100.80 = 1.97 in Map 3 for 51,200 evaluations. In Fig. 5d, we observe that the best speedup for the parallel implementation in CPU is 289.20/84.10 = 3.44 and the parallel implementation in GPU is 289.20/147.70 = 1.96 in Map 4 for 102,400 evaluations.

5 Conclusions

In this work, we have presented a GPU accelerated MemEAPF algorithm for mobile robot path planning. Through the simulation results, we can conclude that the MemEAPF algorithm can be capable of facing complex and bigger path planning problems.

We observed that small population sizes could guide the MemEAPF algorithm to good path planning solutions, but the performance on the GPU implementation is not so good. These results are presented due that the requirement for computationally intensive is not meet, i.e., the time spent on computation not significantly exceeds the time spent on transferring data to and from the GPU memory.

On the other hand, we observe that large population sizes could make the MemEAPF algorithm spend more computation time in finding the solutions, with the benefit that these solutions are better in terms of path length. Also, for large population sizes, the benefits of parallel implementations are evident in terms of performance, for both, CPU implementation with a 3.81 of best speedup and GPU implementation with a 1.99 of best speedup.

Consequently, the GPU implementation can be powerful to accelerate because the path planning problem in this work has been stated as a data-parallel problem, and we can observe best performance results when we satisfy the criteria of computationally intensive and massively parallel.

Acknowledgments This work was supported by the Coordinación de Investigación of CETYS Universidad, and by the Mexico's National Council of Science and Technology (CONACYT).

References

Dao, S.D., K. Abhary, and R. Marian. 2017. A bibliometric analysis of Genetic Algorithms throughout the history. *Computers & Industrial Engineering* 110: 395–403.

Fogel, D.B. 1998. An introduction to evolutionary computation. In *Evolutionary computation: The fossil record*, Wiley-IEEE Press, pp. 1–28.

Holland, J.H. 1992. Adaptation in natural and artificial systems, Cambridge, MA, USA: MIT Press. Second edition. (First edition, Ann Arbor: University of Michigan Press, 1975).

Khatib, O. 1985. Real-time obstacle avoidance for manipulators and mobile robots. In *Proceedings of the IEEE international conference on robotics and automation.*

Liu, C., G. Zhang, H. Liu, M. Gheorghe and F. Ipate. 2010. An improved membrane algorithm for solving time-frequency atom decomposition. In *Membrane computing. WMC 2009.* Lecture Notes in Computer Science, vol 5957, Springer, Berlin, Heidelberg.

Masehian, E., and D. Sedighizadeh. 2007. Classic and heuristic approaches in robot motion planning a chronological review. *International Journal of Mechanical, Aerospace, Industrial, Mechatronic and Manufacturing Engineering* 1 (5): 228–233.

Mitchell, M. 2001. *An introduction to Genetic Algorithms, Cambridge, Massachusetts.* USA: Bradford.

Orozco-Rosas, U., O. Montiel, and R. Sepúlveda. 2018. Parallel bacterial potential field algorithm for path planning in mobile robots: A GPU implementation. In *Fuzzy logic augmentation of neural and optimization algorithms: Theoretical aspects and real applications*, Springer International Publishing, vol. 749, Springer, 2018, pp. 207–222.

Orozco-Rosas, U., K. Picos, O. Montiel and O. Castillo. 2020. Environment recognition for path generation in autonomous mobile robots. In *Hybrid intelligent systems in control, pattern recognition and medicine, studies in computational intelligence*, vol. 827, Springer International Publishing, pp. 273–288.

Orozco-Rosas, U., O. Montiel, and R. Sepúlveda. 2015. Parallel evolutionary artificial potential field for path planning—an implementation on GPU. In *Design of intelligent systems based on fuzzy logic, neural networks and nature-inspired optimization, studies in computational intelligence*, vol. 601, Springer International Publishing, pp. 319–332.

Orozco-Rosas, U., O. Montiel, and R. Sepúlveda. 2017. An optimized GPU implementation for a path planning algorithm based on parallel pseudo-bacterial potential field. In *Nature-inspired design of hybrid intelligent systems, studies in computational intelligence*, vol. 667, Springer International Publishing, 2017, pp. 477–492.

Orozco-Rosas, U., O. Montiel, and R. Sepúlveda. 2019a. Mobile robot path planning using membrane evolutionary artificial potential field. *Applied Soft Computing* 77: 236–251.

Orozco-Rosas, U., K. Picos, and O. Montiel. 2019b. Hybrid path planning algorithm based on membrane pseudo-bacterial potential field for autonomous mobile robots. *IEEE Access* 7: 156787–156803.

Orozco-Rosas, U., K. Picos, and O. Montiel. 2020. Acceleration of path planning computation based on evolutionary artificial potential field for non-static environments. In *Intuitionistic and Type-2 fuzzy logic enhancements in neural and optimization algorithms: Theory and applications, studies in computational intelligence*, vol. 862, Springer International Publishing, pp. 271–297.

Păun, G. 2000. Computing with Membranes. *Journal of Computer and System Sciences* 61 (1): 108–143.

Păun, G., and G. Rozenberg. 2002. A Guide to membrane computing. *Theoretical Computer Science* 287 (1): 73–100.

Wang, X., G. Zhang, J. Zhao, H. Rong, F. Ipate, and R. Lefticaru. 2015. A modified membrane-inspired algorithm based on particle swarm optimization for mobile robot path planning. *International Journal of Computers Communications Control* 10 (5): 732–745.

Zhang, G.-X., M. Gheorghe, and C.-Z. Wu. 2008. A Quantum-Inspired Evolutionary Algorithm Based on P systems for knapsack problem. *Fundamenta Informaticae* 87 (1): 93–116.

Zhang, G., J. Cheng, M. Gheorghe, and Q. Meng. 2013. A hybrid approach based on differential evolution and tissue membrane systems for solving constrained manufacturing parameter optimization problems. *Applied Soft Computing* 13 (3): 1528–1542.

Zhang, G., J. Cheng, and M. Gheorghe. 2014a. Dynamic behavior analysis of membrane-inspired evolutionary algorithms. *International Journal of Computers Communications Control* 9 (2): 227–242.

Zhang, G., M. Gheorghe, L. Pan, and M.J. Pérez-Jiménez. 2014b. Evolutionary membrane computing: A comprehensive survey and new results. *Information Sciences* 279: 528–551.

Optimization of the Internet Shopping Problem with Shipping Costs

Hector Joaquín Fraire Huacuja, Miguel Ángel García Morales, Mario César López Locés, Claudia Guadalupe Gómez Santillán, Laura Cruz Reyes, and María Lucila Morales Rodríguez

Abstract This chapter addresses the Internet Shopping Optimization Problem with Shipping Cost (IShOP) issue. In the related literature, there is only one metaheuristic solution reported. This solution is a cellular processing algorithm that simulates the parallel processing of two or more Search processes through the solution space and is currently considered the best solution of state-of-the-art IShOP. In this paper, we propose a new metaheuristic algorithm based on the memetic algorithm methodology. We carry out a comparative study of the performance of the proposed algorithm against that of the cell processing algorithm. In computational experiments, we use a broad set of standard random instances, and the results show a clear superiority of the proposed algorithm. We apply the non-parametric Wilcoxon hypothesis test to determine the significance of the observed differences in the performance of the assessed algorithms.

Keywords Memetic algorithms · Internet shopping optimization · Metaheuristic algorithms

---

H. J. F. Huacuja (✉) · M. Á. G. Morales · C. G. G. Santillán · L. C. Reyes · M. L. M. Rodríguez
Tecnológico Nacional de México, Instituto Tecnológico de Ciudad Madero, Ciudad Madero, Tamaulipas, Mexico
e-mail: automatas2002@yahoo.com.mx

M. Á. G. Morales
e-mail: soporteclusterlanti@gmail.com

C. G. G. Santillán
e-mail: lauracruzreyes@itcm.edu.mx

L. C. Reyes
e-mail: lmoralesrdz@gmail.com

M. L. M. Rodríguez
e-mail: mariocesar@lopezloc.es

M. C. L. Locés
Graduate Program in Systems Engineering, Universidad Autonoma de Nuevo Leon (UANL), 66455 San Nicolás de los Garza, Nuevo León, Mexico
e-mail: claudia.gomez@itcm.edu.mx

O. Castillo and P. Melin (eds.), *Fuzzy Logic Hybrid Extensions of Neural and Optimization Algorithms: Theory and Applications*, Studies in Computational Intelligence 940,
https://doi.org/10.1007/978-3-030-68776-2_14

1 Introduction

Electronic commerce has become an essential part of society, the implementation of innovate and consistently growing technology has made it inevitable to adapt to this evolution (Musial 2014).

The main advantage is that their offers are available to a broader audience, without most associated costs such as rent, taxes, maintenance, and advertising. Buyers can make purchases from anywhere, at any time, with better prices and a more extensive range of products, if they have Internet access (Lopez-Loces et al. 2016).

In the Internet Shopping Optimization Problem (IShOP) (Musial 2012), we assume that a customer with a shopping list needs to purchase the product in a set of online stores at the lowest possible cost. In Gen and Cheng (2000) the problem was proposed, and they show that it is NP-hard.

In Lopez-Loces et al. (2016) a cell Processing algorithm is proposed that simulates the parallel Processing of two or more Search processes in solution space and is currently considered the best algorithm in the state-of-the-art of IShOP.

In this paper, we propose a memetic algorithm (MAIShOP) that, unlike the algorithm of state-of-the-art, uses a vector to represent the candidate solutions, a mechanism that speeds up the calculation of the objective function and for each generation we update the population leaving only the best solution and recalculating the rest of the community. We design these elements to achieve a significant improvement in the performance of the algorithm. To validate this new approach, we conducted a comparative study of the performance of the proposed algorithm against that of the best algorithm in the state-of-the-art. In computational experiments, we use a broad set of standard instances, and the results show a clear superiority of the memetic algorithm. The Wilcoxon non-parametric hypothesis test is applied.

1.1 Definition of the Problem

A customer needs to buy a set of n products N online, which he can buy in a set of m available stores M. The set N_i contains the products available in-store i, each product $j \in N_i$ has a cost of c_{ij} and a shipping cost d_i.

We add the shipping cost only if the customer buys one or more products from in store i. The problem is to minimize the total cost of buying all the products in N (cost of the products plus shipping).

Formally, the problem consists of determining a partition of the products to be bought in the different stores $X = (X_1, \ldots, X_m)$, so that $X_i \subseteq N_i$ and that all the products $U_{i=1}^{m} X_i = N$ and that the total cost is minimized:

$$F(X) = \sum_{i=1}^{m} \left(\sigma(|X_i|)d_i + \sum_{j \in X_i} c_{ij} \right) \tag{1}$$

where: $|X_i|$ is the cardinality of the container X_i, and $\sigma(x) = 0 if x = 0 and \sigma(X) = 1 if x > 0$.

We calculate the target value of the solution I as follows:

$$F(I) = \sum_{i=1}^{m} \sum_{j=1\&\&I(j)=i}^{n} \left(d_i + c_{ij}\right) \tag{2}$$

If the list of products that the customer will buy consists of N products and there are M stores available, then a solution is represented in a vector of length N, which contains for each product the store where it will be purchased.

2 The General Structure of the Memetic Algorithm

The algorithm uses a vector representation of the solutions that, unlike the state-of-the-art algorithm, reduces the complexity of processing a solution from $O(n^2)$ to $O(n)$. The local Search assigns each product to the store where its purchase is most convenient; this process is carried out intensively in the algorithm. To make the most convenient store more efficient, we incorporated a mechanism that speeds up the objective function. Finally, we modify the traditional structure of the memetic algorithm in such a way that at the end of each generation, we update the population, leaving only the best solution and regenerating the rest of the population. The idea is that the best solution transmits its characteristics to the following generations, and together with the regeneration of the rest of the individuals, we achieve a better balance of the intensification and exploration of the algorithm.

2.1 Selection by Tournament

We randomly select p individuals from the *pop* population, and the fittest is the one that passes to the next generation.

In the memetic algorithm, a binary tournament (with $p = 2$) on the pop population. Each solution participates in two competitions, we select the winning solution, and we add to *NewPop*. In this way, any solution in *pop* can have a maximum of two copies in *NewPop*.

2.2 Crossover Operator

We apply this operator to a pair of $parent_1$ and $parent_2$ solutions, selected sequentially until a percentage of the individuals in the population have been selected (Holland

1975). To generate the solution $child_1$, first, the leading half of $parent_1$ is taken and joined with the second half of $parent_2$. Later, to form the solution $child_2$, the first half of $parent_2$ is taken and joined with the second half of $parent_1$ (Umbakar 2015).

In the crossover operator, we select the point N/2 ó N/2 as the crossover point.

2.3 Mutation Operator

A heuristic mutation method was designed, considering the following procedure.

We select $ps - ps * er$ solutions, and for each one we generate a random number, if this randomly generated value is less than $ps*pm$ we proceed to make all the possible assignment combinations *(product, store)*, we evaluate all these combinations and the best is selected.

2.4 Local Search

This procedure begins by selecting an *IntermediatePop*, where there is a vector of associated values that contains the stores *j* assigned in the product selection sequence X_{old} with their cost, which includes c_{ij} and d_i of the multiset N_i.

Subsequently we verify for each product, from which store we could buy it to reduce the total cost of the selection of products *X*.

When the Search for that store ends, you move to a new product. This process continues until all the *j* stores are reviewed in *X*. Concludes when we assign all the *X's* in *ChildPop.*

2.5 Memetic Algorithm (MAIShOP)

Figure 1 describes the general structure of the MAIShOP algorithm. The algorithm begins by defining the parameters such as the size of the initial population *ps*, percentage of cross *pc*, mutation probability *pm*, rate of elitism *er*, *timelimit* termination criterion, then we generate an initial population (Step 2) using a construction heuristic. In the generational phase (Steps 5–15), we apply the binary tournament operator for selection in Step 6. We copy the best *ps*er* (elitism) solutions from *pop* to *IntermediatePop* in the Step 7. In *ps*pc* of the remaining solutions in *NewPop*, we apply the crossover operator to generate more solutions for *IntermediatePop* and copy the remaining as they are in *IntermediatePop* in the Step 8. We assume that the selection *pc* is such that *ps*pc* is less than or equal to the number of solutions remaining in *NewPop.* In Step 9, we mutated the individuals in *IntermediatePop* except for those selected in Step 6 using elitism. In Step 10, local Search is applied

Input:

timelimit: The time limit in the execution of the algorithm

Output:

BestGlobal: Best overall solution

timesolution: Solution time

1: The following elements are defined: initial population size *ps*, percentage of cross *pc*, probability of mutation *pm*, rate of elitism *er*, *timelimit* termination criterion

2: Generate initial *pop* population of size *ps*

3: Obtain the best local solution *(BestLocal)* among the solutions generated in Step 2.

4: Assign the solution obtained in Step 2 to the overall best *(BestGlobal)*.

5: **while** termination criteria *timelimit* not met **do**

6: Apply binary tournament operator in *pop* to get *NewPop*

7: *ps.er* of the solutions that have a lower cost (elite solutions) from *NewPop* we pass them to *IntermediatePop*

8: Among the resulting individuals in *NewPop*, we apply the crossover operator to *ps.pc* of randomly selected solutions, and we copy the remaining solutions as they are to *IntermediatePop*

9: Apply the mutation operator to *IntermediatePop* with probability *pm* except for those selected in Step 7.

10: Local Search operator is applied to *IntermediatePop* to generate *ChildPop*

11: You get the best solution *(BestLocal)* from *ChildPop*

12: We determine if the target value of the best local solution (BestLocal) is less than the global solution (BestGlobal); if so, we update the global solution. We calculate the execution time of the algorithm until this Step *(timesolution)*.

13: We copy the best local solution *(BestLocal)* to *pop*, and all other individuals are regenerated again.

14: We calculate the algorithm execution time for each generation *(totaltime)*

15: **end while**

16: **return** (BestGlobal, timesolution)

Fig. 1 Memetic algorithm (MAIShOP). *Source* self made

to *IntermediatePop* to improve solutions. In Step 11, we obtain the best local solution (*BestLocal*) from *ChildPop* and compare with the global solution (*BestGlobal*) in Step 12 also we calculate the execution time (*timesolution*) once we evaluate this step, if *BestLocal* has a lower cost then it replaces *BestGlobal*, in Step 13 we copy the best local solution (*BestLocal*) to the initial population and all the other solutions we regenerate again, we calculate the total execution time of the algorithm for each generation in Step 14, and finally, we return the best global solution (*BestGlobal*) with runtime calculation in which we found this solution in Step 16.

3 Computational Experiments

The hardware and software platform where we carried out the experiments includes Intel Core 2 Duo processor at 2.53 Ghz. with 4 GB of RAM, Java version 1.8.0_231.

Table 1 Comparative performance of algorithms. *Source* Self made

Instancia	Pcell (FO)	MAIShOP (FO)	Pcell (Time)	MAIShOP (Time)
3n20m	$63.03_{0.483}\uparrow$	$62.76_{2.57E\text{-}14}$	$0.00049_{0.0006}\uparrow$	$1.2222E\text{-}05_{6.21E\text{-}05}$
4n20m	$78.83_{0.40}\uparrow$	$78.73_{2.45E\text{-}14}$	$0.00185_{0.0012}\uparrow$	$6.6667E\text{-}06_{4.26E\text{-}05}$
5n20m	$102.56_{0.80}$ --	$102.09_{0.0043}$	$0.00273_{0.0014}\uparrow$	$0.00016_{0.0002}$
5n240m	$75.43_{0.77}\downarrow$	$76.32_{1.67}$	$0.27360_{0.0957}\uparrow$	$0.0669_{0.0627}$
5n400m	$69.98_{1.72}$ --	$70.55_{1.68}$	$0.79002_{0.2882}\uparrow$	$0.2425_{0.2139}$
50n240m	$1058.56_{103.06}\uparrow$	$505.47_{17.60}$	$16.4441_{8.0266}\downarrow$	$22.5010_{7.5014}$
50n400m	$951.08_{119.35}\uparrow$	$438.74_{24.09}$	$48.6607_{25.3042}\uparrow$	$33.1063_{3.8351}$
100n240m	$2349.94_{159.48}\uparrow$	$993.70_{38.25}$	$73.8573_{39.7564}\uparrow$	$50.4910_{5.6348}$
100n400m	$2213.90_{173.70}\uparrow$	$801.51_{35.03}$	$181.9521_{95.3004}$ --	$176.1110_{41.5457}$

The name of the instances specifies their size; *m* represents the number of stores and *n* the number of products. We create three sets of cases of different sizes. A collection of small instances, with three subsets: 3n20m, 4n20m, and 5n20m. A group with medium cases, with three subgroups: 5n240m, 5n400m, and 50n240m. A significant instance set with three subsets: 50n400m, 100n240m, and 100n400m. Each subgroup contains 30 cases.

The objective of the computational experiments we carried out was to compare the performance of the memetic algorithm (MAIShOP) and the cell processing algorithm (Pcell). Each instance was solved by carrying out 30 independent runs with each of the algorithms, and the objective value of the best solution found, and the time of said solution was determined. To determine the performance of the algorithm, we calculate the average values of the target value and the solution time; and standard deviations. The Wilcoxon non-parametric hypothesis test was applied to determine the statistical significance of the observed differences in quality and efficiency.

Table 1 shows the results we obtained with each group of 30 instances. For each algorithm, we show the average and standard deviation (subscript) of the target value and the solution time. In the Pcell cells, we indicate if there are significant differences in favor of Pcell (↓) or support of MAIShOP (↑) and if there are no significant differences between the two algorithms (–). We determine the significance with a reliability of 95%. The shaded cells correspond to the algorithm that obtained the best result in quality or efficiency.

4 Conclusions

In this paper, we propose a memetic algorithm (MAIShOP) that incorporates a series of innovative mechanisms to solve the IShOP problem efficiently. To validate this new approach, we conducted a comparative study of the performance of the proposed algorithm and of the best algorithm of the state-of-the-art. The results of the experiments carried out show that the proposed algorithm surpasses the state-of-the-art algorithm in quality and efficiency. In quality improves significantly in 6/9 sets, and in efficiency, improves significantly in 7/9 sets.

Acknowledgements The authors thank the support from CONACYT projects: (a) A1-S-11012-"AnÁlisis de Modelos de NO-Inferioridad para incorporar preferencias en Problemas de Optimización Multicriterio con Dependencias Temporales Imperfectamente Conocidas"; (b) project 3058 from the program Cátedras CONACyT; and, (c) and from project 312397 from Programa de Apoyo para Actividades Científicas, Tecnológicas y de Innovación (PAACTI 2020-1). Also thank the support from TecNM project no. 5797.19-P and Laboratorio Nacional de Tecnologias de Información (LaNTI) del TecNM/Campus ITCM. M.A. García would like to thank CONACYT for the support 658787.

References

Gen, Mitsuo, and Runwei Cheng. 2000. *Genetic algorithms & engineering optimization*. New York: Wiley Series in Engineering Design and Automation. John Wiley & Sons.

Holland, J.H. 1975. *Adaptation in natural and artificial systems*. And Arbor, Michigan: University of Michigan Press.

Lopez-Loces, M.C., J. Musial, J.E. Pecero, H.J. Fraire-Huacuja, J. Blacewicz, and P. Bouvry. 2016. Exact and heuristic approaches to solve the Internet shopping optimization problem with delivery costs. *International Journal of Applied Mathematics and Computer Science* 26 (2): 391–406.

Musial, J. 2012. *Applications of combinatorial optimization for online shopping*. Poznán: NAKOM.

Musial, J., J. Pecero, B. Dorronsoro, and J. Blazewicz. 2014. Internet shopping optimization project (IShOP). *European IST Projects*, 16.

Umbakar, A.J., and P.D. Sheth. 2015. Crossover operators in genetic algorithms: a review. *ICTACT Journal on Soft Computing*, 6(1).

Multiobjective Algorithms Performance When Solving CEC09 Test Instances

Hector Fraire, Eduardo Rodríguez, and Alejandro Santiago

Abstract In this chapter is presented a comparative study of the performance of ten state-of-the-art multiobjective algorithms. The performance metrics used are hypervolume (HV), epsilon indicator ($I_{\varepsilon+}$), inverted generational distance (IGD), and generalized spread (Δ^*). The used instances belong to the set CEC09, which currently are considered very difficult to solve. As far as we know, there is no such performance study with the instances and metrics used here. The Friedman statistical hypothesis tests applied to the results of the computational experiments shows that the algorithms with the best performance regards to HV, IGD, and $I_{\varepsilon+}$ are SPEA2, NSGA-II, AbYSS, and GDE3; and regards to Δ^* the best algorithms are CellDE and OMOPSO.

Keywords Multiobjective optimization · CEC09 test functions · Performance multiobjective metrics

1 Introduction

Many real-life problems need to simultaneous optimize many conflicting objectives. Finding a solution that only optimizes one objective often leads to the worsening of one or more of the other objectives. These problems are called Multiobjective Optimization Problems (MOPs) (Deb 2011).

H. Fraire (✉) · E. Rodríguez
TecNM/Instituto Tecnológico de Cd. Madero, Ciudad Madero, Mexico
e-mail: automatas2002@yahoo.com.mx

E. Rodríguez
e-mail: rodela22@hotmail.com

A. Santiago
Polytechnic University of Altamira, Altamira, Mexico
e-mail: aurelio.santiago@upalt.edu.mx

O. Castillo and P. Melin (eds.), *Fuzzy Logic Hybrid Extensions of Neural and Optimization Algorithms: Theory and Applications*, Studies in Computational Intelligence 940,
https://doi.org/10.1007/978-3-030-68776-2_15

In contrast to a single objective optimization problem, a MOP does not have only a single optimal solution, but a set of solutions known as the Pareto Optimal Set. The plotted solutions in the objective space are known as Pareto Front.

Given the complexity, difficulty, and the high cost associated with the evaluation of objective functions, multiple metaheuristic algorithms have been developed to try to solve MOPs in the most efficient way. The need to maintain multiple different solutions to guarantee the highest number of Pareto fronts elements has made population algorithms a popular choice. Typical algorithms for multiobjective optimization are NSGA-II and SPEA2, which are evolutionary algorithms (EA).

In the literature are reported different indicators and metrics for assessing the quality of the solutions obtained by the algorithms used for solving the MOPs. To measure the performance of an algorithm, tests using benchmarks of synthetic instances are available. These instances are developed to subdue the algorithms to specific features such as scalability, discontinuous fronts, problems that change during execution of the algorithms among others.

The main contribution of this work is a comparative study of the performance of ten state-of-the-art multiobjective algorithms when trying to solve the set of instances CEC09 (Zhang 2008), as far as know does not exist a publication with these instances, which currently are considered very difficult to solve.

The algorithms assessed were: AbYSS (Nebro et al. 2008), CellDE (Durillo 2008), GDE3 (Kukkonen 2005), IBEA (Zitzler 2004), MOCell (Nebro 2007), NSGA-II (Deb et al. 2002), NSGA-III (Yuan 2014), OMOPSO (Sierra 2005), PAES (Knowles and Corne 2000), and SPEA2 (Zitzler 2001). Four different quality indicators were used: Hypervolume (Zitzler and Thiele 1999), Epsilon (Zitzler et al. 2003), IGD (Zhang et al. 2008), and Generalized spread (Zhou 2006).

The structure of The chapter is as follow. Section 2 starts by presenting the definitions related to multiobjective optimization. Section 3 describes the problems that fit the CEC09 set, and we consider only the unconstrained test problems UF1 to UF10. The mathematical functions are presented together with the optimal theoretical fronts of each one of the instances. Section 4 briefly describes the jMetal 5.0 framework (Nebro 2015) and the ten assessed algorithms. Section 5 includes a general description of the metrics used in this work. Section 6 is devoted to the presentation and analysis of the experiments carried out. Finally, in Sect. 7, we include a discussion about the obtained results and future works.

2 Multiobjective Optimization

Now we define the following concepts, multiobjective Optimization problem, Pareto optimality, Pareto dominance, Pareto optimal set, and Pareto front. The following definitions assume, without the loss of generality, the minimization of all the objectives. A Multiobjective Optimization Problem (MOP) can be formally defined as follows.

Definition 1 (*Multiobjective Optimization Problem (MOP)*) A MOP consists in finding a vector $\vec{x}^* = \{x_1^*, x_2^*, \ldots, x_n^*\}$ that optimizes the multiple conflicting objectives of a vector function $\vec{f}(\vec{x}) = [f_1(\vec{x}), f_2(\vec{x}), \ldots, f_k(\vec{x})]$, subject to a set of m inequality constrains $g_i(\vec{x}) \geq 0, i = 1, 2, \ldots, m$ and a set of p equality constraints $h_i(\vec{x}) = 0, i = 1, 2, \ldots, p$. The set of all the values of the variables that satisfy all both type of constraints defines the feasible region Ω, and any point $\vec{x} \in \Omega$ is a feasible solution.

Definition 2 (*Pareto Optimality*) A solution $\vec{x}^* = \{x_1^*, x_2^*, \ldots, x_n^*\}$ is Pareto optimal if for every $\vec{x} \in \Omega$ and $I = \{1, 2, \ldots, k\}$ either $\forall_{i \in I}(f_i(\vec{x}^*) = f_i(\vec{x}))$ or there is at least one $i \in I$ such that $f_i(\vec{x}^*) < f_i(\vec{x})$. $\vec{x}^*$ is Pareto optimal if no feasible vector $\vec{x}$ exists that creates an improvement in minimum one objective without causing a simultaneous worsening in at least one other objective.

Definition 3 (*Pareto Dominance*) A vector $\vec{x} = \{x_1, \ldots, x_k\}$ is said to dominate a vector $\vec{x}' = \{x_1', \ldots, x_k'\}$ (denoted by $\vec{x} \prec \vec{x}'$) if and only if both conditions are fulfilled:

1. $\forall i \in \{1, \ldots, k\}, x_i \leq x_i'$
2. $\exists i \in \{1, \ldots, k\} : x_i < x_i'$

Definition 4 (*Pareto Optimal Set*) For a given MOP, the definition of the Pareto optimal set is $P^* = \left\{\vec{x} \in \Omega | \nexists \vec{x}' \in \Omega : \vec{f}(\vec{x}') \prec \vec{f}(\vec{x})\right\}$.

Definition 5 (*Pareto Front*) For a given MOP and its Pareto optimal set P^*, the Pareto front nis defined as $PF^* = \left\{\vec{f}(\vec{x}), \vec{x} \in P^*\right\}$.

The goal of a multiobjective optimization is to obtain the Pareto front. However, given that a Pareto front can contain many points, only a limited number of them is obtained, which should be as close as possible to the real Pareto front, as well as being uniformly spread (Durillo et al. 2010).

3 CEC09 Test Functions

In real-life applications, most of the time it is not possible to know the elements in the theoretical fronts. The current tendency is to formulate theoretical problems for which it is possible to deduce their theoretical fronts. These problems usually incorporate challenging characteristics for the algorithms.

Proposed in Zhang (2008) the CEC09 test functions attempt to resemble complicated real-life problems. In this chapter, we consider only the unconstrained test instances UF1-UF10. See Tables 1 and 2. For each Table 3 shows their Pareto fronts. Some problems share the same Pareto front (PF^*) but the elements of their Pareto set (P^*) are different. Pareto theoretical optimum fronts is defined in Zhang (2008).

Table 1 Two objective problems in CEC09 set

Problem	Objective Functions	Var Bounds
UF1	$f_1 = x_1 + \frac{2}{\lvert J_1\rvert}\sum_{j\in J_1}\left[x_j - \sin\left(6\pi x_1 + \frac{j\pi}{n}\right)\right]^2$ $f_2 = 1 - \sqrt{x_1} + \frac{2}{\lvert J_2\rvert}\sum_{j\in J_2}\left[x_j - \sin\left(6\pi x_1 + \frac{j\pi}{n}\right)\right]^2$ Where: $J_1 = \{j \mid j \text{ is odd} \wedge 2 \le j \le n\}$ $J_2 = \{j \mid j \text{ is even} \wedge 2 \le j \le n\}$	$0 \le x_1 \le 1$ $-1 \le x_i \le 1$ $\forall i = 2, \ldots, n$
UF2	$f_1 = x_1 + \frac{2}{\lvert J_1\rvert}\sum_{j\in J_1} y_j^2$ $f_2 = 1 - \sqrt{x_1} + \frac{2}{\lvert J_2\rvert}\sum_{j\in J_2} y_j^2$ Where: $J_1 = \{j \mid j \text{ is odd} \wedge 2 \le j \le n\}$ $J_2 = \{j \mid j \text{ is even} \wedge 2 \le j \le n\}$ $y_j = \begin{cases} x_j - \left[0.3x_1^2\cos\left(24\pi x_1 + \frac{4j\pi}{n}\right) + 0.6x_1\right]\cos\left(6\pi x_1 + \frac{j\pi}{n}\right) & j \in J_1 \\ x_j - \left[0.3x_1^2\cos\left(24\pi x_1 + \frac{4j\pi}{n}\right) + 0.6x_1\right]\sin\left(6\pi x_1 + \frac{j\pi}{n}\right) & j \in J_2 \end{cases}$	$0 \le x_1 \le 1$ $-1 \le x_i \le 1$ $\forall i = 2, \ldots, n$
UF3	$f_1 = x_1 + \frac{2}{\lvert J_1\rvert}\left(4\sum_{j\in J_1} y_j^2 - 2\prod_{j\in J_1}\cos\left(\frac{20y_j\pi}{\sqrt{j}}\right) + 2\right)$ $f_2 = 1 - \sqrt{x_1} + \frac{2}{\lvert J_2\rvert}\left(4\sum_{j\in J_2} y_j^2 - 2\prod_{j\in J_2}\cos\left(\frac{20y_j\pi}{\sqrt{j}}\right) + 2\right)$ Where: $J_1 = \{j \mid j \text{ is odd} \wedge 2 \le j \le n\}$ $J_2 = \{j \mid j \text{ is even} \wedge 2 \le j \le n\}$ $y_j = x_j - x_1^{0.5\left(1.0 + \frac{3(j-2)}{n-2}\right)}, j = 2, \ldots, n$	$0 \le x_i \le 1$ $\forall i = 1, \ldots, n$
UF4	$f_1 = x_1 + \frac{2}{\lvert J_1\rvert}\sum_{j\in J_1} h(y_j)$ $f_2 = 1 - x_1^2 + \frac{2}{\lvert J_2\rvert}\sum_{j\in J_2} h(y_j)$ Where: $J_1 = \{j \mid j \text{ is odd} \wedge 2 \le j \le n\}$ $J_2 = \{j \mid j \text{ is even} \wedge 2 \le j \le n\}$ $y_j = x_j - \sin\left(6\pi x_1 + \frac{j\pi}{n}\right), j = 2, \ldots, n$ $h(y_j) = \frac{\lvert y_j\rvert}{1 + e^{2\lvert y_j\rvert}}$	$0 \le x_1 \le 1$ $-2 \le x_i \le 2$ $\forall i = 2, \ldots, n$
UF5	$f_1 = x_1 + \left(\frac{1}{2N} + \varepsilon\right)\lvert\sin(2N\pi x_1)\rvert + \frac{2}{\lvert J_1\rvert}\sum_{j\in J_1} h(y_j)$ $f_2 = 1 - x_1 + \left(\frac{1}{2N} + \varepsilon\right)\lvert\sin(2N\pi x_1)\rvert + \frac{2}{\lvert J_2\rvert}\sum_{j\in J_2} h(y_j)$ Where: $J_1 = \{j \mid j \text{ is odd} \wedge 2 \le j \le n\}$ $J_2 = \{j \mid j \text{ is even} \wedge 2 \le j \le n\}$ $y_j = x_j - \sin\left(6\pi x_1 + \frac{j\pi}{n}\right), j = 2, \ldots, n$ $h(y_j) = 2y_j^2 - \cos(4\pi y_j) + 1$	$0 \le x_1 \le 1$ $-1 \le x_i \le 1$ $\forall i = 2, \ldots, n$
UF6	$f_1 = x_1 + max\left\{0.2\left(\frac{1}{2N} + \varepsilon\right)\sin(2N\pi x_1)\right\} + \frac{2}{\lvert J_1\rvert}\left(4\sum_{j\in J_1} y_j^2 - 2\prod_{j\in J_1}\cos\left(\frac{20y_j\pi}{\sqrt{j}}\right) + 2\right)$ $f_2 = 1 - x_1 max\left\{0.2\left(\frac{1}{2N} + \varepsilon\right)\sin(2N\pi x_1)\right\} + \frac{2}{\lvert J_2\rvert}\left(4\sum_{j\in J_2} y_j^2 - 2\prod_{j\in J_2}\cos\left(\frac{20y_j\pi}{\sqrt{j}}\right) + 2\right)$ Where: $J_1 = \{j \mid j \text{ is odd} \wedge 2 \le j \le n\}$ $J_2 = \{j \mid j \text{ is even} \wedge 2 \le j \le n\}$ $y_j = x_j - \sin\left(6\pi x_1 + \frac{j\pi}{n}\right), j = 2, \ldots, n$	$0 \le x_1 \le 1$ $-1 \le x_i \le 1$ $\forall i = 2, \ldots, n$
UF7	$f_1 = \sqrt[5]{x_1} + \frac{2}{\lvert J_1\rvert}\sum_{j\in J_1} y_j^2$ $f_2 = 1 - \sqrt[5]{x_1} + \frac{2}{\lvert J_2\rvert}\sum_{j\in J_2} y_j^2$ Where: $J_1 = \{j \mid j \text{ is odd} \wedge 2 \le j \le n\}$ $J_2 = \{j \mid j \text{ is even} \wedge 2 \le j \le n\}$ $y_j = x_j - \sin\left(6\pi x_1 + \frac{j\pi}{n}\right), j = 2, \ldots, n$	$0 \le x_1 \le 1$ $-1 \le x_i \le 1$ $\forall i = 2, \ldots, n$

Table 2 Tri objective problems in CEC09 set

Problem	Objective Functions	Var Bounds
UF8	$f_1 = \cos(0.5x_1\pi)\cos(0.5x_2\pi) + \frac{2}{\lvert J_1\rvert}\sum_{j\in J_1}\left(x_j - 2x_2\sin\left(2\pi x_1 + \frac{j\pi}{n}\right)\right)^2$ $f_2 = \cos(0.5x_1\pi)\cos(0.5x_2\pi) + \frac{2}{\lvert J_2\rvert}\sum_{j\in J_2}\left(x_j - 2x_2\sin\left(2\pi x_1 + \frac{j\pi}{n}\right)\right)^2$ $f_3 = \sin(0.5x_1\pi) + \frac{2}{\lvert J_3\rvert}\sum_{j\in J_3}\left(x_j - 2x_2\sin\left(2\pi x_1 + \frac{j\pi}{n}\right)\right)^2$ Where: $J_1 = \{j \mid 3 \le j \le n \wedge j - 1 \text{ is multiple of } 3\}$ $J_2 = \{j \mid 3 \le j \le n \wedge j - 2 \text{ is multiple of } 3\}$ $J_3 = \{j \mid 3 \le j \le n \wedge j \text{ is multiple of } 3\}$	$0 \le x_1 \le 1$ $0 \le x_2 \le 1$ $-2 \le x_i \le 2$ $\forall i = 3, \ldots, n$
UF9	$f_1 = 0.5[max\{0, (1+\varepsilon)(1 - 4(2x_1 - 1)^2)\} + 2x_1]x_2 + \frac{2}{\lvert J_1\rvert}\sum_{j\in J_1}\left(x_j - 2x_2\sin\left(2\pi x_1 + \frac{j\pi}{n}\right)\right)^2$ $f_2 = 0.5[max\{0, (1+\varepsilon)(1 - 4(2x_1 - 1)^2)\} - 2x_1 + 2]x_2 + \frac{2}{\lvert J_2\rvert}\sum_{j\in J_2}\left(x_j - 2x_2\sin\left(2\pi x_1 + \frac{j\pi}{n}\right)\right)^2$ $f_3 = 1 - x_2 + \frac{2}{\lvert J_3\rvert}\sum_{j\in J_3}\left(x_j - 2x_2\sin\left(2\pi x_1 + \frac{j\pi}{n}\right)\right)^2$ Where: $J_1 = \{j \mid 3 \le j \le n \wedge j - 1 \text{ is multiple of } 3\}$ $J_2 = \{j \mid 3 \le j \le n \wedge j - 2 \text{ is multiple of } 3\}$ $J_3 = \{j \mid 3 \le j \le n \wedge j \text{ is multiple of } 3\}$	$0 \le x_1 \le 1$ $0 \le x_2 \le 1$ $-2 \le x_i \le 2$ $\forall i = 3, \ldots, n$
UF10	$f_1 = \cos(0.5x_1\pi)\cos(0.5x_2\pi) + \frac{2}{\lvert J_1\rvert}\sum_{j\in J_1}[4y_j^2 - \cos(8\pi y_j) + 1]$ $f_2 = \cos(0.5x_1\pi)\cos(0.5x_2\pi) + \frac{2}{\lvert J_2\rvert}\sum_{j\in J_2}[4y_j^2 - \cos(8\pi y_j) + 1]$ $f_3 = \sin(0.5x_1\pi) + \frac{2}{\lvert J_3\rvert}\sum_{j\in J_3}[4y_j^2 - \cos(8\pi y_j) + 1]$ Where: $J_1 = \{j \mid 3 \le j \le n \wedge j - 1 \text{ is multiple of } 3\}$ $J_2 = \{j \mid 3 \le j \le n \wedge j - 2 \text{ is multiple of } 3\}$ $J_3 = \{j \mid 3 \le j \le n \wedge j \text{ is multiple of } 3\}$ $y_j = x_j - 2x_2\sin\left(2\pi x_1 + \frac{j\pi}{n}\right), j = 3, \ldots, n$	$0 \le x_1 \le 1$ $0 \le x_2 \le 1$ $-2 \le x_i \le 2$ $\forall i = 3, \ldots, n$

Table 3 Pareto fronts of the instances in the CEC09 set

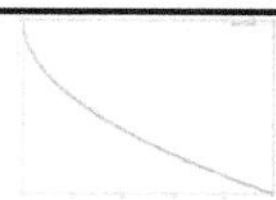 UF1, UF2 & UF4. $f_2 = 1 - f_1^2$ $0 \le f_1 \le 1$	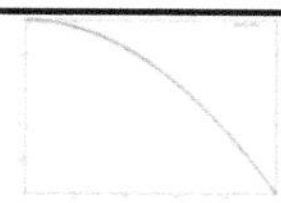 UF3. $f_2 = 1 - \sqrt{f_1}$ $0 \le f_1 \le 1$	UF5. Its PF has $2N + 1$ solutions: $\left(\frac{i}{2N}, 1 - \frac{i}{2N}\right)$ $i = 0, 1, \ldots, 2N$
 UF6. N disconnected parts: $f_2 = 1 - f_1, f_1 \in \bigcup_{i=1}^{N}\left[\frac{2i-1}{2N}, \frac{2i}{2N}\right]$	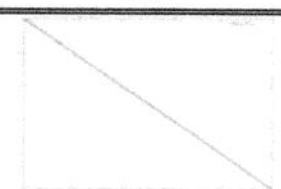 UF7. $f_2 = 1 - f_1$ $0 \le f_1 \le 1$	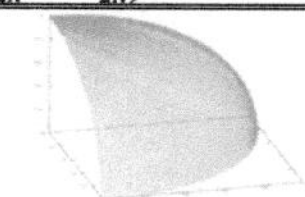 UF8 & UF10. $f_1^2 + f_2^2 + f_3^2 = 1$ $0 \le f_1, f_2, f_3 \le 1$
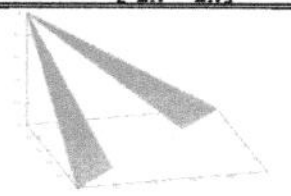 UF9.	The PF has two parts. The first part is: $0 \le f_3 \le 1$ $0 \le f_1 \le \frac{1}{4}(1 - f_3)$ $f_2 = 1 - f_1 - f_3$	The second part is: $0 \le f_3 \le 1$ $\frac{3}{4}(1 - f_3) \le f_1 \le 1$ $f_2 = 1 - f_1 - f_3$

4 Multiobjective Optimization Algorithms

In this section, we briefly described the ten state-of-the-art multiobjective algorithms (sorted alphabetically) used in this work. All experimentation was carried out using the implementation of the algorithms provided in jMetal 5.0 which is a Java framework aimed at a multiobjective optimization with metaheuristics (Nebro 2015).

AbYSS is a Scatter Search hybrid algorithm for multiobjective optimization (Nebro et al. 2008). The algorithm uses an external archive, similar to that used in MOCell algorithm, and also it incorporates mutation and crossover operators from evolutionary algorithms. AbYSS uses Pareto dominance, and an external archive to store the nondominated solutions.

CellDE is a modified version of MOCell algorithm (Durillo 2008). In CellDE, reproductive operators used in differential evolution are implemented instead of the typical genetic crossover and mutation operators. An external archive is used to store the nondominated solutions found during the search process, and the SPEA2 density estimator is applied when the archive becomes full.

GDE3 (Generalized Differential Evolution), is proposed as an extension of differential evolution (DE), for global optimization with an arbitrary number of objectives and constraints (Kukkonen 2005). The algorithm starts with a random initial solution. At each generation, differential evolution operators are used to create an offspring population.

Current population and the newly generated offspring populations are merged and then reduced using non-dominated sorting a diversity preservation technique like those of NSGA-II.

Proposed by Zitzler (2004), IBEA is an evolutionary algorithm that performs binary tournaments for mating selection of individuals and implements environmental selection by iteratively removing the individual with less contribution to the indicator from the population and updating the fitness values of the remaining individuals.

MOCell is a cellular genetic algorithm that includes an external archive to store the non-dominated solutions found during the generations (Nebro 2007). To ensure diversity in the archive, the algorithm utilizes crowding distance, selection based on a binary tournament, and apply genetic crossover and mutation operators. At the end of each iteration, the algorithm selects a fixed number of individuals from the archive to replace the same number of randomly selected solutions from the population.

NSGA-II (Non-dominated Sorting Genetic Algorithm) is a metaheuristic proposed by Deb et al. (2002), it is a genetic algorithm with three key features: (1) It uses an elitist principle, (2) uses an explicit diversity preserving mechanism and (3) emphasizes non-dominated solutions. Each offspring population is obtained by applying classic genetic operators (Selection, crossover, and mutation), then the two populations are sorted according to their rank, and the best solutions are chosen to create a new population using crowding distance.

NSGA-III (Yuan 2014) is a reference-point based on NSGA-II where the maintenance of diversity among population members is aided by supplying and adaptively updating some well-spread reference points. NSGA-III still relies on Pareto-dominance to push the population towards the Pareto front (PF).

OMOPSO (Sierra 2005) particle swarm optimization algorithm for multiobjective optimization problems. Key features: Usage of the crowding distance from NSGA-II to filter the leader solutions, and mutation operators to accelerate the convergence of the swarm.

PAES is an evolutionary algorithm proposed by Knowles and Corne (Knowles and Corne 2000). The algorithm utilizes a $(1 + 1)$ evolution strategy and an external archive of nondominated solutions to ensure diverse solutions in the Pareto optimal set. PAES uses local search from a population and a reference archive of previously found solutions to identify the dominance ranking of the current and candidate solution vectors.

SPEA2 is an improved version of a previous SPEA algorithm with the next features (Zitzler 2001): a fitness assignment strategy, a density estimation technique, and an enhanced archive truncation method.

5 Performance Metrics of Multiobjective Optimization

In this section we desribe the metrics used in this work, which are hypervolume (HV), epsilon ($I_{\varepsilon+}$), inverted generational distance (IGD), and generalized spread (Δ^*). The performance metrics enable the quantification of an algorithm's performance with regards to a specific requirement.

Definition 6 (*Hypervolume (HV)*) The hypervolume is the dominated volume of the front concerning a reference set (Zitzler and Thiele 1999; Fleischer 2003). Let n_{POF} the number of elements of the Pareto front, mathematically, for each solution $i \in n_{POF}$ a hypercube v_i is constructed using a reference point. The hypervolume (Eq. 1) is the union of all the hypercubes.

$$HV = volume\left(\bigcup_{i=1}^{n_{POF}} vi\right) \tag{1}$$

Higher values of the HV performance measure imply more desirable solutions (Durillo et al. 2010). This quality indicator measures both diversity and convergence.

Definition 7 Epsilon indicator ($I_{\varepsilon+}$). Additive Epsilon indicator (Zitzler et al. 2003) is the minimum distance that a Pareto set approximation B needs to translate in each dimension such that another approximation A is weakly dominated. The additive Epsilon indicator is as follows (Eq. 2):

$$I_{\varepsilon+}(A, B) = \min_{\varepsilon}\left\{ \begin{array}{c} \forall \vec{x}_B \in B \exists \vec{x}_A \in A : f_i(\vec{x}_B) - \varepsilon \leq f_i(\vec{x}_A) \\ for\ i \in \{1, \ldots, n\} \end{array} \right\} \tag{2}$$

where $\vec{x}_B$ and $\vec{x}_A$ are variable vectors that form the Pareto front approximations.

Definition 8 (*Inverted Generational Distance (IGD)*) The IGD metric measures both convergence and diversity of found solutions by an algorithm (Zhang et al. 2008). Let POF be a set of uniformly distributed points in the real optimal Pareto front (theoretical front), and let POF' an approximation of the POF. Then IGD is calculated by Eq. 3:

$$IGD = \frac{\sum_{i=1}^{n_{POF}} d_i}{n_{POF}} \tag{3}$$

where n_{POF} is the number of elements in POF and d_i is the Euclidean distance between the *i*-member in the POF and its nearest member in POF'. A low IGD value is achieved when POF' is very close to POF and POF' do not leave aside any section of the whole POF.

Definition 9 Generalized Spread (Δ^*). Zhou (2006) proposed a Generalized Spread Δ^* as an extension of the spread metric Δ previously proposed by Deb et al. (2002). The Generalized Spread Δ^* provides information about how evenly the non-dominated solutions are spaced in the Pareto front and is defined by (Eq. 4):

$$\Delta^*(S,P) = \frac{\sum_{k=1}^{m} d_k^e + \sum_{i=1}^{|n_{POF}|} \left|d_i - \overline{d}\right|}{\sum_{k=1}^{m} d_k^e + |n_{POF}|\overline{d}} \tag{4}$$

where d_i is the distance measure between neighbor solutions, $\overline{d}$ is the mean of the distance measures, and d_k^e is the distance between the extreme solutions of the reference set (P) and the Pareto front (S).

6 Computational Experiments

In this section, we briefly describe the parameter settings of the ten multiobjective algorithms assessed (Table 4), as well as the methodology and the results obtained.

We set the maximal number of function evaluations to 300,000 as suggested by Zhang (2008). The stopping condition for all the algorithms is a number of objective functions evaluations.

We have executed 30 independent runs for each experiment. In this work, an experiment is defined as a pair of an algorithm with a test instance (10 algorithms $\times$ 10 instances = 100 of experiments). Table 5 enlists all the elements of the experimentation.

Tables 6, 7, 8 and 9 show the median value and the interquartile range of each indicator for the 30 independent runs for every experiment. Cells that contain the best and second-best median values for each instance (In a row) are colored, dark grey and lighter grey respectively.

To further analyze this behavior, we include Table 10, which shows four Friedman average rankings for all algorithms (one per indicator).

Due to its low quality, the despicable results appear with an infinity symbol. Taking as an indicator to the hypervolume the results are inferior, not arriving in any of the cases to 7.0 $\times$ E-01. As for additive epsilon, no result exceeds 8.0 $\times$ E-02.

For the IGD indicator it was not possible to obtain a result that exceeds 1.0 $\times$ 10E-04, however it is the best behavior of the indicators, due to the direct relation with the low number of non-dominated solutions found by each algorithm; For example, the best result found by OMOPSO for instance UF2 finds a higher number of non-dominated solutions.

The algorithms struggled to obtain good results in the problems with discontinuous Pareto fronts (Test instances UF5 and UF6). Even the algorithms that obtained the best median values for each one of the indicators proved to be inefficient in this type of instances.

In the work of Jiang et al. (2014), the authors propose a classification of quality indicators using four core performance criteria namely: capacity, convergence, diversity, and convergence-diversity.

Table 4 Parameters used in the algorithms

Parameterization used in AbYSS (Nebro et al. 2008)	
Population size	20 individuals
Reference set	2 References sets 10 each
Archive size	100 individuals (Using Crowding Distance Archive)
Mutation	Polynomial: $p_{mutation} = 1.0/variables$
Parameterization used in Cell DE (Durillo 2008)	
Population size	100 individuals
Archive size	100 individuals (Using Crowding Distance Archive)
Crossover	Differential Evolution ($CR = 0.5$, $F = 0.5$)
Selection of parents	Binary Tournament
Feedback	20
Parameterization used in GDE3 (Kukkonen 2005)	
Population size	100 individuals
Crossover	Differential Evolution ($CR = 0.5$, $F = 0.5$)
Parameterization used in IBEA (Zitzler 2004)	
Population size	100 individuals
Archive size	100 individuals
Selection of parents	Binary Tournament
Crossover	Simulated Binary Crossover
Mutation	Polynomial: $p_{mutation} = 1.0/variables$
Parameterization used in MOCell (Nebro 2007)	
Population size	100 individuals (10×10)
Archive size	100 individuals
Selection of parents	Binary Tournament
Crossover	Simulated Binary Crossover ($p_{cross} = 0.9$)
Mutation	Polynomial: $p_{mutation} = 1.0/variables$
Parameterization used in NSGA-II (Deb et al. 2002)	
Population size	100 individuals
Selection of parents	Binary Tournament (Ranking using Crowding distance)
Recombination	Simulated Binary Crossover ($p_{cross} = 0.9$)
Mutation	Polynomial: $p_{mutation} = 1.0/variables$
Parameterization used in NSGA-III (Yuan 2014)	
Population size	100 individuals
Crossover	Simulated Binary Crossover ($p_{cross} = 0.9$)
Mutation	Polynomial: $p_{mutation} = 1.0/variables$
Selection of parents	Binary Tournament

(continued)

Table 4 (continued)

Parameterization used in AbYSS (Nebro et al. 2008)	
Parameterization used in OMOPSO (Sierra 2005)	
Particles	100 particles
Mutation	Uniform and Non-uniform: $p_{mutation} = 1.0/variables$
Swarm size	100 individuals
Parameterization used in PAES (Knowles and Corne 2000)	
Archive size	100 individuals
Mutation	Polynomial: $p_{mutation} = 1.0/variables$
Parameterization used in SPEA2 (Zitzler 2001)	
Population size	100 individuals
Selection of parents	Binary Tournament
Recombination	Simulated Binary Crossover ($p_{cross} = 0.9$)
Mutation	Polynomial: $p_{mutation} = 1.0/variables$

Table 5 Experimentation elements

Test instances Zhang (2008)		Algorithms		Performance metrics
UF1	UF6	AbYSS (Nebro et al. 2008)	NSGA-II (Deb et al. 2002)	Hypervolume (Zitzler and Thiele 1999) , (Fleischer 2003)
UF2	UF7	CellDE (Durillo 2008)	NSGA-III (Yuan 2014)	Epsilon indicator (Zitzler et al. 2003)
UF3	UF8	GDE3 (Kukkonen 2005)	OMOPSO (Sierra 2005)	IGD (Zhang et al. 2008)
UF4	UF9	IBEA (Zitzler 2004)	PAES (Knowles and Corne 2000)	Generalized spread (Zhou 2006)
UF5	UF10	MOCell (Nebro 2007)	SPEA2 (Zitzler 2001)	

Table 6 Median and interquartile range for CEC09 instances. HV indicator

Problem	ABYSS	CELLDE	GDE3	IBEA	MOCELL
UF1	$\mathbf{5.87E-01_{2.8E-02}}$	$1.52E-01_{1.4E-01}$	$\mathbf{5.68E-01_{1.0E-02}}$	$5.41E-01_{4.6E-02}$	$5.68E-01_{1.1E-02}$
UF2	$\mathbf{6.39E-01_{2.6E-03}}$	$4.75E-01_{2.6E-02}$	$6.34E-01_{1.2E-02}$	$6.26E-01_{9.5E-03}$	$6.23E-01_{5.9E-03}$
UF3	$4.16E-01_{9.8E-02}$	$7.10E-02_{1.9E-02}$	$3.45E-01_{2.5E-02}$	$3.48E-01_{2.8E-02}$	$4.19E-01_{5.3E-02}$
UF4	$2.62E-01_{4.8E-03}$	$1.69E-01_{5.9E-03}$	$\mathbf{2.75E-01_{5.4E-04}}$	$2.62E-01_{5.8E-03}$	$2.58E-01_{7.0E-03}$
UF5	$1.57E-01_{7.4E-02}$	$0.00E00_{0.0E00}$	$0.00E00_{0.0E00}$	$\mathbf{1.90E-01_{9.6E-02}}$	$1.69E-01_{8.3E-02}$
UF6	$2.16E-01_{1.6E-01}$	$0.00E00_{0.0E00}$	$3.40E-04_{2.2E-01}$	$2.23E-01_{1.4E-01}$	$2.13E-01_{9.2E-02}$
UF7	$2.50E-01_{3.7E-02}$	$4.82E-02_{5.9E-02}$	$\mathbf{4.51E-01_{2.2E-03}}$	$2.93E-01_{2.1E-01}$	$3.23E-01_{2.4E-01}$
UF8	$1.76E-01_{1.2E-01}$	$5.72E-04_{4.5E-03}$	$6.05E-02_{4.4E-02}$	$\mathbf{2.09E-01_{4.2E-04}}$	$4.87E-02_{1.0E-01}$
UF9	$4.03E-01_{1.4E-01}$	$2.52E-02_{3.3E-02}$	$4.34E-01_{6.4E-02}$	$\mathbf{5.32E-01_{9.6E-02}}$	$1.54E-01_{1.1E-01}$
UF10	$\mathbf{8.96E-02_{3.3E-02}}$	$0.00E00_{0.0E00}$	$0.00E00_{2.8E-04}$	$4.34E-02_{6.9E-02}$	$0.00E00_{0.0E00}$

Problem	NSGAII	NSGAIII	OMOPSO	PAES	SPEA2
UF1	$5.66E-01_{1.8E-02}$	$4.63E-01_{8.7E-02}$	$5.29E-01_{2.4E-02}$	$4.47E-01_{7.5E-02}$	$5.46E-01_{2.1E-02}$
UF2	$6.30E-01_{9.2E-03}$	$5.81E-01_{1.6E-02}$	$\mathbf{6.35E-01_{4.9E-03}}$	$4.96E-01_{1.3E-01}$	$6.25E-01_{1.1E-02}$
UF3	$\mathbf{4.36E-01_{6.4E-02}}$	$3.38E-01_{3.3E-02}$	$3.40E-01_{4.2E-01}$	$2.72E-01_{7.1E-02}$	$\mathbf{4.57E-01_{5.3E-02}}$
UF4	$\mathbf{2.64E-01_{1.1E-03}}$	$2.04E-01_{1.3E-02}$	$2.37E-01_{4.4E-02}$	$1.48E-01_{5.1E-02}$	$2.63E-01_{6.6E-04}$
UF5	$\mathbf{1.71E-01_{8.6E-02}}$	$1.11E-01_{1.3E-01}$	$0.00E00_{1.4E-02}$	$1.42E-01_{1.5E-01}$	$1.68E-01_{1.0E-01}$
UF6	$\mathbf{2.41E-01_{8.7E-02}}$	$2.02E-01_{1.0E-01}$	$8.37E-02_{9.7E-02}$	$9.84E-02_{1.8E-01}$	$\mathbf{2.34E-01_{8.8E-02}}$
UF7	$3.12E-01_{2.2E-01}$	$2.16E-01_{1.7E-01}$	$\mathbf{4.43E-01_{1.2E-02}}$	$1.92E-01_{9.5E-02}$	$3.86E-01_{1.9E-01}$
UF8	$1.03E-01_{1.2E-01}$	$\mathbf{1.93E-01_{3.7E-03}}$	$0.00E00_{1.6E-01}$	$7.89E-02_{5.6E-02}$	$1.32E-01_{5.2E-02}$
UF9	$3.23E-01_{1.2E-01}$	$4.68E-01_{3.3E-02}$	$1.81E-01_{2.1E-01}$	$2.27E-01_{1.7E-01}$	$\mathbf{4.95E-01_{8.1E-02}}$
UF10	$1.33E-02_{2.5E-02}$	$5.28E-02_{5.9E-02}$	$0.00E00_{0.0E00}$	$\mathbf{6.20E-02_{2.6E-02}}$	$2.42E-02_{5.1E-02}$

Table 7 Median and interquartile range for CEC09 instances. $I_{\varepsilon+}$ indicator

Problem	ABYSS	CELLDE	GDE3	IBEA	MOCELL
UF1	$1.35E-01_{3.4E-02}$	$4.68E-01_{1.2E-01}$	$1.62E-01_{2.0E-02}$	$1.75E-01_{7.6E-02}$	$1.55E-01_{2.1E-02}$
UF2	$1.18E-01_{4.7E-02}$	$2.46E-01_{6.2E-02}$	$7.36E-02_{4.5E-02}$	$1.39E-01_{8.4E-02}$	$1.09E-01_{1.5E-02}$
UF3	$4.11E-01_{1.2E-01}$	$7.00E-01_{1.0E-01}$	$3.20E-01_{1.1E-01}$	$5.09E-01_{7.2E-02}$	$4.10E-01_{1.3E-01}$
UF4	$5.76E-02_{1.0E-02}$	$1.26E-01_{7.0E-03}$	$3.94E-02_{1.0E-03}$	$7.84E-02_{1.9E-02}$	$6.70E-02_{1.8E-02}$
UF5	$4.24E-01_{1.4E-01}$	$2.47E00_{2.3E-01}$	$5.00E-01_{0.0E00}$	$5.65E-01_{2.4E-01}$	$4.77E-01_{2.1E-01}$
UF6	$5.15E-01_{1.7E-01}$	$1.68E00_{2.6E-01}$	$5.00E-01_{2.5E-01}$	$4.69E-01_{2.7E-01}$	$4.76E-01_{3.4E-01}$
UF7	$6.77E-01_{6.3E-02}$	$5.11E-01_{1.6E-01}$	$1.33E-01_{4.0E-03}$	$5.99E-01_{5.5E-01}$	$5.59E-01_{5.9E-01}$
UF8	$6.62E-01_{3.3E-01}$	$9.75E-01_{1.3E-01}$	$7.43E-01_{4.0E-02}$	$7.09E-01_{1.8E-03}$	$7.53E-01_{4.2E-01}$
UF9	$4.82E-01_{5.0E-02}$	$8.69E-01_{1.0E-01}$	$2.45E-01_{4.8E-02}$	$4.41E-01_{2.8E-01}$	$6.45E-01_{2.5E-01}$
UF10	$7.35E-01_{1.4E-01}$	$3.19E00_{4.0E-01}$	$9.91E-01_{1.9E-01}$	$9.48E-01_{1.7E-01}$	$1.03E00_{2.9E-01}$

Problem	NSGAII	NSGAIII	OMOPSO	PAES	SPEA2
UF1	$1.55E-01_{2.1E-02}$	$3.12E-01_{1.2E-01}$	$1.70E-01_{6.1E-02}$	$3.68E-01_{1.2E-01}$	$1.83E-01_{4.2E-02}$
UF2	$1.11E-01_{3.3E-02}$	$1.49E-01_{1.1E-01}$	$6.02E-02_{1.2E-02}$	$3.98E-01_{2.0E-01}$	$1.16E-01_{4.0E-02}$
UF3	$3.81E-01_{1.1E-01}$	$5.23E-01_{1.2E-01}$	$2.13E-01_{3.4E-01}$	$6.27E-01_{1.8E-01}$	$3.68E-01_{9.6E-02}$
UF4	$4.94E-02_{1.7E-03}$	$1.31E-01_{1.7E-02}$	$7.70E-02_{1.8E-02}$	$6.36E-01_{2.6E-01}$	$5.06E-02_{3.4E-03}$
UF5	$4.21E-01_{1.3E-01}$	$7.36E-01_{2.0E-01}$	$6.96E-01_{5.4E-01}$	$8.20E-01_{3.1E-01}$	$4.17E-01_{1.1E-01}$
UF6	$4.22E-01_{2.0E-01}$	$6.05E-01_{3.0E-01}$	$5.31E-01_{3.6E-01}$	$8.72E-01_{2.5E-01}$	$4.19E-01_{2.0E-01}$
UF7	$6.08E-01_{5.8E-01}$	$7.31E-01_{2.9E-01}$	$1.22E-01_{1.9E-02}$	$7.67E-01_{1.7E-01}$	$4.64E-01_{5.4E-01}$
UF8	$7.17E-01_{4.4E-01}$	$7.13E-01_{8.6E-03}$	$3.93E-01_{7.1E-01}$	$8.33E-01_{2.0E-01}$	$7.21E-01_{3.3E-01}$
UF9	$5.08E-01_{2.5E-01}$	$4.80E-01_{6.4E-02}$	$4.69E-01_{5.4E-01}$	$6.88E-01_{2.9E-01}$	$3.23E-01_{2.6E-01}$
UF10	$9.48E-01_{1.1E-01}$	$8.98E-01_{2.2E-01}$	$1.41E00_{4.7E-01}$	$8.49E-01_{1.9E-01}$	$9.28E-01_{1.1E-01}$

Table 8 Median and interquartile range for CEC09 instances. IGD indicator

Problem	ABYSS	CELLDE	GDE3	IBEA	MOCELL
UF1	$2.68E-03_{8.1E-04}$	$1.37E-02_{5.2E-03}$	$2.70E-03_{3.7E-04}$	$4.13E-03_{1.6E-03}$	$3.01E-03_{5.2E-04}$
UF2	$1.82E-03_{1.1E-03}$	$4.84E-03_{5.5E-04}$	$1.04E-03_{4.9E-04}$	$2.95E-03_{2.1E-03}$	$1.61E-03_{4.3E-04}$
UF3	$1.13E-02_{3.9E-03}$	$1.80E-02_{8.3E-04}$	$7.64E-03_{1.1E-03}$	$1.18E-02_{2.7E-04}$	$8.03E-03_{2.5E-03}$
UF4	$1.53E-03_{1.2E-04}$	$3.94E-03_{1.9E-04}$	$1.25E-03_{1.4E-05}$	$1.63E-03_{2.0E-04}$	$1.69E-03_{2.0E-04}$
UF5	$6.42E-02_{1.4E-02}$	$6.53E-01_{4.5E-02}$	$8.72E-02_{0.0E00}$	$9.06E-02_{4.4E-02}$	$7.37E-02_{3.2E-02}$
UF6	$7.18E-03_{6.0E-03}$	$7.25E-02_{2.1E-02}$	$1.58E-02_{9.3E-03}$	$6.61E-03_{1.3E-02}$	$1.02E-02_{1.3E-02}$
UF7	$1.43E-02_{2.2E-03}$	$1.33E-02_{4.1E-03}$	$1.66E-03_{1.6E-04}$	$1.20E-02_{1.3E-02}$	$9.99E-03_{1.4E-02}$
UF8	$2.93E-03_{9.4E-04}$	$5.88E-03_{1.3E-03}$	$3.53E-03_{3.5E-04}$	$5.63E-03_{1.3E-06}$	$3.71E-03_{1.7E-03}$
UF9	$2.97E-03_{1.2E-03}$	$6.78E-03_{1.1E-03}$	$2.00E-03_{4.4E-04}$	$3.66E-03_{2.5E-03}$	$4.38E-03_{1.4E-03}$
UF10	$4.41E-03_{1.1E-03}$	$4.52E-02_{7.4E-03}$	$5.68E-03_{2.2E-03}$	$6.74E-03_{1.2E-03}$	$8.44E-03_{4.5E-03}$

Problem	NSGAII	NSGAIII	OMOPSO	PAES	SPEA2
UF1	$3.38E-03_{7.4E-04}$	$7.43E-03_{3.1E-03}$	$3.08E-03_{1.1E-03}$	$8.36E-03_{2.9E-03}$	$3.73E-03_{1.1E-03}$
UF2	$1.63E-03_{7.5E-04}$	$2.55E-03_{2.8E-03}$	$9.17E-04_{1.3E-04}$	$8.05E-03_{5.2E-03}$	$1.76E-03_{7.8E-04}$
UF3	$8.16E-03_{2.1E-03}$	$1.16E-02_{9.7E-04}$	$7.37E-03_{8.2E-03}$	$1.26E-02_{2.6E-03}$	$7.55E-03_{2.2E-03}$
UF4	$1.46E-03_{2.7E-05}$	$2.74E-03_{4.9E-04}$	$2.26E-03_{1.2E-03}$	$8.97E-03_{4.3E-03}$	$1.51E-03_{1.7E-05}$
UF5	$6.05E-02_{9.4E-03}$	$1.10E-01_{3.2E-02}$	$1.28E-01_{8.7E-02}$	$1.14E-01_{4.9E-02}$	$6.06E-02_{1.7E-02}$
UF6	$6.20E-03_{1.1E-03}$	$1.55E-02_{5.6E-03}$	$1.46E-02_{8.2E-03}$	$1.89E-02_{8.5E-03}$	$5.78E-03_{2.6E-03}$
UF7	$1.23E-02_{1.4E-02}$	$1.61E-02_{9.4E-03}$	$1.55E-03_{2.9E-04}$	$1.73E-02_{5.7E-03}$	$7.65E-03_{1.3E-02}$
UF8	$3.21E-03_{1.6E-03}$	$5.58E-03_{1.8E-04}$	∞	$5.70E-03_{1.2E-03}$	$2.93E-03_{9.6E-04}$
UF9	$3.25E-03_{1.4E-03}$	$2.95E-03_{4.5E-04}$	$4.53E-03$	$4.80E-03_{1.4E-03}$	$1.94E-03_{1.4E-03}$
UF10	$4.87E-03_{1.7E-03}$	$4.30E-03_{1.5E-03}$	$1.82E-02_{7.3E-03}$	$5.50E-03_{1.3E-03}$	$3.65E-03_{3.1E-03}$

Table 9 Median and interquartile range for CEC09 instances. Δ^* indicator

Problem	ABYSS	CELLDE	GDE3	IBEA	MOCELL
UF1	$8.04E-01_{5.2E-02}$	$1.04E-01_{6.8E-02}$	$7.32E-01_{1.6E-02}$	$6.30E-01_{5.1E-02}$	$7.21E-01_{3.8E-02}$
UF2	$1.11E00_{1.5E-01}$	$3.22E-01_{5.2E-02}$	$7.68E-01_{2.1E-01}$	$1.05E00_{1.4E-01}$	$7.73E-01_{1.5E-01}$
UF3	$6.09E-01_{8.6E-02}$	$1.32E-01_{2.8E-02}$	$2.49E-01_{8.6E-02}$	$5.37E-01_{2.5E-02}$	$6.21E-01_{4.5E-02}$
UF4	$1.46E-01_{5.8E-02}$	$1.03E-01_{3.2E-02}$	$7.12E-02_{1.2E-02}$	$2.60E-01_{8.4E-02}$	$1.70E-01_{6.7E-02}$
UF5	$4.56E-01_{1.1E-01}$	$1.00E-01_{1.5E-02}$	$5.52E-01_{0.0E00}$	$5.65E-01_{7.7E-02}$	$5.49E-01_{1.3E-01}$
UF6	$5.88E-01_{1.3E-01}$	$4.65E-02_{3.8E-02}$	$5.03E-01_{8.1E-02}$	$6.07E-01_{1.2E-01}$	$6.23E-01_{1.1E-01}$
UF7	$7.94E-01_{6.6E-02}$	$1.60E-01_{8.5E-02}$	$1.07E00_{2.4E-02}$	$8.42E-01_{2.2E-01}$	$8.88E-01_{2.8E-01}$
UF8	$5.33E-01_{8.1E-02}$	$2.28E-01_{6.5E-02}$	$4.44E-01_{6.8E-02}$	$6.60E-01_{2.0E-03}$	$4.02E-01_{8.0E-02}$
UF9	$4.72E-01_{1.5E-01}$	$2.58E-01_{6.9E-02}$	$1.60E-01_{4.4E-02}$	$8.09E-01_{2.9E-01}$	$3.87E-01_{2.0E-01}$
UF10	$4.27E-01_{1.0E-01}$	$2.43E-02_{6.1E-03}$	$2.22E-01_{9.5E-02}$	$4.62E-01_{1.9E-01}$	$1.60E-01_{9.8E-02}$

Problem	NSGAII	NSGAIII	OMOPSO	PAES	SPEA2
UF1	$7.05E-01_{5.5E-02}$	$6.20E-01_{8.4E-02}$	$5.28E-01_{1.6E-01}$	$5.60E-01_{1.6E-01}$	$6.52E-01_{6.4E-02}$
UF2	$9.75E-01_{1.4E-01}$	$5.98E-01_{4.0E-01}$	$6.25E-01_{3.0E-02}$	$6.73E-01_{3.3E-01}$	$9.39E-01_{1.4E-01}$
UF3	$6.30E-01_{7.7E-02}$	$5.38E-01_{5.4E-02}$	$3.14E-01_{1.2E-01}$	$5.06E-01_{9.1E-02}$	$6.85E-01_{7.7E-02}$
UF4	$9.53E-02_{2.0E-02}$	$2.37E-01_{5.9E-02}$	$1.51E-01_{1.3E-01}$	$5.67E-01_{2.5E-01}$	$7.16E-02_{9.8E-03}$
UF5	$4.68E-01_{1.2E-01}$	$4.93E-01_{2.5E-01}$	$2.51E-01_{1.1E-01}$	$5.37E-01_{9.5E-02}$	$4.67E-01_{1.6E-01}$
UF6	$5.99E-01_{5.6E-02}$	$5.96E-01_{2.9E-01}$	$1.26E-01_{2.0E-01}$	$5.52E-01_{3.9E-01}$	$5.93E-01_{1.1E-01}$
UF7	$8.42E-01_{2.8E-01}$	$7.23E-01_{1.3E-01}$	$7.76E-01_{6.7E-02}$	$6.79E-01_{1.1E-01}$	$9.44E-01_{2.4E-01}$
UF8	$4.79E-01_{8.7E-02}$	$6.67E-01_{5.8E-03}$	∞	$4.63E-01_{1.8E-01}$	$5.36E-01_{5.8E-02}$
UF9	$4.10E-01_{2.2E-01}$	$7.10E-01_{9.6E-02}$	$4.35E-01$	$5.81E-01_{1.3E-01}$	$3.18E-01_{2.9E-01}$
UF10	$3.40E-01_{1.0E-01}$	$5.94E-01_{2.4E-01}$	$8.18E-02_{4.7E-02}$	$4.03E-01_{7.5E-02}$	$5.59E-01_{2.0E-01}$

Table 10 Friedman average rankings

HV		$I_{\varepsilon+}$		*IGD*		Δ^*	
p-value computed: 2.93*E*–10		p-value computed: 2.607*E*–10		p-value computed: 1.000*E*–99		p-value computed: 2.637*E*–10	
Algorithm	Ranking	Algorithm	Ranking	Algorithm	Ranking	Algorithm	Ranking
AbYSS	3.705	GDE3	3.603	GDE3	3.350	CellDE	1.490
SPEA2	3.893	OMOPSO	4.100	SPEA2	3.480	OMOPSO	4.373
IBEA	4.087	SPEA2	4.303	NSGA-II	3.820	GDE3	4.650
NSGA-II	4.112	NSGA-II	4.337	AbYSS	4.210	MOCell	5.737
GDE3	4.963	AbYSS	4.387	MOCell	4.890	NSGA-II	5.920
MOCell	5.325	MOCell	5.273	OMOPSO	5.853	NSGA-III	6.077
NSGA-III	5.622	IBEA	5.277	IBEA	6.077	PAES	6.323
OMOPSO	6.692	NSGA-III	6.547	NSGA-III	6.303	AbYSS	6.597
PAES	7.258	PAES	8.033	PAES	7.940	SPEA2	6.883
CellDE	9.343	CellDE	9.140	CellDE	9.077	IBEA	6.950

Also, experimentation is proposed to investigate the relationship between six different indicators: Generalized distance GD, inverted generalized distance IGD, Epsilon indicator $I_{\varepsilon+}$, Spread metric Δ, generalized spread Δ^* and, hypervolume. Experimental results of their work show that these six metrics have high consistencies when Pareto fronts are convex, whereas they show contradictions on non-convex fronts.

The results obtained in this paper were experimentally similar to those reported by Jiang et al. (2014), even when in this experiment the algorithms were submitted to test instances (CEC09) that have proved to be very challenging. HV, Epsilon, and IGD got similar rankings, and the algorithms performed well, but some of the algorithms with the worst performance in these indicators got the best ranking and performance using generalized spread.

The experimentation was carried out on a computer with a processor AMD Phenom 2 3.20 GHZ, RAM: 8 GB, with Windows 7 service pack 7, all the coding was written on C# using Visual Studio Community.

7 Conclusion and Future Work

Although the SPEA2 algorithm did not get the first place in the ranking of any of the indicators, its performance was very high for the HV, IGD, and $I_{\varepsilon+}$ metrics achieving to be among the three best places of each ranking. The previous indicates that the fronts obtained by SPEA2, are close to the real front and present a good quality diversity of the points representing the solutions obtained. However, the low performance obtained using Δ^* indicator shows, that there are not so many solutions close to the extreme values.

Taking the HV, IGD, and epsilon indicators as the reference, the performance of the CellDE algorithm is surpassed by the rest of the algorithms. However, in the case of the generalized spread Δ^* indicator, the best results were obtained. A generalized idea in the field of optimization is that no single algorithm has the best performance in solving all problems and evaluating its quality with all indicators, the above is supported by the No Free Lunch theorem (Wolpert and Macready 1997).

When evaluating the performance of a new algorithm, that algorithm is compared against the results obtained by the algorithms SPEA2, NSGA-II, AbYSS, and GDE3 considering HV, IGD, and $I_{\varepsilon+}$ as the quality indicators. However, if it is desired to know if the solutions proposed by the algorithm includes the extreme values, the generalized spread indicator is needed.

Among the future works to be developed, it is considered to carry out more exhaustive experimentation that includes all the multiobjective algorithms in jMetal 5.0. The experimentation can be extended to one hundred or more repetitions per experiment. Significant future work is to compare in pairs the algorithms to identify if two or more algorithms are equivalent.

The results obtained for the functions with discontinuous fronts UF5 and UF6 by all algorithms presented in this work had inferior quality. These discontinuous fronts have proven to be challenging for algorithms that do not have specific strategies to deal with them.

Considering the results obtained experimentally in this work and other work focused on the comparison of quality indicators (Durillo et al. 2010; Jiang et al. 2014; Helbig and Engelbrecht 2013). We strongly recommend that when evaluating the performance of a new algorithm, it should be tested against others reported in the state-of-the-art-using more than one quality indicator.

Acknowledgements Authors would like to acknowledge with appreciation and gratitude to CONACYT, TECNM, and PRODEP for their financial support. Also, acknowledge to Laboratorio Nacional de Tecnologías de la Información del Instituto Tecnológico de Ciudad Madero for the access to the cluster. E. Rodríguez and A. Santiago would like to thank CONACYT for the supports 490945 and 360199.

References

Deb, K. 2011. *Multi-objective optimization using evolutionary algorithms: An introduction.* KanGAL Rep., no. 2011003.

Deb, K., A. Pratap, S. Agarwal, and T. Meyarivan. 2002. A fast and elitist multiobjective genetic algorithm: NSGA-II. *IEEE Transactions on Evolutionary Computation* 6 (2): 182–197.

Durillo, J.J., A.J. Nebro, F. Luna, and E. Alba. 2008. Solving three-objective optimization problems using a new hybrid cellular genetic algorithm. In *International conference on parallel problem solving from nature*, pp. 661–670.

Durillo, J.J., A.J. Nebro, C.A.C. Coello, J. García-Nieto, F. Luna, and E. Alba. 2010. A study of multiobjective metaheuristics when solving parameter scalable problems. *IEEE Transactions on Evolutionary Computation* 14 (4): 618–635.
Fleischer, M. 2003. The measure of Pareto optima applications to multi-objective metaheuristics. In *International conference on evolutionary multi-criterion optimization*, pp. 519–533.
Helbig, M., and A.P. Engelbrecht. 2013. Performance measures for dynamic multi-objective optimisation algorithms. *Information Sciences (Ny)* 250: 61–81.
Jiang, S., Y.-S. Ong, J. Zhang, and L. Feng. 2014. Consistencies and contradictions of performance metrics in multiobjective optimization. *IEEE Transactions on Cybernetics* 44 (12): 2391–2404.
Knowles, J.D., and D.W. Corne. 2000. Approximating the nondominated front using the Pareto archived evolution strategy. *Evolutionary Computation* 8 (2): 149–172.
Kukkonen, S., and J. Lampinen. 2005. GDE3: The third evolution step of generalized differential evolution. In *2005 IEEE congress on evolutionary computation*, vol. 1, pp. 443–450.
Nebro, A.J., J.J. Durillo, F. Luna, B. Dorronsoro, and E. Alba. 2007. Design issues in a multi-objective cellular genetic algorithm. In *International conference on evolutionary multi-criterion optimization*, pp. 126–140.
Nebro, A.J., J.J. Durillo, and M. Vergne. 2015. Redesigning the jMetal multi-objective optimization framework. In *Proceedings of the companion publication of the 2015 annual conference on genetic and evolutionary computation*, pp. 1093–1100.
Nebro, A.J., F. Luna, E. Alba, B. Dorronsoro, J.J. Durillo, and A. Beham. 2008. AbYSS: Adapting scatter search to multiobjective optimization. *IEEE Transactions on Evolutionary Computation* 12 (4): 439–457.
Sierra, M.R., and C.A.C. Coello. 2005. Improving PSO-based multi-objective optimization using crowding, mutation and ϵ-dominance. In *International conference on evolutionary multi-criterion optimization*, pp. 505–519.
Wolpert, D.H., and W.G. Macready. 1997. No free lunch theorems for optimization. *IEEE Transactions on Evolutionary Computation* 1 (1): 67–82.
Yuan, Y., H. Xu, and B. Wang. 2014. An improved NSGA-III procedure for evolutionary many-objective optimization. In *Proceedings of the 2014 annual conference on genetic and evolutionary computation*, pp. 661–668.
Zhang, Q., A. Zhou, S. Zhao, P.N. Suganthan, W. Liu, and S. Tiwari. 2008. Multiobjective optimization test instances for the CEC 2009 special session and competition. Univ. Essex, Colchester, UK Nanyang Technol. Univ. Singapore, Spec. Sess. Perform. Assess. multi-objective Optim. algorithms, Tech. Rep., vol. 264.
Zhang, Q., A. Zhou, and Y. Jin. 2008b. RM-MEDA: A regularity model-based multiobjective estimation of distribution algorithm. *IEEE Transactions on Evolutionary Computation* 12 (1): 41–63.
Zhou, A., Y. Jin, Q. Zhang, B. Sendhoff, and E. Tsang. 2006. Combining model-based and genetics-based offspring generation for multi-objective optimization using a convergence criterion. In *2006 IEEE international conference on evolutionary computation*, pp. 892–899.
Zitzler, E., M. Laumanns, L. Thiele, and others. 2001. SPEA2: Improving the strength Pareto evolutionary algorithm. In *Eurogen*, vol. 3242, no. 103, pp. 95–100.
Zitzler, E., and S. Künzli. 2004. Indicator-based selection in multiobjective search. In *International conference on parallel problem solving from nature*, pp. 832–842.
Zitzler, E., and L. Thiele. 1999. Multiobjective evolutionary algorithms: a comparative case study and the strength Pareto approach. *IEEE Transactions on Evolutionary Computation* 3 (4): 257–271.
Zitzler, E., L. Thiele, M. Laumanns, C.M. Fonseca, and V.G. Da Fonseca. 2003. Performance assessment of multiobjective optimizers: An analysis and review. *IEEE Transactions on Evolutionary Computation* 7 (2): 117–132.

Analysis of the Efficient Frontier of the Portfolio Selection Problem Instance of the Mexican Capital Market

Héctor Joaquín Fraire Huacuja, Javier Alberto Rangel González, Juan Frausto Solís, Marco Antonio Aguirre Lam, Lucila Morales Rodríguez, and Juan Martín Carpio Valadez

Abstract Earning profits when investing in a stock exchange and avoiding losses has always been a priority for every investor. This is the main reason why the portfolio selection problem has been of great importance to obtain reasonable compensation between the rate of return and the risk. However, the portfolio problem has been extended by introducing real-world constraints, such as cardinality restriction to limit the number of portfolio assets and quantity constraints to limit the proportion of each portfolio asset within the inferior and superior limits. In this work, nine state-of-the-art multiobjective algorithms were used to solve an instance of the portfolio selection problem for the Mexican Stock Exchange. A comparative experimental study of the efficient frontiers obtained and the behavior of these algorithms when solving the problem instance is reported. Two statistical hypothesis tests were used to support the conclusions in the analysis of the experimental results.

Keywords Multiobjective optimization · Portfolio selection problem · Efficient frontier

---

H. J. F. Huacuja (✉) · J. A. R. González · J. F. Solís · M. A. A. Lam · L. M. Rodríguez
Tecnológico Nacional de México, Instituto Tecnológico de Ciudad Madero, Ciudad Madero, Mexico
e-mail: automatas2002@yahoo.com.mx

J. A. R. González
e-mail: javieralberto64@hotmail.com

J. F. Solís
e-mail: juan.frausto@itcm.edu.mx

M. A. A. Lam
e-mail: marco.itcm@gmail.com

L. M. Rodríguez
e-mail: lmoralesrdz@gmail.com

J. M. C. Valadez
Tecnológico Nacional de México, Instituto Tecnológico de León, León, Mexico
e-mail: jmcarpio61@hotmail.com

O. Castillo and P. Melin (eds.), *Fuzzy Logic Hybrid Extensions of Neural and Optimization Algorithms: Theory and Applications*, Studies in Computational Intelligence 940,
https://doi.org/10.1007/978-3-030-68776-2_16

1 Introduction

Portfolio optimization is one of the most important issues in finance. Since the pioneering work of Markowitz, the problem is formulated as a bi-objective optimization problem, where the return of a portfolio should be maximized, while the risk should be minimized. Markowitz (1952) shows that the risk of an investment portfolio depends on the covariance between the assets (Markowitz 1952). The average return of assets determines the return of a portfolio. The risk depends on the covariance between the assets in the portfolio. Markowitz's model may be written as the following non-linear quadratic programming model:

$$\min \sum_{i \in N} \sum_{j \in N} x_i x_j \sigma_{ij}$$

$$s.t.: \sum_{j \in N} x_j r_j \geq \omega_0 M_0$$

$$\sum_{j \in N} x_j = M_0$$

$$x_j \geq 0,\ j \in N$$

where $N = \{1,\ldots, \mathrm{N}\}$ is the set of assets; σ_{ij} is the covariance between returns i and j; r_j is the average expected return of asset j; ω_0 is the minimum return required by the investor; M_0 is the capital available for investment; and decision variable x_j is the proportion of capital allocated to invest in asset j. The objective function (1) minimizes the portfolio risk, i.e., the covariance between returns. Constraint (2) states that the portfolio return is equal to or greater than the return required by the investor, where such a return is obtained by multiplying the capital available for investment, M_0, and the rate of return that an investor would like to obtain, ω_0. Constraint (3) guarantees that the sum of the capital proportions invested in each asset j is equal to the capital available. The set of constraints (4) establishes the domain of the decision variables.

In Anagnostopoulos and Mamanis (2010) an additional objective which minimizes the number of assets in a portfolio. The non-smoothness of the third objective function generates an additional difficulty in the problem. The third objective function is incorporated by counting the number of non-negative weights in a portfolio. The number of assets in the portfolio should be minimized. The formal definition of the tri-objective portfolio selection problem is as follows:

$$\max \sum_{i=1}^{N} x_i \mu_i$$

$$\min \sum_{i=1}^{N} \sum_{j=1}^{N} x_i x_j \sigma_{ij}$$

$$\min \sum_{i=1}^{N} 1_{x_i > 0}$$

$$\text{s.t.}$$

$$\sum_{i=1}^{N} x_i = 1$$

$$0 \leq x_i \leq 1$$

where the μ, and σ symbolize the expected returns, covariances, and variances of the securities which are the inputs of the portfolio optimization problem. The feasible set of portfolios X is defined by the budget constraint $\left(\sum_{i=1}^{N} x_i = 1\right)$, which requires that all the available capital is invested, and the non-negativity constraints that imply that no short sales are allowed (i.e. $x \geq 0$).

Introducing a third objective into the portfolio optimization model the efficient frontier becomes a surface in the three-dimensional space and finding the exact efficient surface is a very difficult problem.

In Anagnostopoulos and Mamanis (2010) were tested NSGA-II, PESA, and SPEA2 to find an approximation of the efficient frontier, and the quality indicator used was the hypervolume. Unlike that work, in this chapter nine algorithm are tested: CellDE, IBEA, GDE3, MOCell, NSGA-II, NSGA-III, OMOPSO, PAES, SPEA2. The quality indicators used in this work are: hypervolume, ε, generalized spread, inverted generational distance, inverted generational distance plus and overall spread.

2 Multiobjective Algorithms in Comparison

The multi-objective evolutionary algorithms (MOEA) can produce, a complete approximation of the Pareto front in a single execution. An MOEA consists of two main components: (1) a selection mechanism that aims to choose the solutions that represent the best possible exchanges among all the objectives, and (2) a density estimator that avoids the population convergence to a single solution. Several different solutions in a single dimension are generated with these MOEA components. Another important component is elitism, which refers to maintaining the best solutions that have been generated so far. Elitism is usually implemented using an external file in which the best solutions generated in each iteration are stored. In this section, the algorithms tested are described.

3 CellDE

The basic behavior of CellDE is that of a cGA that follows an asynchronous behavior, in the sense that all cells are scanned sequentially (in cGA the synchronous cells are scanned in parallel).

The main difference between CellDE and MOCell arises in the creation of new individuals. Instead of using the operators of classical genetic algorithms (GA) to generate new individuals, CellDE uses the operator used in differential evolution (DE): three different individuals are chosen, and the new offspring solution is obtained according to the differences between them (Durillo et al. 2008).

4 GDE3

The Differential Evolution (DE) algorithm was introduced by Storn and Price (1997) and Price et al. (2006) in 1995. The principles of DE design are simplicity, efficiency and the use of floating-point coding instead of binary numbers. As a typical EA, DE has a random initial population improved by selection, mutation, and crossover operators. A predefined limit for the number of generations provides an appropriate stopping condition for EA.

The third and last version is GDE3 (Kukkonen and Lampinen 2005; Kukkonen and Deb 2006a). In addition to the selection, another part of the basic DE has also been modified. Now, in the case of feasible and non-dominant solutions, both vectors are saved for the next population. The population size is reduced using a non-dominated selection and pruning based on the preservation of the diversity before continuing with the next generation. The pruning technique used in the original GDE3 is based on the crowding distance, which provides a good estimate of overpopulation in the case of two objectives. However, the crowding distance does not approximate the agglomeration of the solutions when the number of targets is more than two (Kukkonen and Deb 2006a). Since the problem provided in Huang et al. (2007) consists of problems with more than two objectives, a more general diversity maintenance technique proposed in Kukkonen and Deb (2006b) was used.

5 IBEA

Indicators based evolutionary algorithm (IBEA) (Zitzler and Künzli 2004) main concept is to formalize the preferences using a generalization of the dominance relation. Based on a binary indicator that describes the preference, an ability is calculated for everyone in the current population. The individual's fitness values are used to promote environmental and mating selection.

6 MOCell

The MOCell (Nebro et al. 2009) algorithm can be considered as the next step towards the use of the canonical cell model of the genetic algorithm (GA) in the multiobjective field. cMOGA (Alba et al. 2007) was the first attempt in this direction analyzing the search engine of the cellular genetic algorithm (cGA) to solve MOP. MOCell differs from cMOGA in the inclusion of a feedback mechanism at the end of each iteration. This allows orienting the search towards undifferentiated solutions that have already been found. Therefore, this new approach not only takes full advantage of the cGA search engine but also the search experience stored in the external file.

7 NSGA-II

The non-dominated sorting-based (NSGA II) was proposed in 2000 by Srinivas and Deb (1994). This algorithm uses elitism and crowded tournament selection. A non-dominated rapid classification approach with computational complexity $O\left(mN^2\right)$ is presented, where m is the number of targets and N is the population size. The crowded tournament selection consists in the choice of individuals from a group using the dominance rank and the crowding distance.

8 NSGA-III

NSGA-III (Yuan et al. 2014) remains similar to the established NSGA-II (Deb et al. 2002) with significant changes in its selection mechanism. NSGA-III begins with the definition of a set of reference points. Then an initial population is randomly generated. In the nth generation, the current population of parents is used to produce a decent population by the use of random selection, simulated binary crossover operator (SBX) and polynomial mutation (Deb and Agrawal 1994).

In the original NSGA-II, the solutions are selected with the highest occupancy distance values. However, the crowding distance measure does not work well for many objective problems (Kukkonen and Deb 2006a). Therefore, the selection mechanism in NSGA-III is modified by performing a more systematic analysis of the members concerning the provided reference points.

9 OMOPSO

Sierra and Coello (2005) propose an approach based on the Pareto domain and the use of a crowding factor for the selection of leaders. For each generation and each particle select a leader. The selection is made through a binary tournament based on the crowding value of the leaders.

This proposal uses two external files: one to store the leaders that are currently used to carry out the flight and another to store the final solutions. The crowding factor is used to filter the list of leaders whenever the maximum limit imposed on that list is exceeded. Only the leaders with the best values of crowding are conserved. Also, the authors propose a scheme in which they subdivide the population into three different subsets. A different mutation operator is applied to each subset.

10 PAES

PAES (Knowles and Corne 1999) algorithm was developed with two main objectives: the first one was that the algorithm should be limited strictly to the local search, it should use only a small change operator (mutation) and move from a current solution to a nearby neighbor. This makes it quite different from the more popular MOGAs that maintain a population of solutions from which selection and descent are carried out. The second objective was that the algorithm is a true Pareto optimizer, treating all undominated solutions as of equal value. However, achieving both goals together is problematic. This is because, when comparing a couple of solutions, neither will dominate the other. This problem is overcome in PAES by maintaining a file of previously uncovered solutions found. This file is then used as a mean to estimate the true dominance ranking of a couple of solutions.

11 SPEA2

Unlike SPEA, SPEA2 uses a fine-grained fitness mapping strategy that incorporates density information (Zitzler et al. 2001). Also, the file size is fixed, each time the number of non-dominated individuals is less than the predefined file size, dominated individuals fill the file. Besides, the grouping technique, which is invoked when the undominated front exceeds the file limit, has been replaced by an alternative truncation method that has similar characteristics but does not lose limit points. Finally, another difference with SPEA is that only file members participate in the mating selection process.

12 Computational Experiments

A series of computational experiments were conducted to determine the approximation of the efficient frontier generated with nine MOEA's. To generate an instance of the portfolio selection problem, the forecasted returns, average return, variance, and covariance were calculated using the assets values in a determined period. The assets of the Mexican Stock Exchange (BMV) were: ALFAA, AMXL,

ASR1, CEMEX, FEMSA, GCARSO, GMEXICO, INVEX, IXE, KOFL, LALA, MEXCHEM, NAFTRACISHRS, PAC, PEÑOLES, PINFRA, TELEVISA. The values of the assets used are reported in Yahoo Finance in the period of 01/12/2016 to 29/12/2017 (13 months). The data of the first 11 months were used in the training phase; the data of the twelfth month were used in the test phase and finally the generated model was used to make a forecast of the thirteenth month. This forecast is generated with a Support Vector Regression (SVR) model that automatically configures its parameters with a GA (GASVR) (Rangel et al. 2019).

Once the BMW instance was generated, the algorithms CellDE, IBEA, GDE3, MOCell, NSGA-II, NSGA-III, OMOPSO, PAES, and SPEA2, were used to solve it. The implementation used for all the algorithms was available in JMetal (Durillo and Nebro 2011). For each algorithm, 30 independent runs were done and calculated the median and interquartile range of the following commonly metrics: Hypervolume, Epsilon, Overall Spread, Generalized Spread, and Inverted Generational Distance.

HV		ε		OS	
FAME2	3.133	FAME2	4.483	OMOPSO	3.5
OMOPSO	8.133	FAME	5.75	FAME2	4.033
FAME	8.133	OMOPSO	9.533	FAME	4.333
NSGAIISS	10.933	NSGAIISS	11.533	MOCELL	5.033
MOCELL	12.466	MOCELL	11.966	CELLDE	5.8
CELLDE	12.733	CELLDE	12.933	GDE3	7.066
GDE3	13.233	GDE3	13.266	NSGAIISS	13.3
NSGAII	15.833	NSGAII	16.066	NSGAII	14.8
SPEA2	16.1	SPEA2	16.366	SPEA2	15.833
NSGAIII	16.6	NSGAIII	16.383	NSGAIII	17
IBEA	17.7	IBEA	17.766	MOEAD	18.7
PAES	18.5	PAES	18.566	IBEA	18.933
MOEAD	19.633	MOEAD	19.666	PAES	19.366

GS		IGD		IGD plus	
IBEA	1.7	FAME	2.33	FAME2	4.066
MOEAD	4	FAME2	3.633	FAME	7.7
NSGAII	4.133	MOCELL	7.1	OMOPSO	9.8
SPEA2	4.6	OMOPSO	10.5	NSGAIISS	11
NSGAIII	5.866	CELLDE	11.466	MOCELL	12.166
MOCELL	7.633	GDE3	12.766	CELLDE	13.333
GDE3	7.733	NSGAIISS	12.766	GDE3	13.533
CELLDE	7.9	NSGAII	15.566	NSGAII	16.1
FAME	8.933	SPEA2	16	SPEA2	16.3

(continued)

(continued)

GS		IGD		IGD plus	
FAME2	9.7	NSGAIII	16.633	NSGAIII	16.466
PAES	9.866	IBEA	17.966	IBEA	17.633
NSGAIISS	17.833	PAES	18.633	PAES	18.6
OMOPSO	19.033	MOEAD	19.6	MOEAD	19.7

13 Conclusions

The evaluations with the metrics have shown that the algorithm with the best results for the BMV is OMOPSO winning 5 first places and 1 third place, followed by MOCell with 1 first place and 4 s places. The computational analysis confirms that the algorithms provide a good approximation of the risk-return border when solving the tri-objective problem. Therefore, the solution of the tri-objective problem with MOEAs generalizes the medium-variance approach by providing additional portfolios to the investor that are not effective in average-variance but have fewer values in the portfolio.

Acknowledgements The authors would like to acknowledge the Consejo Nacional de Ciencia y Tecnología (CONACYT). Besides, they acknowledge the Laboratorio Nacional de Tecnologías de la Informaci6n (LaNTI) of the Instituto Tecnol6gico de Ciudad Madero for the access to the cluster. Also, Javier Alberto Rangel González thanks the scholarship 429340 received from CONACYT in his Ph.D.

References

Alba, E., B. Dorronsoro, F. Luna, A.J. Nebro, P. Bouvry, and L. Hogie. 2007. A cellular multi-objective genetic algorithm for optimal broadcasting strategy in metropolitan MANETs. *Computer Communications* 30 (4): 685–697.

Anagnostopoulos, K.P., and G. Mamanis. 2010. A portfolio optimization model with three objectives and discrete variables. *Computers & Operations Research* 37 (7): 1285–1297.

Deb, K., and R.B. Agrawal. 1994. Simulated binary crossover for continuous search space. *Complex System* 9 (3): 1–15.

Deb, K., A. Pratap, S. Agarwal, and T. Meyarivan. 2002. A fast and elitist multiobjective genetic algorithm: NSGA-II. *IEEE Transactions on Evolutionary Computation* 6 (2): 182–197.

Durillo, J.J., A.J. Nebro, F. Luna, and E. Alba. 2008. Solving three-objective optimization problems using a new hybrid cellular genetic algorithm. In *International Conference on Parallel Problem Solving from Nature*, pp. 661–670.

Durillo, J.J., and A.J. Nebro. 2011. jMetal: A Java framework for multi-objective optimization. *Advanced Engineering Software* 42 (10): 760–771.

Huang, V.L., et al. 2007. Problem definitions for performance assessment of multi-objective optimization algorithms.

Knowles, J., and D. Corne. 1999. The pareto archived evolution strategy: A new baseline algorithm for pareto multiobjective optimisation. In *Proceedings of the 1999 Congress on* Evolutionary *Computation, 1999. CEC99*, vol. 1, pp. 98–105.

Kukkonen, S., and J. Lampinen. 2005. GDE3: The third evolution step of generalized differential evolution. In *The 2005 IEEE Congress on Evolutionary Computation, 2005*, vol. 1, pp. 443–450.

Kukkonen, S., and K. Deb. 2006a. Improved pruning of non-dominated solutions based on crowding distance for bi-objective optimization problems. In *CEC 2006. IEEE Congress on Evolutionary Computation, 2006*, pp. 1179–1186.

Kukkonen, S., and K. Deb. 2006b. A fast and effective method for pruning of non-dominated solutions in many-objective problems. In *Parallel problem solving from nature-PPSN IX*, 553–562. Springer.

Markowitz, H. 1952. Portfolio selection. *Journal of Finance* 7 (1): 77–91.

Nebro, A.J., J.J. Durillo, F. Luna, B. Dorronsoro, and E. Alba. 2009. Mocell: A cellular genetic algorithm for multiobjective optimization. *International Journal of Intelligent Systems* 24 (7): 726–746.

Price, K., R.M. Storn, and J.A. Lampinen. 2006. *Differential evolution: a practical approach to global optimization*. Springer Science & Business Media.

Rangel, R.A.P., J.A.R. González, J.F. Solís, H.J.F. Huacuja, J.J.G. Barbosa. 2019. Fuzzy GA-SVR for Mexican stock exchange's financial time series forecast with online parameter tuning. *International Journal of Combinatorial Optimization Problems and Informatics* 10(1): 40–50.

Sierra, M.R., and C.A.C. Coello. 2005. Improving PSO-based multi-objective optimization using crowding, mutation and –dominance. In *International Conference on Evolutionary Multi-Criterion Optimization*, pp. 505–519.

Srinivas, N., and K. Deb. 1994. Muiltiobjective optimization using nondominated sorting in genetic algorithms. *Evolutionary Computation* 2 (3): 221–248.

Storn, R., and K. Price. 1997. Differential evolution–a simple and efficient heuristic for global optimization over continuous spaces. *Journal of Global Optimization* 11 (4): 341–359.

Yuan, Y., H. Xu, and B. Wang. 2014. An improved NSGA-III procedure for evolutionary many-objective optimization. In *Proceedings of the 2014 Annual Conference on Genetic and Evolutionary Computation*, pp. 661–668.

Zitzler, E., M. Laumanns, and L. Thiele. 2001. SPEA2: Improving the strength Pareto evolutionary algorithm.

Zitzler, E., and S. Künzli. 2004. Indicator-based selection in multiobjective search. In *International Conference on Parallel Problem Solving from Nature*, pp. 832–842.

Multi-objective Portfolio Optimization Problem with Trapezoidal Fuzzy Parameters

Claudia Guadalupe Gómez-Santillán, Alejandro Estrada Padilla, Héctor Fraire-Huacuja, Laura Cruz-Reyes, Nelson Rangel-Valdez, and María Lucila Morales-Rodríguez

Abstract In this paper, we approach the Multi-Objective Portfolio Optimization Problem with trapezoidal fuzzy parameters. As to the best of our knowledge, there are not reports of this version of the problem. In this work, a formulation of the problem and a solution algorithm is presented for the first time. Traditionally, this kind of algorithm uses the crowding distance density estimator, therefore, we propose substituting this estimator for the Spatial Spread Deviation to improve the distribution of the solutions in the Pareto fronts. We apply a defuzzification process that permits to measure the algorithm performance using the commonly used real metrics. The computational experiments use a set of problem instances and the metrics of hyper-volume and generalized spread. The results obtained are encouraging as they confirm the feasibility of the proposed approach.

Keywords Multi-objective optimization · Multi-objective portfolio optimization problem · Trapezoidal fuzzy numbers · Density estimators

C. G. Gómez-Santillán · A. E. Padilla · H. Fraire-Huacuja (✉) · L. Cruz-Reyes · N. Rangel-Valdez · M. L. Morales-Rodríguez
Tecnológico Nacional de México, Instituto Tecnológico de Ciudad Madero, Ciudad Madero, Tamaulipas, Mexico
e-mail: automatas2002@yahoo.com.mx

C. G. Gómez-Santillán
e-mail: claudia.gomez@itcm.edu.mx

A. E. Padilla
e-mail: aestrada1993@hotmail.com

L. Cruz-Reyes
e-mail: lauracruzreyes@itcm.edu.mx

N. Rangel-Valdez
e-mail: nelson.rangel@itcm.edu.mx

M. L. Morales-Rodríguez
e-mail: lmoralesrdz@gmail.com

O. Castillo and P. Melin (eds.), *Fuzzy Logic Hybrid Extensions of Neural and Optimization Algorithms: Theory and Applications*, Studies in Computational Intelligence 940,
https://doi.org/10.1007/978-3-030-68776-2_17

1 Introduction

Organizations often approach portfolio optimization problems. In many practical cases, the decision-maker faces uncertainty relating to future uncertain states of nature that cause variability in project benefits, in resources to be consumed by the project and resources available to support the portfolio, this often carries uncertainty, due to cognitive limitation of human beings, a great quantity of deal of the information of interest. In this context, we require a tool for describing and representing uncertainty associated with problems of real-life decision-making. The portfolio optimization problem (POP) consists in selecting a subset of projects that can be carried out with the available resources and maximize the generated benefits. The POP has already been addressed in different research works considering that the available resources and the benefits derived from the projects have precise values (Salo 2011).

However, in many cases this assumption is not always fulfilled, because these parameters are frequently only imprecisely or approximately known. The uncertainty in the values of the parameters can be derived from different aspects such as arbitrariness, imprecision and poor determination of the values (Balderas-Jaramillo and Fausto 2018). A tool to model the uncertainty of the parameter values in the portfolio optimization problem is to use fuzzy parameters.

The main contributions of this work are: a novel formulation of the problem, a novel algorithm to solve it and a strategy to measure the performance of the algorithms with commonly used real metrics.

2 Elements of Fuzzy Theory

2.1 *Fuzzy Sets*

Let X a collection of objects x then a fuzzy set A defined over X is a set of ordered pairs: $A = \{(x, \mu_A(x))/x \in X\}$ where $\mu_A(x)$ is called the membership function or grade of membership of x in A which maps X to the membership real subspace M (Sadeh 1965). The range of the membership function is a subset of the nonnegative real numbers whose supremum is finite. Elements with a zero degree of membership are normally not listed.

2.2 *Generalized Fuzzy Numbers*

A generalized fuzzy number $\boldsymbol{A}$ is any fuzzy subset of the real line R, whose membership function $\mu_A(x)$ satisfies the following conditions (Vahidi and Rezvani 2013):

1. $\mu_A(x)$is a continous function from R to the colsed interval [0, 1]
2. $\mu_A(x) = 0, -\infty < x < a$
3. $\mu_A(x) = L(x)$, is strictly increasing on [a, b]
4. $\mu_A(x) = w$, for $b < x < \alpha$
5. $\mu_A(x) = R(x)$is strictly decreasing on $[\alpha, \beta]$
6. $\mu_A(x) = 0$, for $\beta < x < \infty$

where $0 < w < 1, a, b, \alpha, \beta$ are real numbers.

We denote this type of generalized fuzzy number as $A = (a, b, \alpha, \beta, w)_{LR}$. When $w = 1$ the generalized fuzzy number is denoted as $A = (a, b, \alpha, \beta)_{LR}$. When $L(x)$ and $R(x)$ are straight lines, then A is a trapezoidal fuzzy number and it is denoted as $A = (a, b, \alpha, \beta)$.

2.3 Addition Operator

Given two trapezoidal numbers $A_1 = (a_1, b_1, \alpha_1, \beta_1)$and $A_2 = (a_2, b_2, \alpha_2, \beta_2)$ then $A_1 + A_2 = (a_1 + a_2, b_1 + b_2, \alpha_1 + \alpha_2, \beta_1 + \beta_2)$.

2.4 Graded Mean Integration (GMI)

Graded mean integration is a defuzzification method used to compare two generalized fuzzy numbers. We compare the numbers based on their defuzzified values. The number with higher defuzzified value is larger. The formula to calculate the graded mean integration of a trapezoidal number A is given by:

$$P(A) = \left(\int_0^w h\left(\frac{L^{-1}(h) + R^{-1}(h)}{2}\right)dh\right) / \int_0^w h\,dh$$

For a trapezoidal fuzzy number $A = (a, b, \alpha, \beta)$, it reduces to $P(A) = (3a + 3b + \beta - \alpha)/6$.

2.5 Order Relation in the Set of the Trapezoidal Fuzzy Numbers

Given to trapezoidal fuzzy numbers $\boldsymbol{A}_1$ *and* $\boldsymbol{A}_2$, then:

(a) $\boldsymbol{A}_1 < \boldsymbol{A}_2\, if\ only\ if\ P(\boldsymbol{A}_1) < P(\boldsymbol{A}_2)$
(b) $\boldsymbol{A}_1 > \boldsymbol{A}_2\, if\ only\ if\ P(\boldsymbol{A}_1) > P(\boldsymbol{A}_2)$
(c) $\boldsymbol{A}_1 = \boldsymbol{A}_2\, if\ only\ if\ P(\boldsymbol{A}_1) = P(\boldsymbol{A}_2)$

2.6 Pareto Dominance

Given the following fuzzy vectors: $\hat{x} = (\boldsymbol{x}_1, \boldsymbol{x}_2, \ldots\ldots, \boldsymbol{x}_n)$ e $\hat{y} = (\boldsymbol{y}_1, \boldsymbol{y}_2, \ldots\ldots, \boldsymbol{y}_n)$ where x_i and y_i are trapezoidal fuzzy numbers, then we say that $\hat{x}$ dominates to $\hat{y}$, if only if $\boldsymbol{x}_i \geq \boldsymbol{y}_i$ for all $i = 1, 2, \ldots, n$ and $\boldsymbol{x}_i > y_i$ for some $i = 1, 2, \ldots, n$ (Yao et al. 2011).

3 Multi-objective Portfolio Optimization Problem with Trapezoidal Fuzzy Parameters

Let n be the number of projects to consider, $\boldsymbol{C}$ the total available budget, O the number of objectives, $\boldsymbol{c}_i$ the cost of the project i, $\boldsymbol{b}_{ij}$ the produced benefit with the execution of the project i in objective j, K the number of areas to consider, M the number of regions, $\boldsymbol{A}_k^{min}$ and $\boldsymbol{A}_k^{max}$ the lower and upper limits in the available budget for the area k, $\boldsymbol{R}_m^{min}$ and $\boldsymbol{R}_m^{max}$ the lower and upper limits in the available budget for the region m. The arrays a_i *and* b_i contains the area and region assigned to the project i. $\hat{x} = (x_1, x_2, \ldots\ldots, x_n)$ is a binary vector that specifies the selected projects included in the portfolio. If $x_i = 1$ then the project i is selected, else the project i is not selected. Now we define the multi-objective portfolio optimization problem with trapezoidal fuzzy parameters as following:

$$\text{Maximize } \hat{z} = (\boldsymbol{z}_1, \boldsymbol{z}_2, \ldots\ldots, \boldsymbol{z}_O) \tag{2}$$

where

$$z_j = \sum_{i=1}^{n} \boldsymbol{b}_{ij} x_i \quad j = 1, 2, \ldots O \tag{3}$$

Subject to the following constraints:

$$\sum_{i=1}^{n} \boldsymbol{c}_i x_i \leq \boldsymbol{C} \tag{4}$$

$$\boldsymbol{A}_k^{min} \leq \sum_{i=1, a_i=k}^{n} \boldsymbol{c}_i x_i \leq \boldsymbol{A}_k^{max} \quad k = 1, 2, \ldots, K \tag{5}$$

$$\boldsymbol{R}_k^{min} \leq \sum_{i=1, b_i=k}^{n} \boldsymbol{c}_i x_i \leq \boldsymbol{R}_k^{max} \quad k = 1, 2, \ldots\ldots M \tag{6}$$

$$x_i \varepsilon \{0, 1\} \quad \text{for every} \quad i = 1, 2, \ldots\ldots, n \tag{7}$$

In this formulation, all the parameters and variables that appear in ***bold*** and *italic* are trapezoidal fuzzy numbers.

The objective function tries to maximize the contributions of each objective (2). We calculate each objective adding all the contributions of the selected projects in the binary vector (3). The constraint (4) set that the sum of the costs required for all the selected projects do not exceed the available budget. The set of constraints (5) set that the sum of the costs of the projects for each area, is in the range of the available budget for that area. The set of constraints (6) set that the sum of the costs of the projects for each region, is in the range of the available budget for that region. The set of constraints (7) make sure that all the values in x_i are binary.

4 Proposal Algorithm T-NSGA-II

In this section, we present the design of all the components included of the proposal algorithm. The algorithm is an adaptation of the classic Deb algorithm NSGA-II (Deb 2000) to work with the trapezoidal fuzzy numbers considered in the formulation of the multi-objective portfolio optimization problem. The input to the algorithm is an instance of the problem that contains the values of all the trapezoidal fuzzy parameters. The output is an approximate Pareto front for the instance. To improve the reproducibility of the algorithm we describe what structure is used to represent the solutions, how to evaluate the solutions, the genetic operators and how to build the initial solution. Also we describe how to sort the population, how to make the non-dominated sorting process, how to implement the density estimators and finally we present the pseudocode of the algorithm.

4.1 Representation of the Solutions

As we can see in the formulation of the problem, it is the space of the binary vectors $\boldsymbol{S} = \{0, 1\}^{\boldsymbol{n}}$. A solution is a binary vector that set the selected projects in the portfolio. In the algorithm, a candidate solution is represented in a binary vector structure.

4.2 Evaluating the Solutions

The objective function of all the solutions in the population must be calculated. To make this process for a solution, the accumulators of each objective are initialized at 0 and all the projects from that solution are explored. If a project has a value of 1, the profits that the project contributes to each objective are added. Since the values of the profits are trapezoidal fuzzy numbers, the addition is done with the "addition" operator described previously. Finally, the GMI of the accumulated trapezoidal fuzzy values obtained are calculated.

4.3 One-Point Crossover Operator

Two solutions from the population are chosen to apply the one-point crossover operator (Umbarkar and Sheth 1995) and obtain two new solutions that will keep some elements from the original solutions, and they will be added to the crossover population. The one-point crossover operator consists in generating a random number between 0 and (n–1), where n is the number of projects and, therefore, the size of the solution vector, this number will work as the point or the position where the two parental solutions will be divided. The left side of the point of the first solution and the right side of the point of the second solution will be used to create the first new solution, while the left side of the point of the second solution and the right side of the point of the first solution will be used to create the second new solution. The number of crossovers that are done is a defined parameter and the two parental solutions are chosen randomly from the population.

4.4 Uniform Mutation Operator

A solution is chosen to apply uniform mutation on it (Reeves 1975) and generate a new solution that will be added to the mutation population. The uniform mutation operator consists in exploring each of the elements of a solution, and for each element, a random number between 0 and 1 is generated. If this random number is of less value than *mut* (a given parameter), then that element switches its value. The number of new mutated solutions that are generated is a defined parameter and the solutions that undergo this process are randomly chosen from the population.

4.5 Initial Population

A predefined number of randomly generated solutions are created to have an initial population. The size of the solution depends on the number of projects the instance has. When a new random solution is generated, the costs of the projects it uses are added together. Since the costs of the projects are trapezoidal fuzzy numbers, the addition operator is used, then its GMI is calculated. If this sum is of lesser value than the GMI of the instance's budget, then that solution is added to the population, but if this was not the case, then that solution is discarded and a new solution is generated. This is done to make sure that all the solutions in the population are feasible.

4.6 Population Sorting

This process consists in sorting the population's solutions and it is composed of two parts, an "elitist" part, which makes sure the best solutions are kept, and a "diversification" part, which makes sure the solutions are different enough between them so the main algorithm doesn't get stuck in local optima. The elitist part is also known as "no-dominated sorting", and it consists in separating the population in "fronts" or sets of solutions, making sure that the best solutions are always in the first front, the next best solutions in the second front, and so on. The diversification part is also known as crowding distance, and it consists in sorting the solutions in each front respect to this indicator.

4.7 No-Dominated Sorting

This process is divided in two parts. The first part consists in the construction of the first front, to do this, comparisons are made with the objective values of all the solutions in the population (the objective value of one solution is compared with the objective values of the rest of the solutions) to make sure if one solution dominates the other. As the objective function of a solution is a trapezoidal fuzzy numbers vector, the Pareto dominance used is the previously defined for this kind of fuzzy numbers. During the comparison process, for each solution what is saved is the following: the front ranking, the number of solutions that dominate that solution and the set of solutions that solution dominates. At the end of the comparison process, the set of solutions that aren't dominated by any other solution is added to the first front.

The second part of the process consists in the construction of the rest of the fronts. To do this, each of the solutions of the actual front is explored. For each solution, the counter that indicates the number of solutions that dominate that solution is subtracted by one, if the counter reaches 0, that solution is added to the new front. This process is repeated with the rest of the solutions in the current front, and at the end the new front of solutions will be finished. This process is repeated until there are no more fronts to be made.

4.8 Calculating the Crowding Distance (Deb et al. 2000)

This process consists in rearranging each of the solutions in all the fronts by their crowding distance (CD), so the solutions that are scattered far away from the rest of the other solutions are located at the beginning of the front. For each objective, the solutions are rearranged according to their objective values (converted into GMI), because the only important thing are the distances, it doesn't matter if the solutions are sorted from lower to higher objective values, or viceversa, the only important thing is that the same criteria is used during the whole process. Then, the first and last solutions of the front after the sorting will be given a CD value of infinite (∞), and for the rest of the solutions their CD is calculated by the following formula:

$$d_{I_j^m} = d_{I_j^m} + \frac{f_{m^{j+1}}^{I^m} - f_{m^{j-1}}^{I^m}}{f_m^{max} - f_m^{min}}$$

where d is the crowding distance, I is the solution's position in the whole population in general, j is the solution position within the front where the CD is being calculated, f is the objective value and m is the current objective. Once this process has been done with the first objective, the same process is repeated with all the other objectives, the CD of each solution are accumulator values that keep growing.

Once this process is done with all the objectives, the front solutions are rearranged by their CD, from higher to lower values.

4.9 Calculating the Spatial Spread Deviation (SSD) (Santiago et al. 2019)

This is another density estimator, it's objective is to rearrange the solutions in a front so they are not spread by a wide margin. This is done by calculating for each solution a value called SSD using normalized distances between the solutions in the front, at the end, the solutions are sorted from the lowest to highest value by it's SSD. This is done because SSD punishes solutions according to their standard deviation and their proximity to their closest k-neighbors. The formulas to calculate SSD for each solution i are (Fig. 1).

Where $D(\mathrm{i}, \mathrm{j})$ is the distance from solution i to solution j. $D_{\max}$ is the biggest distance between all the solutions and $D_{\min}$ is the closest distance between all the solutions. K is the set of k neighbors closest to solution i. SSD_0 is the initial value of *SSD*, which is-*INF* if the solution is at one of the ends of the front when the normalized values of the GMI of the objective values are calculated.

$$temp1(i) = \frac{1}{n-1}\sqrt{\sum\nolimits_{j=1}^{n}\left(D(i,j) - (D_{max} - D_{min})\right)^2}\ \forall\, i \neq j$$

$$temp2(i) = \sum\nolimits_{j\varepsilon K}\frac{(D_{max} - D_{min})}{D(i,j)}$$

$$SSD(i) = SSD_0(i) + temp1(i) + temp2(i)$$

Fig. 1 SSD calculations

4.10 Pseudocode of the T-NSGA-II Algorithm

The T-NSGA-II is based in the structure of the classic multi-objective algorithm NSGA-II proposed by Deb (2000). As previously described, the processes were modified to work with trapezoidal fuzzy numbers and the problem formulation proposed. In this section, we present the structure of the algorithm T-NSGA-II.

Algorithm 1. T-NSGA-II

Instance lecture
Create the initial population ***pop***
Calculate the objective functions
Non-dominated Sorting
Calculate Spatial Spread Deviation/Crowding distance
pop sorting due to fronts and Spatial Spread Deviation/CD
Main loop, until stopping condition is met
  Create ***popc*** using crossover operator
  Create ***popm*** using Mutation operator
  Join ***popc*** and ***popm*** to create ***popj***
  Evaluate solutions in ***popj*** and *put feasibles* in ***popf***
  Add ***popf*** to ***pop***, and calculate objective functions
  Non-dominated sorting
  Calculate Spatial Spread Deviation/Crowding distance
  pop sorting due to the front ranking and Spatial Spread Deviation/CD
  Truncate ***pop*** to keep a population of original size
  Non-dominated sorting
  Calculate Spatial Spread Deviation/Crowding distance
  pop sorting due to front ranking and Spatial Spread Deviation/CD
End of main loop
Print First Front

---

5 Proposed Strategy to Assess the Performance of Multi-objective Algorithms in the Fuzzy Trapezoidal Numbers Domain

For small instances, we can find the optimal Pareto Front using an exhaustive algorithm that explores all the solution space. However, for instances relevant in real applications we cannot apply this kind of methods. Therefore, in these cases we use genetic algorithms like NSGA-II, which produces approximated Pareto Fronts (Deb 2000). In the previous section, we described an implementation of this algorithm that produces approximated Pareto Fronts for the portfolio optimization problem with trapezoidal fuzzy parameters. In space of the objectives, these fronts are sets of vectors of trapezoidal fuzzy numbers. Suppose that we have the optimal Pareto front and an approximation of the optimal front, how we can assess the quality of the approximation respect to the optimal front? To answer this question, we apply a map from the trapezoidal fuzzy numbers domain F to the set of real numbers R. This mapping transform a Pareto front in a set of real vectors. The main idea of this strategy is to defuzzify the fuzzy numbers. The map that we propose in this work is $\delta : F \to R$ such that $\delta(\boldsymbol{A}) = P(\boldsymbol{A})$. For each trapezoidal fuzzy number, the map associates the graded mean integration value. A remarkable property of this map is that if $A \subset F^n$, then $\delta(A) \subset R^n$. This makes possible to apply the defuzzification process in case of multiple objectives. Once the front is mapped, we can use the metrics commonly used to determine the quality and the spread of the solutions in the front.

To improve the spread of the solution in the generated fronts, in the algorithm T-NSGA-II we propose substituting the crowding distance (Deb 2000) for the spatial spread deviation density estimator (Santiago 2019).

6 Computational Experiments

The software and hardware platforms that were used for these experiments include: an Intel Core i5 1.6 GHz processor, RAM 4 GB, and IntelliJ IDEA CE IDE.

The purpose of this experiment is to evaluate the usage impact of density estimator SSD in the performance of T-NSGA-II algorithm. The algorithms that are compared are T-NSGA-II-CD which is the adaptation of classic algorithm NSGA-II to solve the portfolio optimization problem with Trapezoidal Fuzzy Numbers. This algorithm uses crowding distance as its density estimator. The other algorithm is the proposed T-NSGA-II-SSD which substitutes the crowding distance for the spatial spread deviation. Table 1 show the main parameters of the algorithms and the used values. A series of preliminarily experiments were done to determine that values.

To measure the performance of both algorithms, two metrics are used: hypervolume (While et al. 2012) and generalized spread (Zhou 2006). The Hypervolume is the volume of a n-dimensional space dominated by the solutions of a reference set A, if such that space is big, then that means that the set is close to the Pareto Front. Therefore, it is desirable that the values are high. Generalized Spread calculates the average of the distances of the points of the reference set to their nearest neighbor. If

Table 1 Algorithm parameters

Iterations in main loop	100
Population size	50
Number of projects	25
Crossover %	0.7
Mutation %	0.4

this indicator has low values, then that means that the solutions in the reference set are well spread.

For the experiment, 13 instances for the portfolio optimization problem with trapezoidal fuzzy parameters were created randomly, each instance was solved 30 times with each algorithm, all the objective values of the front are already transformed from Trapezoidal Fuzzy Numbers into real numbers with the GMI formula. The performance metrics were calculated for each of the 30 fronts of each instance, all the values were sorted in ascending form and the median and the interquartile range were calculated. These measures of central tendency and data dispersion are used because they are less sensitive than the mean and variance when outliers appear. A Wilcoxon nonparametric hypothesis test was conducted to determine if the results have statistically significant differences.

Tables 2 and 3 shows the result of the Wilcoxon hypothesis test done with the 30 Hypervolumen (Generalized Spread) measures for each instance. The first column shows the name of the instances, the second column shows the results of the T-NSGA-II-CD algorithm with the corresponding metric, where the first value shows the median, and the smaller value to the right shows the interquartile range. Similarly, the third column shows the results for the T-NSGA-II-SSD algorithm. The cells that are shaded show the algorithms with the best performance for the correspond metric. When do not exists significant differences, both algorithms are marked The fourth column displays the *P-value* obtained in the Wilcoxon test. Finally, the last column displays if for the considered instance the observed differences in the metric values are statistically significant with a confidence of 95%.

The results show that the T-NSGA-II-CD algorithm obtained better results with the hypervolume metric in 10 of 13 of the instances. Meanwhile T-NSGA-II-SSD algorithm obtained better results with the generalized spread metric in all the instances. This means that the SSD density estimator improves the spreading of the solutions

Table 2 Algorithms performance evaluation with the hypervolume metric

Instance	T-NSGA-II-CD	T-NSGA-II-SSD	Wilcoxon *P-value*	Significative dif. *P-value*<0.001?
o2p25_0G	$0.4848_{0.2248}$	$0.4405_{0.2169}$	0.00001	*Yes*
o2p25_1G	$0.3961_{0.0855}$	$0.3045_{0.2227}$	0.0018	*Yes*
o2p25_2G	$0.3101_{0.0668}$	$0.2258_{0.2263}$	0.00194	*Yes*
o2p25_3G	$0.3006_{0.0735}$	$0.2279_{0.1980}$	0.00362	*Yes*
o2p25_4G	$0.3673_{0.1093}$	$0.3552_{0.1623}$	0.17068	*No*
o2p25_5G	$0.2679_{0.0450}$	$0.1937_{0.1612}$	0.00001	*Yes*
o2p25_6G	$0.3876_{0.0744}$	$0.2998_{0.2066}$	0.00001	*Yes*
o2p25_7G	$0.2973_{0.3620}$	$0.2732_{0.3036}$	0.00001	*Yes*
o2p25_8G	$0.2676_{0.0616}$	$0.2349_{0.1628}$	0.24604	*No*
o2p25_9G	$0.3991_{0.0970}$	$0.2697_{0.2326}$	0.0001	*Yes*
o2p25_10G	$0.3220_{0.0992}$	$0.2603_{0.1758}$	0.0001	*Yes*
o2p25_11G	$0.4449_{0.1150}$	$0.2335_{0.2770}$	0.0001	*Yes*
o2p25_12G	$0.3076_{0.1106}$	$0.3023_{0.2433}$	0.13622	*No*

Table 3 Algorithms performance evaluation with the generalized spread metric

Instance	T-NSGA-II-CD	T-NSGA-II-SSD	Wilcoxon *P-value*	Significative dif. *P-value*<0.001?
o2p25_0G	$0.5565_{0.1683}$	$0.3605_{0.2383}$	0.00001	*Yes*
o2p25_1G	$0.6353_{0.1901}$	$0.4921_{0.1161}$	0.00001	*Yes*
o2p25_2G	$0.5897_{0.1770}$	$0.3913_{0.1368}$	0.00001	*Yes*
o2p25_3G	$0.6881_{0.2583}$	$0.5037_{0.1696}$	0.00001	*Yes*
o2p25_4G	$0.6033_{0.2359}$	$0.4950_{0.1425}$	0.00001	*Yes*
o2p25_5G	$0.6828_{0.1869}$	$0.4671_{0.1465}$	0.00001	*Yes*
o2p25_6G	$0.5890_{0.1509}$	$0.4345_{0.1747}$	0.00001	*Yes*
o2p25_7G	$0.6310_{0.4323}$	$0.4882_{0.2981}$	0.00001	*Yes*
o2p25_8G	$0.7526_{0.1873}$	$0.4359_{0.1122}$	0.00001	*Yes*
o2p25_9G	$0.7250_{0.3545}$	$0.4769_{0.1162}$	0.00001	*Yes*
o2p25_10G	$0.5460_{0.2298}$	$0.4672_{0.1367}$	0.00001	*Yes*
o2p25_11G	$0.5949_{0.2515}$	$0.4695_{0.2008}$	0.00014	*Yes*
o2p25_12G	$0.5784_{0.2946}$	$0.4747_{0.1580}$	0.00001	*Yes*

in the fronts. We think that when the algorithm spread the solutions in the front, it is reducing the number of the solutions in the front and in consequence increase the difference regards to the reference front. This can be a reason for the low performance regards to the hypervolume metric.

7 Conclusions

In this work, we approach the Multi-Objective Portfolio Optimization Problem with trapezoidal fuzzy parameters. As to the best of our knowledge, there are not research reports about this version of the problem. In this paper for the first time we present a formulation of the problem. Also, we contribute with the first multi-objective algorithm T-NSGA-II to solve the problem. Traditionally this type of algorithm uses the crowding distance density estimator, therefore, we propose substituting this estimator for the Spatial Spread Deviation to improve the distribution of the solutions in the Pareto fronts. We apply an innovative defuzzification process that permits to measure the algorithm performance using the commonly used real metrics. The computational experiments use a set of problem instances and the metrics of hypervolume and generalized spread. The results obtained show that the Spatial Spread Deviation improve the distribution of the solutions in the front.

We are working now to improve the performance of the T-NSGA-II-SSD algorithm regards the hypervolume metric.

Acknowledgements The authors thank the support from CONACYT projects: (a) A1-S-11012-"Análisis de Modelos de NO-Inferioridad para incorporar preferencias en Problemas de Optimización Multicriterio con Dependencias Temporales Imperfectamente Conocidas"; (b) project 3058 from the program Cátedras CONACyT; and, (c) and from project 312397 from Programa de

Apoyo para Actividades Científicas, Tecnológicas y de Innovación (PAACTI 2020-1). Also thank the support from TecNM project no. 5797.19-P and Laboratorio Nacional de Tecnologias de Información (LaNTI) del TecNM/Campus ITCM. A. Estrada would like to thank CONACYT for the support 740442.

References

Balderas-Jaramillo, Fausto A. 2018. Modelando la imprecisión del problema de cartera de proyectos con filosofía gris. Instituto Tecnológico de Tijuana.

Deb, K., S. Agrawal, A. Pratap, and T. Meyarivan. 2000. A fast elitist non-dominated sorting genetic algorithm for multi-objective optimization: NSGA-II. In *Proceedings of the 6th international conference on parallel problem solving from nature*, vol. 1917. Lecture Notes in Computer Science. Berlin: Springer.

Reeves, C. 1975. Chapter 3 Genetic algorithms Part A : Background.

Salo, A., J. Keisler, and A. Morton. 2011. *Portfolio decision analysis. Improved methods for resource allocation.* International Series in Operations Research & Management Science, 162, 3–27. New York: Springer.

Santiago, A., B. Dorronsoro, A.J. Nebro, J.J. Durillo, O. Castillo, and H.J. Fraire. (2019). A novel multi-objective evolutionary algorithm with fuzzy logic based adaptive selection of operators: FAME. *Information Sciences* 471: 233–251. Elsevier Inc. (January 2019).

Umbarkar, A.J., and P.D. Sheth. 1995. Crossover operators in genetic algorithms: A review. *ICTAC Journal on Soft Computing* 6 (1): 1083–1092.

Vahidi, J., and S. Rezvani. 2013. Arithmetic operations on trapezoidal fuzzy numbers. *Nonlinear Analysis and Applications. ISPACS.* 2013: 1–8.

While, L., L. Bradstreet, and L. Barone. 2012. A fast way of calculating exact hypervolumes. *IEEE Transactions on Evolutionary Computation* 16 (1): 86–95.

Yao, S., Z. Jiang, N. Li, H. Zhang, and N. Geng. 2011. A multi-objective dynamic scheduling approach using multiple attribute decision making in semiconductor manufacturing. *International Journal of Production Economics* 130 (1): 125–133.

Sadeh, L.A. 1965. *Fuzzy sets. Information and control*, vol. 8, No. 3, pp. 338–353. Elsevier.

Zhou, A., Y. Jin, Q. Zhang, B. Sendhoff, and E. Tsang. 2006. Combining model-based and genetics-based offspring generation for multi-objective optimization using a convergence criterion. In *Congress on evolutionary computation*, 3234–3241. IEEE.

A Study on the Use of Hyper-heuristics Based on Meta-Heuristics for Dynamic Optimization

Teodoro Macias-Escobar, Laura Cruz-Reyes, and Bernabé Dorronsoro

Abstract The study of dynamic multi-objective optimization problems (DMOP) is an area that has recently been receiving increased attention from researchers. Within the literature, various alternatives have been proposed to solve DMOPs, among them are the dynamic multi-objective evolutionary algorithms (DMOEA), which use stochastic methods to obtain solutions close to the optimum. With the constant proposal of new DMOPs with different challenges and properties, as well as DMOEAs to solve them, the issue of determining which alternatives are adequate for each problem arises. Hyper-heuristics are methodologies that use multiple heuristics to solve a problem. This allows them to effectively cover a wider spectrum of characteristics of optimization problems. This advantage also involves DMOPs, since a suitable hyper-heuristic can satisfactorily solve a greater number of problems compared to DMOEAs used individually. This paper presents a guide, as well as a checklist to support researchers in the design of hyper-heuristics to solve DMOPs using DMOEAs as their heuristics. This work also presents two case studies which include state-of-the-art proposals that follow each step of the proposed guide, the obtained results were efficient and satisfactory, which shows the effectiveness of this guide.

Keywords Hyper-heuristics · Dynamic optimization · Dynamic optimization problems · Meta-heuristics

---

T. Macias-Escobar
Tecnológico Nacional de México, Instituto Tecnológico de Tijuana, Tijuana, Mexico
e-mail: teodoro_macias@hotmail.com

L. Cruz-Reyes (✉)
Tecnológico Nacional de México, Instituto Tecnológico de Ciudad Madero, Ciudad Madero, Mexico
e-mail: lauracruzreyes@itcm.edu.mx

T. Macias-Escobar · B. Dorronsoro
Universidad de Cádiz, Cadiz, Spain
e-mail: bernabe.dorronsoro@uca.es

O. Castillo and P. Melin (eds.), *Fuzzy Logic Hybrid Extensions of Neural and Optimization Algorithms: Theory and Applications*, Studies in Computational Intelligence 940,
https://doi.org/10.1007/978-3-030-68776-2_18

1 Introduction

A common situation in daily life is to find scenarios in which there are multiple objectives to satisfy. The world is under constant change, this of course, can cause considerable alterations in the environment that surrounds the problem. This type of problem is defined as dynamic optimization problems (DOP). When a DOP has two or more objectives to solve, it is called a multi-objective dynamic optimization problem (DMOP).

Over the years, many researchers have proposed several alternatives, methods and strategies to solve DMOPs. Heuristics are stochastic approaches that seek to obtain one or more high-quality solutions in an acceptable computational time that, although not optimal, are good enough to be considered acceptable solutions.

Meta-heuristics and hyper-heuristics are heuristics with a more general view. These are methods which seek to solve a wider range of problems, with different characteristics, objectives, requirements, and limitations. Hyper-heuristics have become very popular in the last decade to solve optimization problems, and they are being considered as a powerful alternative to solve DOPs.

With the increasing number of DMOPs, as well as heuristics and meta-heuristics to solve them, it is desirable to find a strategy that allows defining which are the most suitable alternatives for each problem. Hyper-heuristics seek to take advantage of the strengths of heuristics and meta-heuristics to cover possible weaknesses, seeking to obtain better solutions than these could obtain individually.

Within the literature there are several works that analyze and classify several hyper-heuristics used to solve static optimization problems (Burke et al. 2010, 2013, 2018) and DOPs (Macias-Escobar et al. 2020). Several of the works reviewed within these investigations involve hyper-heuristics that use meta-heuristics within their process.

This work seeks to identify patterns in the different proposed hyper-heuristics that use meta-heuristics to solve optimization problems and present a guide to researchers to facilitate the design and development of future hyper-heuristics to solve DMOPs. The intention is to define a set of points to check and steps to follow to avoid common inconveniences when working with these strategies. This seeks to allow researchers and practitioners to develop effective hyper-heuristics more quickly.

The content of the rest of this work is organized as follows. Section 2 presents a set of relevant definitions that are important to know before starting to design hyper-heuristics. Section 3 explains critical criteria to consider when analyzing a DMOP. Section 4 presents some of the most common approaches used by hyper-heuristics to solve DOPs. This section classifies several previous works based on previously proposed classifications. Section 5 proposes a guide to design hyper-heuristics supported by meta-heuristics to solve DMOPs. It also proposes a checklist to analyze on each step of the design of said hyper-heuristics. Section 6 presents two case studies which follow the proposed guide and checklist. Finally, Sect. 7 presents the conclusions and future work of this study.

2 Background and Definitions

This section presents the set of concepts and definitions necessary to understand before seeking to design and develop a hyper-heuristic which uses meta-heuristics within its process to solve DMOPs. The concepts are related to dynamic optimization and multi-objective problem solving.

2.1 Dynamic Multi-objective Optimization Problem

A dynamic multi-objective optimization problem (DMOP) is a problem which requires to define the value of a set of decision variables to obtain the best possible result for a set of objectives that suffers changes over time while satisfying a set of constraints (Azzouz et al. 2017). Considering a minimization problem, the formal definition of DMOP can be defined as:

$$\begin{aligned} \min \ & F(\vec{x},t)=\{f_1(\vec{x},t), f_2(\vec{x},t), \ldots, f_m(\vec{x},t)\} \\ s.t \ & g(\vec{x},t) > 0, h(\vec{x},t) = 0 \end{aligned} \tag{1}$$

Let $\vec{x}$ be a vector of decision variables, and F the set of m objective functions to be optimized during a static time step t. The inequality and equality constraints are represented by g and h, respectively.

As previously mentioned, the objectives of a DMOP are usually in conflict with each other. This means that it is very likely that there is no single optimal solution to the problem, but rather that there is a set of non-dominated solutions that are not surpassed by any other. This group of solutions is called the Pareto optimal solution set (POS). The formal definition of POS is

$$POS_t=\{\vec{x} | \nexists f(\vec{y},t) \le f(\vec{x},t), \vec{y} \in V\} \tag{2}$$

where V is the feasible solution search space. These solutions define the Dynamic Pareto Optimal Front (POF), which is the objective values of the POS for time step t, Eq. (3) shows the formal definition of POF

$$POF_t=\{f(\vec{x},t) | \nexists f(\vec{y},t) \le f(\vec{x},t), \vec{y} \in V\} \tag{3}$$

While several proposals to classify DMOPs have been proposed before, the most widely known and accepted classification system is based on considering if the changes of each time step alter the current POS and POF (Farina et al. 2004). Table 1 presents the four different types of DMOP defined in that classification.

Table 1 Types of DMOP

	POF changes	POF does not change
POS changes	Type II	Type I
POS does not change	Type III	Type IV

2.2 Dynamic Multi-objective Evolutionary Algorithm

In the literature there are several proposed methods to solve DMOPs. Among those, the dynamic multi-objective evolutionary algorithms (DMOEA) are an alternative based on the use of stochastic techniques to solve DMOPs. A DMOEA is a dynamic version of MOEA which includes a change detection method and a change adaptation method.

Various DMOEAs have proven to produce good-quality solutions for DMOPs with two or three objectives (Deb and Karthik 2007). DMOEAs are implicitly required to fulfill the two purposes of multi-objective optimization, convergence and diversity towards the POF. Due to this, several methods have been proposed, which attempt to solve a DMOP using different approaches. These approaches are classified by Azzouz et al. (2017) as follows:

Diversity-based approaches: DMOEAs seek to maintain the diversity of solutions by altering the content of the population to a certain degree. A common strategy for this approach is to replace a subset of the population for new, randomly-spread solutions (Deb and Karthik 2007).

Change Prediction-based approaches: DMOEAs predict the location of new optimal solutions based on previously collected information, detecting possible movement patterns. An example of this approach is presented in Liu (2010), where the DMOEA takes information from two previous time steps to determine the position of the reinitialized population when a change is detected.

Memory-based approaches: DMOEAs guide the search direction of future generations on the basis of information collected by previous generations. The work of Goh and Tan (2009) presents an example of this approach, by using an external population that stores the relevant information regarding the obtained Pareto fronts from previous time steps and using it whenever an environmental change occurs.

Parallel approaches: DMOEAs apply parallelization techniques to solve a problem quickly and efficiently. In Zheng (2007), the population is divided into $m + 1$ subsets where m is the number of objectives.

Conversion-based approaches: DMOEAs split a DMOP in periods, where each period is treated as a static optimization problem. An example of this approach can be seen in Liu and Wang (2006), where each subperiod is solved using a static EA.

2.3 Hyper-heuristic

The constant increase in the interest on DMOPs, as well as proposed alternatives to solve them, present a new area of research. This area focuses on identifying the best alternatives to solve a set of DMOPs. It is necessary to remember the "No Free Lunch" (NFL) theorem (Wolpert and Macready 1997), which establishes that an algorithm that is efficient for a certain set of problems will be inefficient for a different set.

Applying NFL for DMOPs and DMOEAs means that it is impossible for a single DMOEA to be able to solve all existent DMOPs efficiently and with a higher quality that the rest of the DMOEAs present in the literature, as there will be at least one DMOP where another DMOEA outperforms it.

While it is theoretically impossible to overcome NFL completely, there are approaches that seek to cover a wider array of optimization problems. An alternative to overcome NFL limitations, for the most practical situations, is the use of hyper-heuristics to combine heuristics (Poli 2008).

Hyper-heuristics are defined as methodologies that select or generate heuristics, named low-level heuristics (LLH), to solve problems (Burke et al. 2010). One of the main benefits of hyper-heuristics is to take advantage of the strengths of certain heuristics to cover the weaknesses of others, allowing the hyper-heuristic to obtain better results in comparison to what each heuristic could obtain individually.

It is important to note that hyper-heuristics do not act directly with the solution search space, as heuristics and meta-heuristics do. Instead, they explore a *heuristic search space* to define which will be the heuristics that will act on the solution search space.

A classification of hyper-heuristic is proposed in Burke et al. (2010) based on two criteria, the source of feedback during the learning stage of the hyper-heuristic and the nature of the heuristic search space.

Based on the source of the feedback, this classification proposes three categories:

No learning: Feedback is not used during the process. Only the information obtained during the current state of the hyper-heuristic is used to define which LLHs will be used for certain time.

Offline learning: A set of training instances is established prior the execution of the hyper-heuristic. All the information obtained by those instances is compiled in order to define which LLH will be used for instances not yet seen.

Online learning: The learning process is performed while an optimization problem is being solved. This allows the hyper-heuristic to use the current information along with the data obtained on previous stages of the problem to define which LLH is the most suitable to use under the current environment.

Meanwhile, hyper-heuristics have two categories when classified by the nature of the heuristic search space.

Selection-based hyper-heuristics: Methodologies select from a set of predesigned LLHs those that produce higher quality solutions under the current environment to use for a certain period.

Generation-based hyper-heuristics: Methodologies generate new heuristics using elements from heuristics within the search space.

The LLHs from both selection and generation hyper-heuristics can have two different focuses. *Perturbation* of current solutions or *generation* of new solutions.

This work focuses on selection-based hyper-heuristics that use perturbative LLHs. This means that the LLHs within the heuristic search space are completely usable from selection, without the need of further modification. The main responsibility of this type of hyper-heuristics, as mentioned previously, is to find which heuristic (or meta-heuristic) in the search space is the most suitable to apply in the current period.

This type of hyper-heuristics is composed by two phases. First, a *heuristic selection*, which performs the selection of the most suitable LLH according to the hyper-heuristic being used. And second, *move acceptance*, which uses a criterion to determine if the solution obtained by the selected LLH can replace the current solution (Bilgin et al. 2006).

There is no limit regarding the complexity of LLHs. This means that they can be as simple or complex as needed. The only condition is that they are focused on working directly in the solution search space. Therefore, the use of meta-heuristics as LLHs, which can be as complex as the hyper-heuristic itself, is possible.

2.4 Indicators to Evaluate DMOEAs Performance Over DMOPs

All indicators used to assess the performance of DMOEAs when solving DMOPs must consider the environmental changes for their evaluation. The most common way is to divide the evaluation of the indicator by each static period. This section presents a list of some of the most commonly used performance metrics to evaluate DMOEAs.

Ratio of non-dominated solutions (*RNI*) (Tan et al. 2002). Obtains a ratio regarding the obtained non-dominated solutions (*PF*) with respect to the size of the obtained population P at a time step t.

$$RNI_t = \frac{|PF_t|}{|P_t|} \tag{4}$$

Hyper-volume ratio (*HVR*) (van Veldhuizen 1999). It measures the ratio between the POF and the Pareto front (*PF*) obtained by an algorithm by comparing the hyper-volume obtained by both using an equal reference point.

$$HVR_t = \frac{HV(PF_t)}{HV(POF_t)} \tag{5}$$

Variable space generational distance (*VD*) (Goh and Tan 2009). Measures the minimum distance between each solution from the POF with respect to *PF*. Euclidean distance $d(v, POF_t)$ is used to evaluate the closeness of v, which represents a solution from PF, to its closest solution from POF.

$$VD_t = \frac{\sqrt{|PF_t| \sum_{v \in PF_t} d(v, POF_t)^2}}{|PF_t|} \tag{6}$$

Inverted generational distance (IGD) (Sierra and Coello 2005). A variation of generational distance which reverses the roles of PF and POF. This metric measures the minimum Euclidean distance between each of the solutions from the POF with respect to PF.

$$IGD_t = \frac{\sum_{v \in POF_t} d(v, PF_t)^2}{|POF_t|} \tag{7}$$

Another way to gather information to evaluate algorithm performance is through the use of fitness landscape analysis (FLA) methods. Within the literature there are proposals to use FLA methods under dynamic environments (Richter 2013). The use of population evolvability (Wang et al. 2017) has recently been proposed to solve DMOPs, since it can obtain information from both the problem and the algorithm that is solving it. This allows to obtain more information to carry out processes such as the performance evaluation of DMOEAs, as well as supporting the LLH selection process within a hyper-heuristic.

Population evolvability indicates the ability of a population to evolve in a better population based on two factors: (i) the probability of the population to evolve positively and (ii) the degree of improvement of the evolution. Equation (8) shows how this method is evaluated.

$$evp(P_i) = \begin{cases} \dfrac{\sum_{P_{ij} \in N^+(P_i)} \frac{\left|f^b(P_i) - f^b\left(P_{ij}\right)\right| / NP}{\sigma(f(P_i))}}{|N(P_i)|}, & |N^+(P_i)|)0 \\ 0, & |N^+(P_i)| = 0 \end{cases} \tag{8}$$

Let P_i be the current population, $f^{\,b}(P)$ represents the best fitness value obtained in a population P, $\sigma(f(P_i))$ is the standard deviation of the fitness values of the population P_i, NP is the size of P_i, $N(P_i)$ is a set of one-step offspring populations of P_i defining its neighborhood, $N^+(P_i)$ is a subset of $N(P_i)$ containing only neighbors with better fitness than P_i. For minimization problems this is denoted as $N^+(P_i) = \{P_{ij} \mid P_{ij} \in N(P_i), f^{\,b}(P_{ij}) < f^{\,b}(P_i)\}$ in which $j = 1,\ldots,| N^+(P_i)|$. As it can be seen, the range of evp is $[0, +\infty)$. Higher values mean that the algorithm has a better capability to produce solutions of better quality as it evolves the current population.

3 Relevant Properties to Consider from DMOPs

Based on the information and concepts defined in the previous section, it is understandable to assume that DMOPs can have varying levels of complexity. This depends on how the properties and main elements of each problem are defined. This section briefly explains the characteristics that have the most relevance and impact in defining the difficulty of a DMOP.

3.1 Objective Function

Clearly, it is of utmost importance to take into consideration how the objective function of a DOP is defined. Within the objective function there are three factors that determine the difficulty and complexity of a DOP:

Number of objectives: In an optimization problem, the number of objectives is a criterion that serves as a starting point to determine the possible difficulty and complexity of a problem. Single objective DOPs such as the Moving Peaks Benchmark (Branke 1999) are an example of how a DOP can be presented. Within the literature, several two and three-objective DMOPs have also been proposed, such as the FDA set (Farina et al. 2004), dMOP set (Goh and Tan 2009) or DMZDT set (Wang and Li 2010). Those benchmark instances present DMOPs of different types (POF and POS may change after some time) and POFs of different shapes, such as convex, non-convex, non-continuous or with a non-uniform distribution in the search space. There are also DMOPs with four or more objectives, called dynamic many objective optimization problems (DMaOP). These problems require different strategies to obtain satisfactory solutions and, while relevant to mention, will not be considered for this work.

Difficulty of each objective: Each objective can have a certain degree of complexity. This mainly depends on the number of decision variables involved, the difficulty of the equation and how the dynamism of the problem is handled. For example, the FDA4 and FDA5 instances do not have a preset number of objectives to solve. Instead, the m number of objectives to solve has to be manually defined. The authors of this benchmark also recommend using $m + 9$ decision variables. Therefore, a bigger value on m means more objectives to solve, a bigger decision variable vector and a more complex objective function as each objective requires uses a certain subset of variables to calculate its value.

Objective function dynamism: A change in the environment can mean a modification in each of the objective values. One of the most common approaches to incorporate dynamism into a DOP, and therefore a DMOP, is by inserting a time variable t in the objective function. This variable changes over time. The FDA, dMOP and DMZDT benchmarks use this variable to feature dynamism.

3.2 Decision Variables

Decision variables are critical in an optimization problem, as they determine the value of each objective. There are three factors to consider when trying to determine how complex are the decision variables in the DMOP.

Number of decision variables: As the size of the decision variables vector increases, so does the number of dimensions of the solution search space. This can increase the difficulty of finding a set of solutions that belong in the POS. For DMOPs such as the instances from the FDA or dMOP benchmarks, the authors propose a predetermined number of decision variables. For example, FDA4 and FDA5 handle $m + 9$ decision variables, where m is the number of objectives.

Relevance of each variable: This property is completely related to the objective function handled by the DMOP. Each decision variable should, in theory, be related to at least one of the objectives of the DMOP. However, depending on the objective function, each variable can be assigned to one or multiple objectives and have different levels of importance. A clear example of this is FDA1, where the first decision variable directly represents the value of the first objective. This means that the value of this variable is critical while searching for good quality solutions.

Decision variables dynamism: In the same way as the objective function, various aspects of the decision variables can change over time. Of these changes, two take on the main relevance. The first being a modification in the values of each decision variable. This can be caused by a variable susceptible to time or as a consequence of a change adaptation method used by the algorithm to solve a DMOP. An example of this can be seen when working with DNSGA-II (Deb and Karthik 2007). The change adaptation method of DNSGA-II-B, a version of DNSGA-II presented in that work, replaces a subset of the population by mutating the values of the decision variables vector. This replacement is used when solving FDA2 and a case study based on simulations of a real-life problem. Also, the size of the vector of decision variables can change, which requires a reevaluation of each value of the objective function. The dynamic vehicle routing problem (DVRP), presented in works such as Ghannadpour et al. (2014) can have new destinations to cover introduced as times passes. This presents new variables to consider as the updated route has now to consider those new additions.

3.3 Constraints

A DMOP is limited by equality and inequality constraints. There are three criteria to consider in the constraints to understand how difficult a DMOP is to solve.

Number of constraints: Due to the nature of decision variables, each of them must have a constraint to define its range. In addition to those, some DMOPs handle additional constraints, such as the set of DCTP problems (Azzouz et al. 2015), which increases the unfeasible area in the objective search space for a DMOP.

Strictness of each constraint: Each constraint generates an infeasible area within the objective search space. It is reasonable to think that if the restrictions limit the feasible area too much or if the feasible areas are split into different regions of the search space, the difficulty of the DMOP may be increased. Considering once again the DCTP set, those benchmark functions present several challenges regarding this criterion. For example, there are cases where the only feasible area of the search space is near the POF.

Constraint dynamism: In the same way as the other two elements previously mentioned, constraints are also susceptible to changes in the environment. There are cases, such as the DCTP set, in which the constraints are altered with time.

4 Known Hyper-heuristic Approaches Towards Solving DOPs

Within the literature there are multiple proposals for hyper-heuristics to solve DOPs. Both focused on the generation and the selection of LLH. As previously mentioned, this work focuses entirely on the design and development of selection-based hyper-heuristics. Based on this consideration and following the classification established by Burke et al. (2010), as well as the research carried out in Macias-Escobar et al. (2020) it is possible to realize that most of these studies use online learning feedback methods (Baykasoğlu and Ozsoydan 2017; Cowling et al. 2000; Kiraz et al. 2013; Sabar et al. 2015; Topcuoglu et al. 2014; van der Stockt and Engelbrecht 2015). Even so, there are works that use offline learning (Uludag et al. 2012; Uludağ et al. 2013) and even no learning strategies, using only the information obtained at the time the LLH selection is performed (Kiraz et al. 2013; Köle et al. 2012; Ozcan et al. 2009; Topcuoglu et al. 2014).

In addition to the classification established in (Burke et al. 2010). The survey regarding dynamic hyper-heuristics previously mentioned also takes into consideration the complexity of the LLHs used. Most of the dynamic hyper-heuristics proposed use simple heuristics, specifically focused on a type of problem to solve a small array of instances (Chen et al. 2017; Garrido and Riff 2010; Gökçe et al. 2017; Kiraz et al. 2013; Ozcan et al. 2009; Topcuoglu et al. 2014; Uludag et al. 2012; Uludağ et al. 2013). However, there are multiple works that propose the use of meta-heuristics, which might have a higher degree of complexity than problem-specific heuristics, since they can be applied to a larger vector of instances (Baykasoğlu and Ozsoydan 2017; van der Stockt and Engelbrecht 2018; Wang et al. 2009). The results obtained by these hyper-heuristics are highly promising and open the possibility of using more complex algorithms as LLHs with the possibility of finding higher quality results in a computationally reasonable time.

As indicated in the survey presented in Macias-Escobar et al. (2020), the use of selection-based hyper-heuristics to solve DMOPs is an area that has not received much exploration. Recently, two proposals which seek to make an initial exploration in this research area have been proposed. The first of these works (Macias-Escobar et al. 2019) seeks to incorporate multiple DMOEAs within a selection-based hyper-heuristic supported by population evolvability to solve DMOPs. The results yielded promising conclusions, as the proposed hyper-heuristic is capable of performing better than its LLHs when working individually. The second work presented by these authors (Macias-Escobar et al. 2020) seeks to deepen the use of hyper-heuristics in DMOPs further by incorporating strategies that allow adding preferences of the decision maker (DM). The results reveal that the application of a hyper-heuristic to solve DMOPs with preferences of a DM is feasible. However, each of the hyper-heuristic properties must be carefully defined and designed to obtain effective results.

5 Proposed Checklist and Design Guide for Dynamic Hyper-heuristics

As mentioned at the end of Sect. 4. The effectiveness of using hyper-heuristics when solving DMOPs is susceptible to the properties with which it is configured. A hyper-heuristic that uses the wrong approach or unsuitable LLHs may not turn out to be able to obtain satisfactory results. This section proposes a checklist and a guide for the adequate design of hyper-heuristics capable of solving DMOPs in an efficient and satisfactory way.

Step 1 Identify the desired scope of the hyper-heuristic: The initial step to design a hyper-heuristic is to recognize the properties of the problems to be solved. In this case, since it is seeking to solve DMOPs, the properties mentioned in Sect. 3 should be reviewed. Special care should be taken in identifying the number of objectives that the DMOPs to solve have. Strategies used for optimization problems with up to three objectives are different with respect to problems with four objectives or more, also called many-objective optimization problems. In the same way, it is also very important to recognize which are the elements of each DMOP in which changes occur after a certain time. The points to check in this step are:

(1) How many objectives to solve do each DMOP has?
(2) What is the size of the decision variables vector for each DMOP?
(3) How strict are the constraints of each DMOP?
(4) Where is the dynamism of each DMOP applied?
(5) What are the dynamic properties of each DMOP (shape of the POF, change frequency, change severity)?

Step 2 Design the selection-based hyper-heuristic structure: Once the DMOPs to be solved have been identified, it is recommended to start the design of the basic structure of the hyper-heuristic. This work focuses on selection-based hyper-heuristics. As seen previously in Sect. 2.3, there are two main stages of this type of methodologies.

The first stage is the *heuristic selection*, this stage is very important, since it will determine which of the possible heuristics is the most appropriate to use for a certain period. It is imperative that the hyper-heuristic is able to make a correct selection in order to obtain good results. The points to consider for this stage are:

(1) Does the selection method use information from previous periods (is adaptive)?
(2) How many LLHs can be chosen in each selection? There are cases in which the selection can have permutations of LLHs to be applied in a period.
(3) What is the complexity of the selection method? The selection method can be from a simple randomly-based selections to complex strategies such as the double ranking Choice Function (Maashi et al. 2014).

The second stage is *move acceptance*. This step is also crucial, as it determines if the solution obtained in one period is acceptable to replace the current solution. The following points should be considered:

(1) Is the acceptance criterion deterministic or non-deterministic?
(2) How strict will the criterion be to accept the solution?
(3) Is the complexity of the acceptance criterion adequate for current DMOPs?

Step 3 Select performance indicators and metrics that support the hyper-heuristic: Both the heuristic selection and move acceptance stages require some type of valuation to determine which are the appropriate alternatives. Once the methods to be used in both stages have been set, it is necessary to define how the results obtained by each LLH will be measured. The following points should be considered:

(1) Are the indicators appropriate for the type of problem? In this particular case, it is necessary to know if the indicators are adapted to evaluate heuristics in multi-objective and dynamic environments. Some examples can be seen in Sect. 2.4.
(2) Is the performance of LLHs evaluated considering convergence and diversity?
(3) How much relevant information does each indicator provide?
(4) What is the complexity of each indicator? A very complex indicator must be used as little as possible to avoid additional computational cost. However, this must not affect the quality of the evaluation.

Step 4 Determine the DMOEAs to use as LLHs: After defining the selection method to use and the indicators to support the selection, it must be determined which are the LLHs to use. Care must be taken when selecting these heuristics, as LLHs will explore the solution search space. A set of suitable LLHs will not only be capable of obtaining good results but may even improve the performance of what each one can obtain individually. It is necessary to check the following points:

(1) Are the LLHs suitable for the DMOPs to be solved? For example, if the DMOP has four or more objectives, DMOEAs like DNSGA-II may not be suitable as they have strategies focused on problems with up to three objectives.
(2) What is the individual performance of each LLH in the DMOPs? It should be noted that a poorly performing LLH in a DMOP may affect the performance of hyper-heuristic.

(3) What are the strengths and weaknesses of each LLH?

Step 5 Determine the conditions of the experimentation: Having defined the design of the hyper-heuristic to be used, it is necessary to correctly state how the experimentation process is going to be carried out to determine if the designed and developed hyper-heuristic is effective. The points to consider in this step are similar in many respects to any other work presented for dynamic optimization problems:

(1) Which are the DMOPs to be solved?
(2) How is the change severity and change frequency defined? One or multiple combinations of these two elements can be used to test the capability of the hyper-heuristic. This means setting DMOPs under environments with constant or random changes which can have high or low impact on the POS.
(3) Are DMOPs sufficient to extract relevant information? The guide presented in Helbig and Engelbrecht (2014) establishes a set of considerations to check this point.
(4) What are the state-of-the-art algorithms that the proposed hyper-heuristic will be compared to? Usually the hyper-heuristic is compared with their individually implemented LLHs. However, it is also recommended, if possible, to use state-of-the-art algorithms to complement the comparison.
(5) What is the number of runs per algorithm in each DMOP? It is desirable to have an acceptable sample to perform a solution analysis and statistical tests.
(6) Is the environment where the experiments are carried out neutral? There should not be any form of bias towards any algorithm and the environment where the algorithms are executed must be identical for each one.

Step 6 Determine the evaluation criteria for results: Once the experiments are carried out, the obtained results must be evaluated in such a way that strong and well supported conclusions can be generated. For this, it is necessary to determine a set of indicators that allow the performance of each tested algorithm to be fairly evaluated. The following points should be considered:

(1) Are the indicators suitable for this type of problem?
(2) What are the DMOP's characteristics covered by each performance indicator?
(3) Is it possible to verify a statistically significant difference between the results of the hyper-heuristic and the other algorithms tested?
(4) Is the information obtained sufficient to generate solid conclusions from the experiment?

Step 7 Generate conclusions based on the results obtained: After evaluating the results of each algorithm tested in the experimentation, it is necessary to generate conclusions that cover all possible areas. It is recommended that the conclusions of the experiment and analysis follow the next points:

(1) Based on each of the evaluated characteristics (convergence, diversity, adaptation to dynamism).
(2) Based on each of the DMOPs evaluated.

(3) Based on the ability to adapt to problems with different levels of change frequency and change severity.
(4) Based on the comparison of the results obtained by the proposed hyper-heuristic against each of the state-of-the-art algorithms and LLHs used.

6 Case Studies Using the Proposed Guide and Checklist

In this section, we analyze two previously presented works that use the guide and checklist presented in Sect. 5 to design a selection-based hyper-heuristic to solve DMOPs while using DMOEAs as LLHs.

6.1 *Case Study 1*

The guide and checklist proposed in this work is used in the design of a selection-based hyper-heuristic in Macias-Escobar et al. (2019). The Dynamic Population-Evolvability based Multi-objective Hyper-Heuristic (DPEM_HH), seeks to solve DMOPs using DMOEAs as LLHs, in addition to that it relies on population evolvability, a FLA method, to carry out the selection of LLHs.

DPEM_HH follows step 1 of the guide by defining and identifying the characteristics of the DMOPs to solve. In that work, DMOPs from the FDA and DMZDT benchmarks are used, each of them being bi-objective while having decision variables vectors of varying size. Also, these instances do not require additional constraints and the dynamism occurs within the objective function in the form of a time dependent variable. For step 2, DPEM_HH uses the Choice Function method based on the work of (Maashi et al. 2014) and adapted to dynamic optimization as the LLH selection method. This method is adaptive, taking information from previous periods and requires a set of indicators to carry out the selection process. For the solution acceptance, the "All Moves" method (Cowling et al. 2000) is used, which is deterministic since it accepts any new solution obtained by the selected LLH.

For step 3, DPEM_HH uses as indicators and measures to support the LLH selection process the maximum spread metric (MS) (Goh and Tan 2007) as well as HVR, IGD and RNI. The population evolvability in Eq. (8) presented in Sect. 2 is also used. This allows DPEM_HH to obtain information regarding the quality of convergence and diversity of each LLH, as well as its ability to evolve into better solutions. For step 4, DPEM_HH uses three different versions of DNSGA-II, which use different change adaptation methods. DNSGA-II presents good performance in bi-objective DMOPs under the considered properties of the problems selected for the experimentation.

In step 5, the characteristics of the experimentation stage are determined. In this case, common elements from other works related to dynamic optimization were maintained, such as using DMOPs of different types and with varied forms of POF, with a regular change severity and frequency by having 10 changes scheduled each of

Fig. 1 HVR comparison between DPEM_HH and DNSGA-II-B

them after 25 generations. The selected LLHs are used as algorithms for performance comparison against the hyper-heuristic, as those also belong into the state-of-the-art. All experiments were run on an equal and non-biased environment.

The results of the experiments were evaluated according to the recommendations of step 6 of our proposed guide. Using as performance indicators MS, HVR and IGD, supported by the Friedman aligned ranks tests (Hodges and Lehmann 1962) to identify statistically significant differences. With this, the quality of each algorithm is evaluated based on its convergence and diversity with respect to the POF. Following step 7, comparisons of DPEM_HH are made regarding convergence, diversity and capacity to adapt to change with respect to each of the LLHs executed individually.

The results obtained by DPEM_HH were satisfactory, as DPEM_HH was able to equal or exceed in performance every LLHs used individually for most of the instances tested. Figure 1 presents a comparison regarding HVR between DPEM_HH and DNSGAII-B, one of its LLHs, for each DMOP tested. As it can be seen DPEM_HH is equal or outperforms DNSGAII-B in most of cases, with FDA3 being the only exception. This shows the capability of DPEM_HH to obtain better solutions than one of its LLHs while performing individually.

6.2 *Case Study 2*

Another case in which the guide presented in Sect. 5 applies is in Macias-Escobar et al. (2020). In this work, a hyper-heuristic named Dynamic Hyper-Heuristic with Plane Separation (DHH-PS) is proposed. DHH-PS follows a strategy similar to DPEM_HH. However, this work also seeks to solve dynamic optimization problems that incorporate DM preferences. Following step 1 of the guide, the authors define the DMOPs to be solved. In this case, several bi-objective problems are chosen from the FDA and dMOP benchmarks, all of them focusing on a certain area of their POF based on preferences defined by the DM. This increases the difficulty of the problem and adds extra constraints in the search for solutions.

For step 2, DHH-PS follows a structure similar to DPEM_HH, the Choice Function method based on a double ranking is used as the LLH selection method. For the move acceptance criterion, the "All Moves" method is also used. The performance indicators used to support the LLH selection, following the recommendations of step 3, are RNI, VD, IGD and HVR, which allow evaluating the performance of each LLH regarding convergence and diversity against the POF.

For step 4, two variations of the DNSGA-II are used for DHH-PS, as well as a dynamic version of the GDE3 algorithm (Kukkonen and Lampinen 2005), the three LLHS are adapted to include Plane Separation (PS), a novel reference-point based preference incorporation method. These algorithms presented a good capability to solve preferential DMOPs in previous individual experiments, for which they were selected. The conditions of the experimentation were determined following the step 5 of the proposed guide, using DMOPs with different properties and offering different challenges, as well as defining a regular change frequency and chance severity, with 10 changes each occurring in a period of 25 generations. Also, two DM preference settings are used, each one focused on the maximum or minimum extremes of the values obtained from the first objective.

For step 6, the VD, IGD and HVR metrics are used as performance indicators, as well as a statistical study based on the Friedman aligned ranks test to identify statistically significant differences between algorithms. Those performance metrics offer enough information to identify the quality of the results obtained regarding convergence as well as diversity with respect to a POF. Also, following step 7 of the guide, comparisons of the hyper-heuristic results were made with respect to each of the LLHs used and a dynamic version of RNSGA-II (Deb and Sundar 2006), a well-known algorithm used to solve optimization problems under a set of preferences defined by a DM, named DRNSGA-II. The results were promising, since DHH-PS was superior in most cases with respect to DRNSGA-II. DHH-PS was also capable to reach equal or better performance than the LLHs applied individually for some instances. These experimental tests of a hyper-heuristic in a dynamic and preferential environment establish a proof of the feasibility of its application in this type of problems and mark a starting point for future studies in this field.

Figure 2 shows a comparison regarding HVR between DHH-PS, DNSGA-II-A-PS, a version of DNSGA-II-A that uses PS on its process, which is used an LLH, and DRNSGA-II. All tested DMOPs in this scenario have preferences incorporated which aim towards the minimum values of the first objective. DHH-PS is capable to perform equally or better than DNSGA-II-A-PS and DRNSA-II for most of the DMOPs tested, with the exception of dMOP3.

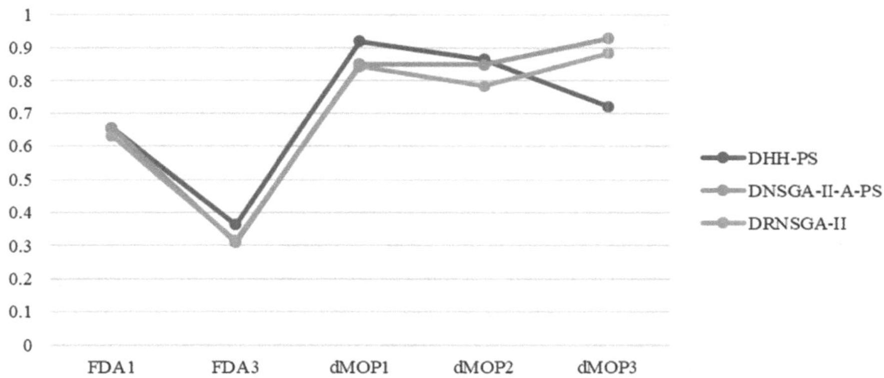

Fig. 2 HVR comparison between DHH-PS, DNSGA-II-A-PS and DRNSGA-II

7 Conclusions and Future Work

The design and development of hyper-heuristics to solve dynamic optimization problems is a complicated process. There are many factors to consider which must be reviewed with caution, since improperly choosing elements such as LLHs, heuristic selection and move acceptance methods or even performance indicators can lead to undesirable results.

This work seeks to present a guide to researchers to facilitate this process, so that they can focus their investigation on critical elements that determine the effectiveness of a hyper-heuristic, such as the adequate selection and definition of the problems to be solved, the heuristic selection and move acceptance methods of a hyper-heuristic, the LLHs to be used and the performance indicators. The guide presented in this work is focused on the resolution of DMOPs. However, it can also be easily adapted and used for the design of hyper-heuristics that seek to solve DOPs with a single objective or with many objectives (four or more).

The case studies presented in this work where the proposed guide is applied were able to develop competent and feasible hyper-heuristics. This clearly shows that this guide is effective in guiding the design of hyper-heuristics to solve DMOP. Within the scope of the possible future work is the application of this guide in the design of hyper-heuristics to solve DMOPs with four or more objectives which have a set of preferences established by a DM.

Acknowledgements This work was supported by the following projects: CONACyT PAACTI2020-1 under contract 312397; CONACyT ICB 17-18 under contract A1-S-11012; Cátedras CONACyT under contract 3058; CONACyT National Grant System under contract 465554; Spanish MINECO and ERDF under contract RTI2018-100754-B-I00; Junta de Andalucía and ERDF for contract P18-FR-2399; and ERDF for project FEDER-UCA18-108393. Also, thanks to the support from Laboratorio Nacional de Tecnologías de Información (LaNTI) from TecNM/Campus ITCM.

References

Azzouz, R., S. Bechikh and L. Ben Said. 2015, July. Multi-objective optimization with dynamic constraints and objectives: new challenges for evolutionary algorithms. In *Proceedings of the 2015 Annual Conference on Genetic and Evolutionary Computation*, pp. 615–622.

Azzouz, R., S. Bechikh and L.B. Said. 2017. Dynamic multi-objective optimization using evolutionary algorithms: a survey. In *Recent advances in evolutionary multi-objective optimization*, 31–70. Cham:Springer.

Baykasoğlu, A., and F.B. Ozsoydan. 2017. Evolutionary and population-based methods versus constructive search strategies in dynamic combinatorial optimization. *Information Sciences* 420: 159–183.

Branke, J. 1999, July. Memory enhanced evolutionary algorithms for changing optimization problems. In *Proceedings of the 1999 Congress on Evolutionary Computation-CEC99* (Cat. No. 99TH8406) 3, 1875–1882. IEEE.

Bilgin, B., E. Özcan, and E.E. Korkmaz. 2006, August. An experimental study on hyper-heuristics and exam timetabling. In *International Conference on the Practice and Theory of Automated Timetabling*, 394–412. Berlin: Springer.

Burke, E.K., M. Hyde, G. Kendall, G. Ochoa, E. Özcan, and J.R. Woodward. 2010. A classification of hyper-heuristic approaches. In *Handbook of metaheuristics*, 449–468. Boston, MA: Springer.

Burke, E.K., M. Gendreau, M. Hyde, G. Kendall, G. Ochoa, E. Özcan, and R. Qu. 2013. Hyper-heuristics: A survey of the state of the art. *Journal of the Operational Research Society* 64 (12): 1695–1724.

Burke, E.K., M.R. Hyde, G. Kendall, G. Ochoa, E. Özcan, and J.R. Woodward. 2018. A classification of hyper-heuristic approaches: revisited. *Handbook of Metaheuristics* 272: 453.

Chen, Y., P. Cowling, F. Polack, S. Remde, and P. Mourdjis. 2017. Dynamic optimisation of preventative and corrective maintenance schedules for a large scale urban drainage system. *European Journal of Operational Research* 257 (2): 494–510.

Cowling, P., G. Kendall, and E. Soubeiga. 2000, August. A hyperheuristic approach to scheduling a sales summit. In *International Conference on the Practice and Theory of Automated Timetabling*, 176–190. Berlin: Springer.

Deb, K., and J. Sundar. 2006, July. Reference point based multi-objective optimization using evolutionary algorithms. In *Proceedings of the 8th annual conference on Genetic and evolutionary computation*, 635–642.

Deb, K., U.B. Rao, and S. Karthik. 2007, March. Dynamic multi-objective optimization and decision-making using modified NSGA-II: a case study on hydro-thermal power scheduling. In *International conference on evolutionary multi-criterion optimization*, 803–817. Berlin: Springer.

Farina, M., K. Deb, and P. Amato. 2004. Dynamic multiobjective optimization problems: test cases, approximations, and applications. *IEEE Transactions on Evolutionary Computation* 8 (5): 425–442.

Ghannadpour, S.F., S. Noori, R. Tavakkoli-Moghaddam, and K. Ghoseiri. 2014. A multi-objective dynamic vehicle routing problem with fuzzy time windows: Model, solution and application. *Applied Soft Computing* 14: 504–527.

Garrido, P., and M.C. Riff. 2010. DVRP: a hard dynamic combinatorial optimisation problem tackled by an evolutionary hyper-heuristic. *Journal of Heuristics* 16 (6): 795–834.

Goh, C.K., and K.C. Tan. 2007. An investigation on noisy environments in evolutionary multiobjective optimization. *IEEE Transactions on Evolutionary Computation* 11 (3): 354–381.

Goh, C.K., and K.C. Tan. 2009. A competitive-cooperative coevolutionary paradigm for dy-namic multiobjective optimization. *IEEE Transactions on Evolutionary Computation* 13 (1): 103–127.

Gökçe, M.A., B. Beygo, and T. Ekmekçi. 2017. A Hyperheuristic approach for dynamic multilevel capacitated lot sizing with linked lot sizes for APS implementations. *Journal of Yaşar University* 12 (45): 1–13.

Helbig, M., and A.P. Engelbrecht. 2014. Benchmarks for dynamic multi-objective optimisation algorithms. *ACM Computing Surveys (CSUR)* 46 (3): 1–39.

Hodges, J.L., and E.L. Lehmann. 1962. Rank methods for combination of independent experiments in analysis of variance. In *Annals of mathematical statistics.*

Kiraz, B., Etaner-Uyar, A. Ş., & Özcan, E. (2013, April). An ant-based selection hyper-heuristic for dynamic environments. In *European conference on the applications of evolutionary computation*, 626–635. Berlin: Springer.

Kiraz, B., A.Ş. Etaner-Uyar, and E. Özcan. 2013. Selection hyper-heuristics in dynamic environments. *Journal of the Operational Research Society* 64 (12): 1753–1769.

Köle, M., A.Ş. Etaner-Uyar, B. Kiraz, and E. Özcan. 2012, September. Heuristics for car setup optimisation in torcs. In *2012 12th UK workshop on computational intelligence (UKCI)*, 1–8. IEEE.

Kukkonen, S., and J. Lampinen. 2005, September. GDE3: The third evolution step of generalized differential evolution. In *2005 IEEE congress on evolutionary computation*, 1, 443–450. IEEE.

Liu, C.A., and Y. Wang. 2006, September. New evolutionary algorithm for dynamic multiobjective optimization problems. In *International conference on natural computation*, pp 889–892. Berlin: Springer.

Liu, C.A. 2010, June. New dynamic multiobjective evolutionary algorithm with core estimation of distribution. In 2010 *international conference on electrical and control engineering*, 1345–1348. IEEE.

Maashi, M., E. Özcan, and G. Kendall. 2014. A multi-objective hyper-heuristic based on choice function. *Expert Systems with Applications* 41 (9): 4475–4493.

Macias-Escobar, T., L. Cruz-Reyes, B. Dorronsoro, H. Fraire-Huacuja, N. Rangel-Valdez, and C. Gómez-Santillán. 2019. Application of population evolvability in a hyper-heuristic for dynamic multi-objective optimization. *Technological and Economic Development of Economy* 25 (5): 951–978.

Macias-Escobar, T., B. Dorronsoro, L. Cruz-Reyes, N. Rangel-Valdez, and C. Gómez-Santillán. 2020. A survey of hyper-heuristics for dynamic optimization problems. In *Intuitionistic and type-2 fuzzy logic enhancements in neural and optimization algorithms: Theory and applications*, 463–477. Cham: Springer.

Macias-Escobar, T., L. Cruz-Reyes, H. Fraire, and B. Dorronsoro. 2020b. Plane Separation: A method to solve dynamic multi-objective optimization problems with incorporated preferences. *Future Generation Computer Systems* 110: 864–875.

Ozcan, E., S.E. Uyar, and E. Burke. 2009, July. A greedy hyper-heuristic in dynamic environments. In *Proceedings of the 11th annual conference companion on genetic and evolutionary computation conference: Late breaking papers*, 2201–2204. ACM.

Poli, R. 2008. *Some Ideas about No-Free Lunch for Hyper-Heuristics*. Technical Report, Department of Computing and Electronic Systems, University of Essex, Essex.

Richter, H. 2013. Dynamic fitness landscape analysis. In *Evolutionary computation for dynamic optimization problems*, 269–297. Berlin: Springer.

Sabar, N.R., M. Ayob, G. Kendall, and R. Qu. 2015. A dynamic multiarmed bandit-gene expression programming hyper-heuristic for combinatorial optimization problems. *IEEE Transactions on Cybernetics* 45 (2): 217–228.

Sierra, M.R., and C.A.C. Coello. 2005, March. Improving PSO-based multi-objective optimization using crowding, mutation and$\in$-dominance. In *International conference on evolutionary multi-criterion optimization*, 505–519. Berlin: Springer.

Tan, K.C., T.H. Lee, and E.F. Khor. 2002. Evolutionary algorithms for multi-objective optimization: Performance assessments and comparisons. *Artificial Intelligence Review* 17 (4): 251–290.

Topcuoglu, H.R., A. Ucar, and L. Altin. 2014. A hyper-heuristic based framework for dynamic optimization problems. *Applied Soft Computing* 19: 236–251.

Uludag, G., B. Kiraz, A.S. Etaner-Uyar, and E. Ozcan. 2012, September. Heuristic selection in a multi-phase hybrid approach for dynamic environments. In *UKCI*, 1–8.

Uludağ, G., B. Kiraz, A.Ş. Etaner-Uyar, and E. Özcan. 2013. A hybrid multi-population framework for dynamic environments combining online and offline learning. *Soft Computing* 17 (12): 2327–2348.

van der Stockt, S., and A.P. Engelbrecht. 2015, May. Analysis of global information sharing in hyper-heuristics for different dynamic environments. In *2015 IEEE Congress on Evolutionary computation (CEC)*, 822–829. IEEE.

van der Stockt, S.A., and A.P. Engelbrecht. 2018. Analysis of selection hyper-heuristics for population-based meta-heuristics in real-valued dynamic optimization. *Swarm and Evolutionary Computation*.

van Veldhuizen, D. A. (1999). *Multiobjective evolutionary algorithms: classifications, analyses, and new innovations* (No. AFIT/DS/ENG/99-01). Air Force Institute of Technology, Wright-Patterson AFB OH, School of Engineering.

Wolpert, D.H., and W.G. Macready. 1997. No free lunch theorems for optimization. *IEEE Transactions on Evolutionary Computation* 1 (1): 67–82.

Wang, H., D. Wang, and S. Yang. 2009. A memetic algorithm with adaptive hill climbing strategy for dynamic optimization problems. *Soft Computing* 13 (8–9): 763–780.

Wang, Y., and B. Li. 2010. Multi-strategy ensemble evolutionary algorithm for dynamic multi-objective optimization. *Memetic Computing* 2 (1): 3–24.

Wang, M., B. Li, G. Zhang, and X. Yao. 2017. Population evolvability: Dynamic fitness landscape analysis for population-based metaheuristic algorithms. *IEEE Transactions on Evolutionary Computation*.

Zheng, B. 2007, August. A new dynamic multi-objective optimization evolutionary algorithm. In *Third international conference on natural computation (ICNC 2007)*, 5, 565–570. IEEE.

On the Adequacy of a Takagi–Sugeno–Kang Protocol as an Empirical Identification Tool for Sigmoidal Allometries in Geometrical Space

Cecilia Leal-Ramírez and Héctor Echavarría-Heras

Abstract Examining sigmoidal allometries in geometrical space can be carried away by direct nonlinear regression or generalized additive modeling approaches. Nevertheless, producing consistent estimates of breakpoints characterizing phases composing sigmoidal heterogeneity could be problematic. Here, we explain how the paradigm of weighted multiple–phase allometries embraced by the mixture structure of the total output of a first-order Takagi–Sugeno–Kang fuzzy model can carry on this task in a direct, intuitive and efficient way. Present calibration tasks relied on log-transformed amniote testes mass allometry data. The considered TSK fuzzy model approach not only offers a way to back the assumption that analyzed testes mass allometry is sigmoidal in geometrical space but beyond this, it provided meaningful estimates for transition among involved phases. Results confirm previously raised views on the superior capabilities of the addressed fuzzy approach to validating prior subjective knowledge in allometry.

Keywords Sigmoidal allometry · Heterogeneity · Takagi–Sugeno–Kang fuzzy model · Breakpoint estimation

1 Introduction

Allometry or biological scaling has traditionally been one of the most addressed subjects of theoretical examination in biology. Initially, allometry meant linkage of trait and whole body sizes for an organism. Rooted in interpretations by Snell (1892), and Thompson (1992), the concept advanced into a study subject with the seminal work of (Huxley 1932). He embedded it in the theory of constant relative growth by two body parts and formulated though the scaling equation (Huxley 1932),

---

C. Leal-Ramírez · H. Echavarría-Heras (✉)
Centro de Investigación Científica y de Educación Superior de Ensenada, Carretera Ensenada-Tijuana, No 3918, Zona Playitas, 22860 Ensenada, BC, Mexico
e-mail: heheras@icloud.com

C. Leal-Ramírez
e-mail: cleal@cicese.mx

O. Castillo and P. Melin (eds.), *Fuzzy Logic Hybrid Extensions of Neural and Optimization Algorithms: Theory and Applications*, Studies in Computational Intelligence 940,
https://doi.org/10.1007/978-3-030-68776-2_19

$$y = \beta x^{\alpha} \tag{1}$$

where y and x are measureable traits, the parameter α designated as the allometric exponent and β is recognized as the normalization constant. The afore given model also identifies as the equation of simple allometry. It is of widespread use in research problems in biology (Huxley 1932; Lu et al. 2016; Savage et al. 2004; Myhrvold 2016; West and Brown 2005), biomedical sciences (Mayhew 2009; Paul et al. 2014; Moore et al. 2011; Eleveld et al. 2017; Kwak et al. 2016) economics (Champernowne 1953; Samaniego and Moses 2008; Wang et al. 2014; Coccia 2018; William 1979), earth and planetary sciences (Neukum et al. 1994; Maritan et al. 2002; Liu et al. 2018; Wolinsky et al. 2010; Bull 1975; Newman 2007; Naschie 2004; Ji-Huan and Jun-Fang 2009; Dreyer 2001; Pouliquen 1999) resource management and conservation (Zeng and Tang 2011a; De Robertis and Williams 2008; Rodríguez et al. 2017; Ofstad et al. 2016; Sutherland et al. 2000; Echavarria-Heras et al. 2018; Solana-Arellano et al. 2014; Montesinos-López et al. 2018), among other fields.

A prevalent device to obtain estimates of the parameters α and β relies on the logarithmic transformation of the original data to convey a linear regression model in log-scales. Afterward, retransforming identifies the projected two-parameter power function of Eq. (1) in the arithmetical scales. We call this here Log Transformation Method (hereafter LM). Despite the pervasiveness of an LM slant, views assert it produces unreliable results (Packard 2017a, b, 2013, 2009; Packard and Birchard 2008), stresses on the significance of graphical analysis in allometry. Plot assessment could readily explain LM failure. Revision of spreading in tied geometrical scale plot could reveal nonlinear dependence of the log-transformed response on its descriptor. This curvature upset commonly discusses non-log linear allometry (Packard 2012; Strauss and Huxley 1993), and could manifest through complex forms, with observations tracking smooth curves in log scales (Gould 1966; Echavarria-Heras et al. 2019a; MacLeod and MacLeod 2009). Polynomial regression could offer a cumbersome device to approximate devised shapes (Kolokotrones et al. 2010; Lemaître et al. 2014; Glazier et al. 2013; Tidière et al. 2017; Echavarria-Heras et al. 2019). This approach endorses that a single form for the mean response function holds over the whole range of the descriptor. Thus for instance, a biphasic heterogeneity in the log transformed response, as contemplated in Huxley's original theoretical perspective could not be embraced (Huxley 1932). Likewise, plot exploration could envisage an allometric relationship having two or more log-linear segments (Macleod 2010, 2014; Packard 2016). And this suggests that an affine model structure deems necessary to accommodate the anticipated complexity. A conventional identification route could be modeling the conditional distribution of the response given the descriptor as a mixture of linear sub-models. This way, the covariate domain can be clustered and associated with local regression models concurrently fitted. However, identification of breakpoints for transition among phases may depend on the subjectivity of the practitioner (Bervian et al. 2006; Forbes and López 1989; Sommerton 1980). Highly standardized piecewise regression methods could provide a seemingly automatic transition point estimation (Forbes and López 1989; Beckman and

Cook 1979; Ertel and Fowlkes 1976; Tsuboi et al. 2018; Ramírez-Ramírez et al. 2019; Muggeo 2003). Nevertheless, this technique depends on non-linear regression, thereby requiring initial values for estimates. Local maxima kind complications may upsurge. Inferences in estimated parameters could also be troublesome.

A fuzzy set theory approach offers a direct and intuitive non-probabilistic method to manage uncertainty (Dechnik-Vázquez et al. 2019; Zimmerman 1991; Zadeh 1965). Furthermore, construction of a Takagi–Sugeno fuzzy model (Takagi and Sugeno 1985; Sugeno and Kang 1988) encompasses splitting the covariate domain into fuzzy subdomains. Each acquired cluster or partition region associates with a fuzzy IF–THEN rule (Bezdek and Pal 1992). Each rule specifies a local model as an output. Membership functions of the fuzzy partitioning sets are assumed to have a known parametric form identified through subtractive clustering techniques (Castro et al. 2016; Chiu 1994). Normalizing membership functions produce weights for the local models. Weights interpret as the activating strengths of the rules. Products of weights and corresponding local models combine to create a regression scheme known as the Takagi–Sugeno–Kang fuzzy model's output. Identifying the parameters involved in the local models is achieved through recursive least squares techniques (Jang et al. 1997; Wang and Mendel 1992). Consequently, like Cohn et al. (1997) submit, a TSK protocol endorses a fuzzy parallel to a mixture regression model. Moreover, Azeem et al. (2000) assert that a Generalized Fuzzy Model (GFM) structure hosts the Mamdani (1977) and Larsen (1980) or the additive TSK construct as specific characterizations. Gan et al. (2005) established that the conditional mean of a Gaussian Mixture Model and a GFM defuzzified output match mathematically. Hence, the probability density function of a Gaussian mixture can be adapted to put up the TSK paradigm. Besides, Ying (1998) established that the overall output of a TSK fuzzy model could produce a uniform and decidedly precise approximation to any continuous function. Then, it is sound assuming that a TSK construct could capably address the identification of allometries in logarithmic scales, regardless of how complex they manifest. Echavarria-Heras et al. (2019) illustrate the masterful performance of a TSK device while identifying biphasic allometries in log scales. This examination also exemplifies the equivalence of named fuzzy slant and conventional-crisp mixture regression approaches while analyzing real data. Moreover, Echavarria-Heras et al. (2018) demonstrate how the TSK arrangement can perform direct scales allometric law identification. Besides, Echavarria-Heras et al. (2020) explain how a TSK scheme contributes to elucidating competing allometric interpretations at a qualitative glance. They demonstrate how this fuzzy analytical arrangement clarifies the existence of breakpoint allometries as envisioned in Huxley (1932), a paradigm questioned by a contemporary examination based on conventional statistical methods (Packard and Huxley 2012). Here we embrace an evaluation of the capabilities of the aforementioned fuzzy model on efficient identification of sigmoidal allometries. The performance of focusing arrangement on automatic detection of breakpoints composing inherent heterogeneity turned outstanding, even with complexity masked by high dispersion as in presently analyzed data. Since these uncertainty inducing effects are commonplace in most of the allometric examination settings, present results strengthen views in Echavarría-Heras et al. (2020) that project the addressed TSK scheme as a versatile identification device in biological scaling.

This paper organizes such that Sect. 2 explains the formalities of the model of complex allometry in direct scales. There, we explain associated multiplicative error regression scheme and its additive error consequent in geometrical scales. At hand we also elucidate the steps backing the presently addressed TSK fuzzy paradigm. This composes the log transformed formula of the allometric response as a linear combination of weighted local regression models of the linear form offered by the traditional analysis method of allometry. The results section compares the performances of the present TSK approach and the traditional alternative based in additive modelling referred in Macleod (2014) as a suitable device to identify sigmoidal allometries. A discussion section elaborates on the potential advantages, and limitations that a TSK slant bear for identification of sigmoidal allometries in geometrical space.

2 Methods

2.1 Model of Complex Allometry

For present aims, we assume that the allometric response y and its covariate x vary one to one in sets Y and X with both domains confined to R^+. Additionally, we conceive that the linkage between y and x expresses through the equation

$$y = w(x, \ \boldsymbol{p}) \tag{2}$$

with $w(x, \boldsymbol{p}) : X \rightarrow Y$ called the complex allometry function defined in terms of parameter set $\boldsymbol{p} = (p_1, \ldots, p_n)$. Since, present assay concerns to identification of $w(x, \boldsymbol{p})$ through log transformation methods, we firstly envision the regression model in direct scales

$$y = w(x, \ \boldsymbol{p})\psi(\varepsilon), \tag{3}$$

with a multiplicative error $\psi(\epsilon) = \exp(\epsilon)$, being ϵ a ϕ-distributed random variable with zero mean and variance σ^2, that is, $\epsilon \sim \phi(0, \sigma)$. Concurrently, the error term $\psi(\epsilon)$ acquires a log-ϕ distribution with log mean zero and log deviation σ. Accordingly, we consider a log transformation $u = lnx$ and $v = lny$.This sets domains U and V in geometrical space for u and v respectively. As a result, Eq. (3) becomes the log scales regression model

$$v = z(u, \ \boldsymbol{\pi}) + \varepsilon, \tag{4}$$

where the anticipated-nonlinear systematic part $z(u, \boldsymbol{\pi})$ formally defines like

$$z(u,\ \boldsymbol{\pi}) = ln(w(e^u,\ \boldsymbol{p})), \tag{5}$$

being $\boldsymbol{\pi}$ the concomitant log transformation-borne parameter set, and with additive error term $\epsilon = ln(\psi)$ that distributes as specified above. Associated mean response symbolizes

$$E_\phi(v|u) = z(u,\ \boldsymbol{\pi}). \tag{6}$$

2.2 TSK Fuzzy Model

Echavarría-Heras et al. (2020) interpreted the overall TSK output as a seeming paradigm for weighted multiple log-linear phase allometry. We here abide by this idea when aiming to identify possible sigmoidal shapes characterizing $z(u, \boldsymbol{\pi})$. Putin up to this requires a first step obtaining a fuzzy partition of the input domain U. For that aim, we contemplate membership functions $\mu_{\Phi_i}(u, \boldsymbol{\theta}_i)$ for $i = 1, 2, \ldots, q$, and having a Gaussian form i.e.

$$\mu_{\Phi i}(u,\ \theta_i) = exp\left\{-\frac{1}{2}\left[\left(\frac{u - a_i}{b_i}\right)^2\right]\right\} \tag{7}$$

with θ_i being the parameter pair $(a_{i,}, b_i)$. Then, we obtain $\vartheta^i(u, \boldsymbol{\theta}_i)$ recognized as the ith firing strength and that calculates through

$$\vartheta^i(u, \theta_i) = \frac{\mu_{\Phi i}(u,\ \theta_i).}{\sum_1^q \mu_{\Phi i}(u,\ \theta_i).}. \tag{8}$$

Next, we dispose linear consequents $f^i(u, \boldsymbol{c}_i)$ defined in terms of parameter pairs $\boldsymbol{c}_i = (\alpha_i, \beta_i)$, namely

$$f^i(u, \boldsymbol{c}_i) = ln\beta_i + \alpha_i u. \tag{9}$$

This builds a proxy for $z(u, \boldsymbol{\pi})$ expanded by the combined TSK output that expresses in the form

$$z(u, \boldsymbol{\pi}) = \sum\nolimits_1^q \vartheta^i(u, \boldsymbol{\theta}_i) f^i(u, \boldsymbol{c}_i). \tag{10}$$

In what follows, we will refer the parameter q as the heterogeneity index as it determines the number of phases composing the analysed allometry. Also, the comprehensive parameter set of fitted membership functions will be denoted by means of the symbol $\boldsymbol{\theta}$, that is, $\boldsymbol{\theta} = \bigcup_i^q \{\theta_i\}$ and correspondingly $\boldsymbol{c}$ shall represent the set containing all parameters o f consequent functions.

2.3 Data

MacLeod and MacLeod (2009) and Macleod (2014) analyzed general amniote testes mass composing log-transformed body mass and log-transformed percentage testes mass during the breeding season. For present aims, we acquired a related proxy data set resulting from electronically scanning points displayed in Fig. 1 of Macleod (2014). So achieved data distinguishes as present data or indistinctly through the McLWD shorthand. Additional information provided in MacLeod and MacLeod (2009) assert that the inflection points of an Additive Modeling smoother (AM) fitted on the general amniote testes mass data allowed a partition into three size ranges. These conform into species <30 g, species among 30 g and 10,000 g, and species >10,000 g. So intended ranges match the arrangement of linear segments portrayed in Fig. 1 in Macleod (2014). Joining piecewise linear trajectory refers here through McLPLA for short. Thresholds for transition between phases associating to the McLPLA path provide reference values to asses adequacy of candidate breakpoints composing the multiple phase loglinear allometry pattern deriving from present methods.

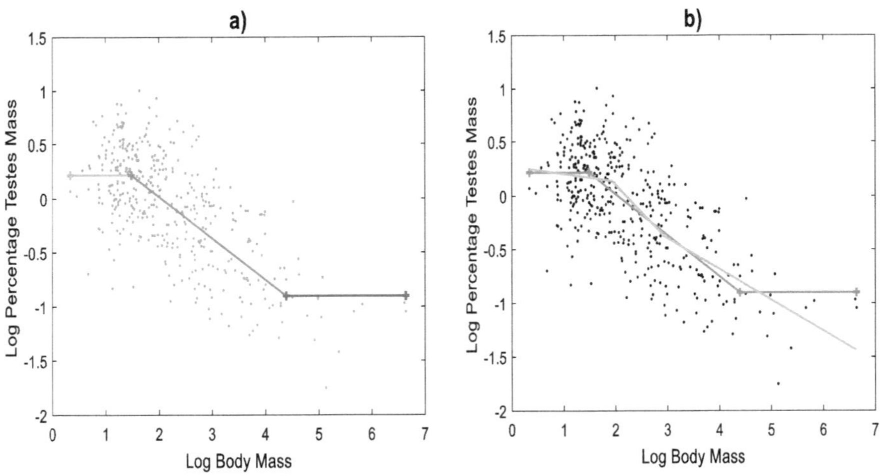

Fig. 1 Dots in shown plots represent proxy values (McLWD) for the data set analysed in MacLeod and MacLeod (2009) and Macleod (2014), composing pairs of log-transformed body mass and log-transformed percentage testes mass ($n = 395$). Continuous lines in panel **a** represent what we refer to as the McLPLA trajectory, that mimic the piecewise-linear arrangement in Fig. 1 of Macleod (2014). Panel **b** displays what we recognize as the HRPLA trajectory resulting from McLWD proxies (green lines). Concurring red lines represent the McLPLA, trajectory

2.4 Reproducibility Assessment

Assessment of reproducibility here mainly relies on analyzing Lin's Concordance Correlation Coefficient (CCC), signified by means of ρ (Lin 1989). Agreement will be understood as meager whenever$\rho < 0.90$, modest for $0.90 \leq \rho < 0.95$, good for $0.95 \leq \rho < 0.99$, or outstanding for $\rho \geq 0.99$ (McBride 2005). Assessments also relies on AIC values (Akaike 1974), and in other goodness-of-fit statistics, such as the standard error of estimate (SEE), and the mean prediction error (MPE) (Zeng and Tang 2011a, b; Zeng et al. 2017a, b).

2.5 TSK Identification Procedures

One approach to estimation of the parameters of the TSK model depends on Subtractive Clustering techniques (SC) (Castro et al. 2016; Chiu 1994) that yields the value of the parameter q. This arrays the number of inference rules R_i, for $i = 1.2.,,,.q$, each one associating to a local model $f^i(u, \boldsymbol{c}_i)$ composing the TSK representation of $z(u, \boldsymbol{\pi})$. The SC step also characterizes the membership functions $\mu_{\Phi_i}(u, \boldsymbol{\theta}_i)$ as given by Eq. (7). This identifies the normalized firing strength functions $\vartheta^i(u, \boldsymbol{\theta}_i)$, that is, the weighting factors of the linear consequents $f^i(u, \boldsymbol{c}_i)$ stablished in Eq. (10). Finally, estimates of the parameters in the consequents draw off from a Recursive Least Squares (RLS) routine (Jang et al. 1997; Wang and Mendel 1992) or from a maximum likelihood approach (Kalbfleisch 1985). Projected breakpoint values resulted by applying the firing strength intersection criteria by Echavarría-Heras et al. (2019, 2020). Here, identification of the TSK model relied in the genfis2 function of Matlab version R2016b, which generates a Sugeno-type Fuzzy Inference System (FIS) structure. This method relies on a subtractive grouping technique to extract a set of rules for the FIS that models the trends in data. The grouping procedure first determines q, the number of rules and membership functions called antecedents of the FIS. Then, the genfis2 code implements least squares estimation to identify the linear models that characterize the consequent function of each rule. At the end of the method, the generated FIS contains a set of fuzzy rules to cover the feature space. The $radii$ parameter in genfis2 function specifies the range of influence of a cluster center on each of the dimensions of the data. Since complex allometry pattern to be analysed involves three phases (Macleod 2014), we accordingly set the $radii$ value in such a way that resulting FIS composed three membership functions, providing that reproducibility index values resulted in highest values.

2.6 Piecewise-Linear Schemes

In order to explore suitability of the three-phase loglinear arrangement shown in Fig. 1 of Macleod (2014), we fit alternate piecewise-linear composites to associated McLWD surrogates. In what follows, the symbols u_{mb1} and u_{mb2} will represent the required first and second break points in corresponding order.

(a) We called on a Broken Line Regression Protocol (BLRP) that relies on the segmented package in R. The estimation method is discussed in Muggeo (2003). Case studies that explain implementation are available from Muggeo (2008). In present application of the procedure, we considered an initial pair (u_{0b1}, u_{0b2}) of first and second break points. The output break point pair denotes through (u_{mb1}, u_{mb2}). Resulting piecewise linear arrangement intended as a model for the covariate–response linkage underlying in the McLWD proxies. Corresponding path will be subsequently referred as a Broken Line Regression Trajectory (BLRT).

(b) We conceived an intuitive based broken line path drawing procedure that we will further on recognize as the Highest Reproducibility Piecewise-Linear Procedure (HRPLP). This departs from an initial 4-tuple of knot points: $(u_{01}, u_{0b1}, u_{0b2}, u_{04})$ in the McLWD set satisfying the order relationship $u_{01} < u_{0b1} < u_{0b2} < u_{04}$. We particularly choose u_{01} and u_{04} corresponding one to one, to the first and the last u covariate values in the McLWD set. Similarly, $(v_{01}, v_{0b1}, v_{0b2}, v_{04})$ identifies the 4-tuple of response values associating to the initial knot set. Then, we draw a linear segment beginning at the point (u_{01}, v_{01}) and ending at (u_{0b1}, v_{0b1}), a second one joining (u_{0b1}, v_{0b1}) and (u_{0b2}, v_{0b2}), and a last one connecting (u_{0b2}, v_{0b2}) and (u_{04}, v_{04}). Next, we calculate and record the CCC value of resulting initial piecewise linear arrangement. In a second step, we randomly choose a subsequent knot set of covariate values $(u_{11}, u_{1b1}, u_{1b2}, u_{14})$ such that $u_{01} \leq u_{11} < u_{1b1} < u_{1b2} < u_{14} \leq u_{04}$. Acquisition of conforming piecewise linear arrangement and also assessment of matching CCC value repeat. The HRPLP routine so on iterates as to calculate the CCC values of all possible 4-knot-set-determined piecewise-linear trajectories that can be acquired from the McLWD points. The 4-knot-set leading to the highest CCC value denotes through $(u_{m1}, u_{mb1}, u_{mb2}, u_{m4})$. Alongside, we distinguish the resulting piecewise composite as a Highest Reproducibility Piecewise-Linear Arrangement (HRPLA). Particularly, according to adopted notation convention the internal knots u_{mb1} and u_{mb2} identify as the first and second breakpoints that characterize built HRPLA.

3 Results

Table 1 presents the values of the CCC, AIC, SEE and MPE reproducibility statistics. In what follows, for the sake of easy of presentation for method performance assessment aims we only refer to CCC and AIC. A piece-wise linear composite as it interprets from Fig. 1 in Macleod (2014) reproduces here as in Fig. 1a. In further matters, we recognize this path as McLPLA, which furnishes a shorthand for MacLeod's Piecewise Linear Arrangement. Reproducibility strength of this trajectory when projecting response values on the McLWD surrogates resulted in a CCC = 0.6222 value. An electronic scanning procedure aimed to acquire reference knot values $(u_{r1}, u_{rb1}, u_{rb2,}, u_{r4})$ from the McLPLA delivered (0.3362, 1.4844, 4.3905, 6.6372) in a one to one correspondence. On further developments, we understand u_{rb1} and u_{rb2} one to one as the break points for transition from a first to a second and from this one to a third allometric phase, as contemplated in Macleod (2014). Similarly, the pair (u_{mb1}, u_{mb2}) will generically designate corresponding estimates gotten by the different methods addressed here. Running a BLRP we got break points estimates (u_{mb1}, u_{mb2}) placed at (1.560, 5.362) respectively. Comparing with reference values $(u_{rb1}, u_{rb2,})$ we can be aware of absolute differences of $\Delta_{mb1} = 0.0756$ and $\Delta_{mb2} = 0.9715$. We can make a distinction of a barely agreement between the second reference break point values and that resulting from the BLRT. Associated BLRT trajectory displays a reproducibility strength as corresponds to a CCC = 0.5697 value. We can also ascertain adequate matching in CCC values for McLPLA and BLRT approaches, implying similar reproducibility power. Furthermore, when testing the strength of either one of these trajectories to reproduce the other we obtained a CCC = 0.9066 agreement index value. We arranged the above given McPLA reference values as the initial knot set in a HRPLP. This conveyed a HRPLA trajectory as shown in Fig. 1b with linked knot values calculated at (0.3363, 1.9363, 2.9363, 6.6372). Paralleling estimated and reference break point values led to absolute differences of $\Delta_{mb1} = 0.0756$ and $\Delta_{mb2} = 0.9715$ (Table 2). The reproducibility strength of HRPLA when projecting McLWD proxy values

Table 1 Reproducibility strength of the different methods fitted on the McLWD proxies. BLRP stands for Broken Line Regression Protocol, HRPLP refers to Highest Reproducibility Piecewise-Linear Procedure, TSK(r) signifies Takagi–Sugeno–Kang fuzzy model fitted using $raddi = r$, AM stands for Additive Modeling protocol fitted by MacLeod and MacLeod (2009). AIC stands for Akaike Information Criterion, SEE means Standard Error of Estimate and MPE, Mean Prediction Error

Model	CCC	AIC	SEE	MPE
BLRP	0.5697	569.55	0.4913	−13.2926
HRPLP	0.6733	389.77	0.3933	−44.8990
TSK (0.30)	0.6420	330.32	0.4727	0.3648
TSK (0.25)	0.6420	334.27	0.3657	−41.7542
TSK (0.16)	0.6494	344.21	0.3668	−41.8703

Table 2 Breakpoints u_{mb1} and u_{mb2} as identified by the different methods fitted on McLWD proxies. Correspondingly Δu_{mb1} and Δu_{mb2} represent one to one their deviations from reference breakpoints u_{rb1} and u_{rb2}. Breakpoints for BLRP derived from implementation of the Segmented package in R. Those for HRPLP gotten accordingly to scheme explained in Sect. 2.5. Finally breakpoints for the TSK method resulted from the Echavarria-Heras et al. (2020) criterion, TSK(r) signifies Takagi–Sugeno–Kang protocol fitted using $raddi = r$

Model	u_{mb1}	u_{mb2}	Δu_{mb1}	Δu_{mb2}
McPLA	1.4844	4.3905	0	0
BLRP	1.560	5.362	0.0756	0.9715
HRPLP	1.9363	2.9363	0.4519	1.4542
TSK (0.30)	2.1858	3.3805	0.7014	1.0100
TSK (0.25)	1.4578	3.1770	0.0266	1.2135
TSK (0.16)	1.4756	3.5487	0.0088	0.8418

corresponded to a CCC = 0.6733 value. Comparing to the McLPLA counterpart (reed lines), we can be aware that although the HRPLA trajectory displays a slight gain in reproducibility strength, conforming break point placement show noticeable differences relative to reference McLPLA ones (Table 2).

A $radii = r$ fit of the Takagi–Sugeno–Kang paradigm on a specified data set will be afterwards recognized by means of TSK(r) for short. Similarly, corresponding mean response $z(u, \pi)$ will be afterwards recognized by means of a TSK(r)T shorthand. Figure 2a displays the spread about TSK(0.3)T the form of $z(u, \pi)$ curve (black lines) obtained by a $radii = 0.3$, fit of the TSK model on the McLWD surrogate. Shown shape of TSK(0.3)T resembles sigmoidal allometry as inferred by MacLeod

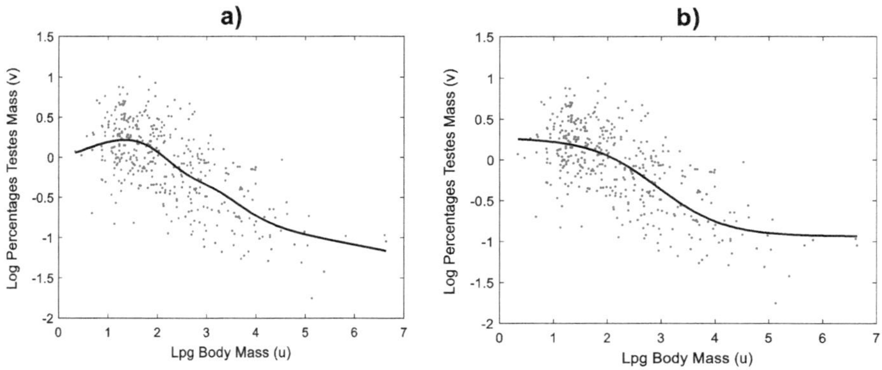

Fig. 2 Panel **a** Spread about TSK(0.3)T the form of the mean response $z(u, \pi)$ curve (black lines) obtained by a $radii = 0.3$, TSK model fitted on McLWD proxies (see text for details). Shape of TSK(0.3)T resembles sigmoidal allometry. Panel **b** Continuous lines represent the McLAMT, the sigmoidal mean response curve $z(u, \pi)$ fitted by MacLeod and MacLeod (2009) using an additive modeling protocol

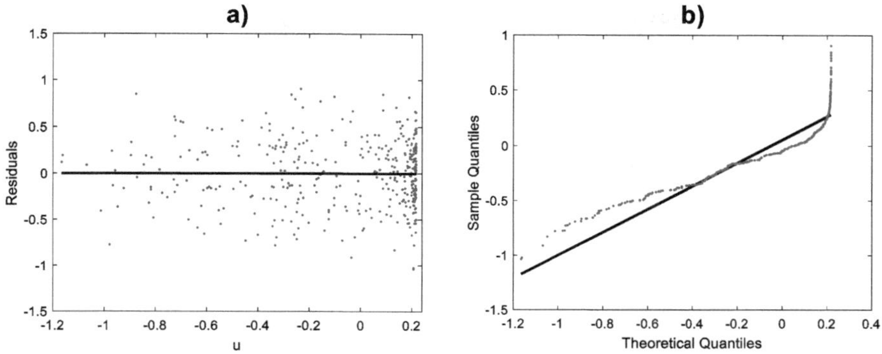

Fig. 3 Panel **a** Residual plot associating to TSK(0.3)T the form of the mean response $z(u, \boldsymbol{\pi})$ function resulting from a $radii = 0.3$, fit of the TSK model to McLWD proxies (see text for details). Residual spread suggests a lack of normality. Panel **b** linked Normal-QQ plot. Deviations from reference line as well as heavy tail shown at right extreme of distribution suggests a lack of normality of residuals. A quest for characterization of a suitable error structure deems necessary

and MacLeod (2009) and Macleod (2014). Calculated concordance correlation coefficient resulted in CCC = 0.6420 with a corresponding AIC index value of AIC = 330.32 (Table 1). Figure 2b portraits the sigmoidal mean response curve $z(u, \boldsymbol{\pi})$ fitted by MacLeod and MacLeod (2009) on concurring data and using Additive Modeling (AM). Resulting trajectory will be subsequently acknowledged by means the symbol McLAMT. This trajectory associated to CCC = 0.6324 and AIC = 328.1436 values. We may be aware that both fits produce similar $z(u, \boldsymbol{\pi})$ shapes and that albeit slight differences in reproducibility measure values favor the TSK(0.3)T scheme, we may say that both trajectories associate equivalent projection features. Furthermore, when testing the strength on what TSK(0.3)T reproduce points composing the McLAMT trajectory we obtained a CCC = 0.6324 agreement index value. The TSK residual and QQ-plots appear in Fig. 3a and b respectively. Figure 4 pertains to corresponding plots for the fit of the sigmoidal-AM counterpart. Examining these plots reveals that an assumption of a structure conforming to the normality of errors is incompatible with resulting spread in both cases.

The TSK acquired mean response curve $z(u, \boldsymbol{\pi})$ displaying in Fig. 1 (panel **a**) suggests a sigmoidal-like pattern of complex allometry that as ascertained by MacLeod and MacLeod (2009) and Macleod (2014) is consistent with data analyzed. Moreover, calculated CCC values imply that the reproducibility strengths of the TSK(0.3)T and McLAMT trajectories are equivalent. We now explain the performance the TSK slant at identifying candidate values for breakpoints u_{mb1} and u_{mb2} directly obtained from the McLWD proxies. Carrying away a $radii = 0.3$ fit of the said paradigm on available data identified a $q = 3$ value for the heterogeneity index. Therefore, present fuzzy scheme backs heterogeneity as composed by three allometric phases, that as pointed out in Macleod (2014) deem necessary to interpret complexity appropriately. The firing strength intersection criterion proposed by Echavarría-Heras et al. (2020), estimated a first breakpoint point at $u_{mb1} = 2.1858$

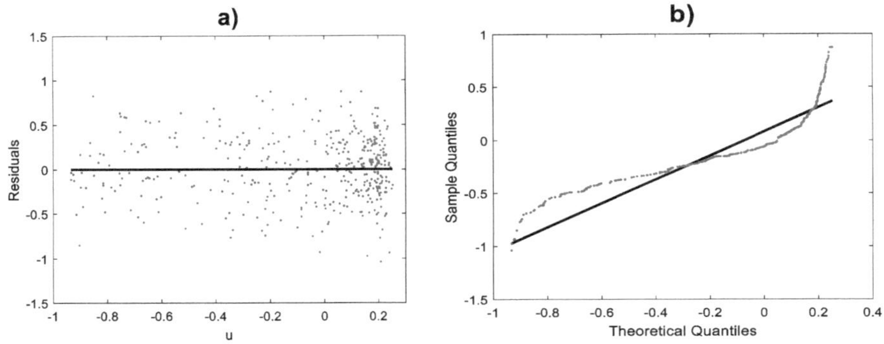

Fig. 4 Plot of residuals gotten from response values in McLWD and their projections through McLAMT, the AM-sigmoidal-shape form of the mean response $z(u, \boldsymbol{\pi})$ fitted by MacLeod and MacLeod (2009). Residual spread suggests a lack of normality. Panel **b** displays the linked Normal-QQ plot. As it infers from a similar pattern in Fig. 3b, deviations from the reference line and heavy tail shown at the right extreme of distribution back the perception of a lack of normality of residuals. Addressing a different error structure for enhanced dependability of results deems pertinent

separating the first and second phases. Concurrently, a second breakpoint identified at $u_{mb2} = 3.3805$ interprets as a transition threshold from the second to the third phase (Fig. 5a). Comparing with reference values ($u_{rb1} = 1.4844$, $u_{rb2} = 4.3905$) inferred from the HRPLA, we can be aware of absolute differences of $\Delta_{mb1} = 0.7014$ and $\Delta_{mb2} = 1.010$ (Table 2). Table 3 includes linked parameter sets $\boldsymbol{\theta}$ and $\boldsymbol{c}$. Figure 5a displays plots of firing strength functions $\vartheta^i(u, \boldsymbol{\theta}_i)$ for $i = 1, 2, 3$. First linear consequent function resulted in $f^1(u, \boldsymbol{c}_1) = 0.2719u - 0.0273$ while second and third ones were $f^2(u, \boldsymbol{c}_2) = 0.3108u - 1.2803$ and $f^3(u, \boldsymbol{c}_3) = -0.1167u - 0.3915$). Figure 5b provides plots of consequent functions $f^i(u, \boldsymbol{c}_i)$ for $i = 1, 2, 3$.

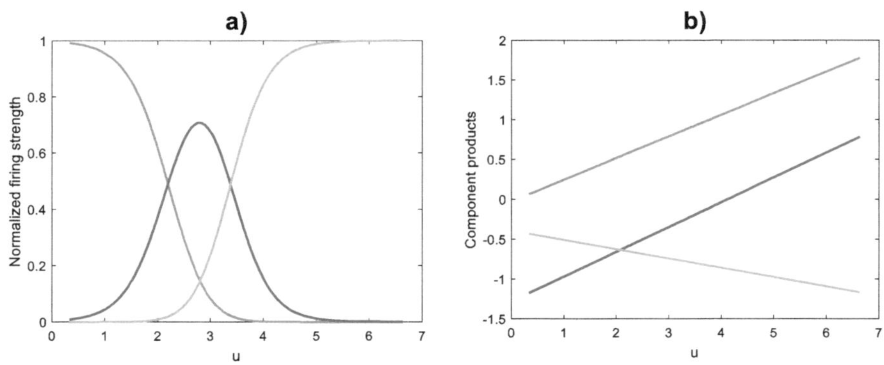

Fig. 5 Panel **a**, normalized firing strengths $\vartheta^i(u, \boldsymbol{\theta}_i)$ for $i = 1, 2, 3$, produced by a $radii = 0.3$, $q = 3$, fit of the TSK model of combined output given by Eq. (10) and fitted on Macleod (2014) data (McLWD). Intersection of firing strength curves yield estimated breakpoints at $u_{mb1} = 2.19845$ and $u_{mb2} = 3.3805$. Panel **b**, related consequent functions s $f^i(u, \boldsymbol{c}_i)$

Table 3 Reproducibility measure statistics for TSK(r) fits of radii $r = 0.37$, $r = 0.25$ and $r = 0.17$ performed on simulated points (u, v) conforming random replicates about the McLPLA trajectory (n = 395) (see text for details)

Model	CCC	AIC	SEE	MPE
BLRP	0.9930	−1.259e + 03	0.0488	−2.4792
HRPLP	0.9928	−1.243e + 03	0.0494	−2.4619
TSK (0.37)	0.9928	−1255.15	0.0490	−2.4438
TSK (0.25)	0.9933	−1277.79	0.0475	−2.3688
TSK (0.17)	0.9938	−1302.08	0.0459	−2.2858

As explained by Echavarría-Heras et al. (2020), the TSK scheme can be used to explore the fitness of subjective knowledge based assertions over breakpoint location. Here, we address the capabilities of this paradigm to explore feasibility of breakpoint placement as conjectured by Macleod (2014), that is $u_{rb1} = 1.4844$, and $u_{rb2} = 4.3905$. A $radii = 0.25$ fit of the TSK fuzzy model on Macleod (2014) data proxies McLWD (TSK(0.25)) resulted in a $q = 4$ value for the heterogeneity index. Associating mean response curve TSK(0.25)T displays in Fig. 6a (black lines). Accompanying, green, yellow and red segments reproduce the arrangement drawn in Fig. 1 by Macleod (2014). According to the criterion in Echavarría-Heras et al. (2020) first TSK breakpoint u_{mb1} coincides with the covariate value on which the firing strengths $\vartheta^1(u, \boldsymbol{\theta}_1)$ and $\vartheta^2(u, \boldsymbol{\theta}_2)$ intersect. Moreover, intersection of the $\vartheta^3(u, \boldsymbol{\theta}_3)$ and $\vartheta^4(u, \boldsymbol{\theta}_4)$ firing strengths closely matches u_{rb2} value thereby delivering an estimated value for u_{mb2}. Dashed purple colored vertical lines mark the resulting TSK(0.25) estimated break points $u_{mb1} = 1.4578$ and $u_{mb2} = 3.1770$. Absolute deviations from reference values turned out to be $\Delta_{mb1} = 0.0266$ and $\Delta_{mb2} = 1.2135$.

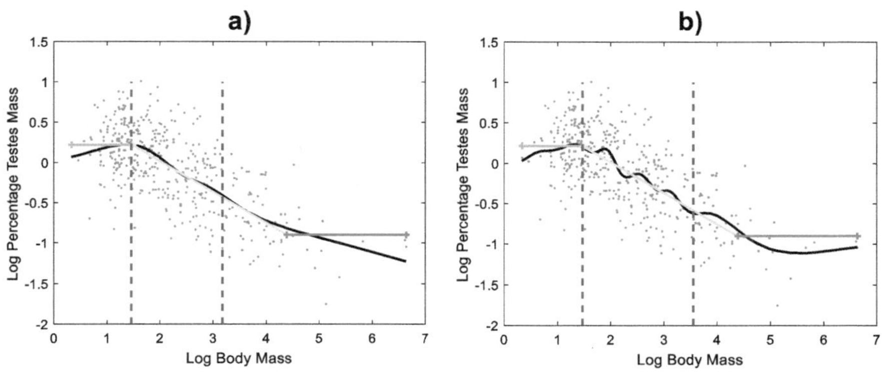

Fig. 6 Panel **a**, break point arrangement resulting from a $radii = 0.25$ fit of the TSK fuzzy model on the McLWD surrogates. Green, yellow and red segments correspond to the McLPLA. Dashed blue colored vertical lines mark estimated breakpoints. Panel **b** portraits response TSK (0.16) T with vertical lines labeling resulting breakpoints, joining green yellow and red lines represent the McLPLA. We can be aware that breakpoint estimates are closer to reference values

Then, the TSK(0.25) estimate for the first breakpoint reasonably matches breakpoint positioning inferred by intersection of segments suggested by Macleod (2014). Reproducibility indices, for TSK (0.25)T got AIC = 334.27 and CCC = 0.6420, values. These figures are respectively, higher and lower than those obtained by fitting the McLWD proxies to the TSK model using $radii = 0.3$ (Table 1). Figure 6b portraits plot resulting from a TSK(0.16) fit which produces $u_{mb1} = 1.4756$, and $u_{mb2} = 3.5487$. Absolute deviations from reference now become be $\Delta_{mb1} = 0.0088$, and $\Delta_{mb2} = 0.8418$. As shown in this last plot decreasing the of the $radii$ parameter triggers an interpolation-like mode for the TSK(0.16)T mean response but still resembling a sigmoidal trend. Concomitantly breakpoint estimates get closer to reference values that those gotten from TSK(0.25). Moreover, calculating the modulii of vectors of reference-estimates deviations, we can ascertain that overall agreement of reference and TSK(0.16) estimated breakpoints is higher than that resulting from a BLRP.

Dispersion of McLWD proxies is significant enough as to influence reproducibility strength of TSK projections in a sensible way. In order to illustrate extent of the goodness of fit loss that associates to referred spread pattern, we performed fits of the presently addressed TSK model to simulated (u, v) pairs conforming normally distributed replicates $\left(\mu = 0, \sigma^2 = 0.1\right)$ around points conforming the McLPLA. Trajectory Exploration includes assessment of the deviations in detected breakpoint placement relative to reference ones as conjectured from Fig. 1 in Macleod (2014), that is, $u_{rb1} = 1.4844$, and $u_{rb2} = 4.3905$. Table 3 presents linked reproducibility measure statistics and Table 4 presents breakpoint deviations relative to reference values. Comparing with reproducibility index values in Table 1 calculated for the TSK fits on McLWD proxies ($var(v) = 0.249$) we can be aware of remarkable improvements in goodness of fit gained from a moderate reduction in variability in simulated McLPLA replicates ($var(v) = 0.100$). We can also learn that breakpoint values estimated from the simulated McLPLA replicates got closer to reference ones.

Table 4 Breakpoints u_{mb1} and u_{mb2} values as identified by the different methods fitted on simulated proxies(see text for details). Correspondingly Δu_{mb1} and Δu_{mb2} represent one to one their deviations from reference breakpoints u_{rb1} and u_{rb2}. Breakpoints for BLRP derived from implementation of the segmented package in R. Those for HRPLP gotten accordingly to scheme explained in Sect. 2.6b. Finally breakpoints for the TSK method resulted from the Echavarria-Heras et al. (Echavarria-Heras et al. 2020) criterion, TSK(r) signifies Takagi–Sugeno–Kang protocol fitted using $raddi = r$

Model	u_{mb1}	u_{mb2}	Δu_{mb1}	Δu_{mb2}
McPLA	1.4844	4.3905	0	0
BLRP	1.505	4.355	0.0206	0.0355
HRPLP	1.5462	4.3462	0.0618	0.0443
TSK (0.37)	2.2358	3.7165	0.7514	0.674
TSK (0.25)	1.6833	3.3960	0.1989	0.9945
TSK (0.17)	1.5065	4.0038	0.0221	0.3867

Moreover, BLRP and HRPLP now display equivalent reproducibility strength but were indeed surpassed by the TSK alternates in goodness of fit (Table 3). As expected the BLRP and HRPLP delivered remarkably consistent estimates of breakpoint positioning. Nevertheless, the uncertainty ranges for BLRP breakpoint estimates do not exclude values comparable to the ones estimated by the TSK alternates. Figure 7 portraits the relative increase in accuracy of the BLRP and HRPLP broken line slants. Similarly, Fig. 8 gives an appraisal of the gain in reproducibility power achieved by the TSK(r)T mean response. Finally, the comparing TSK(r) fits performed on

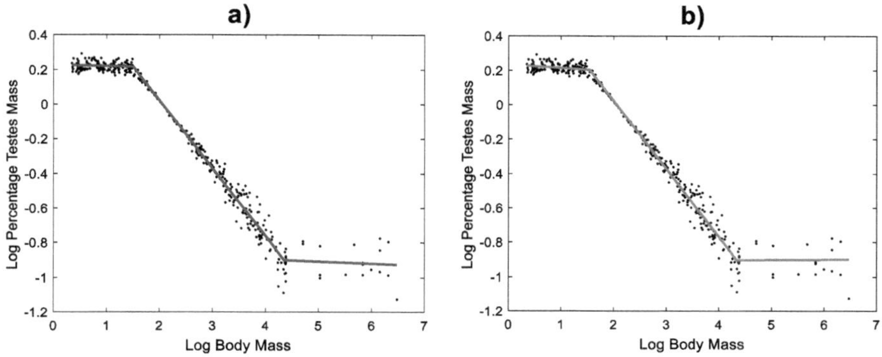

Fig. 7 Panel **a** Mean response trajectory obtained by fitting the BRLP method on simulated replicates of points over the McPLA path (blue line). Panel **b** Mean response curve obtained by fitting the HRPLP method on simulated replicates of points over the McPLA path (red line). In both plots black dots represent said simulated replicates produced by assuming a uniform distribution of a zero mean and a 0.1 variance

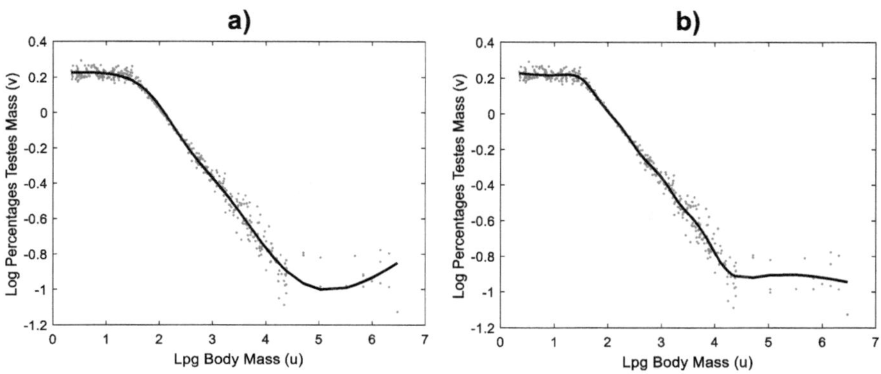

Fig. 8 Panel **a** 3-phase mean response curve obtained by a TSK (0.37) fit on simulated replicates of points placing over the McLPLA trajectory (black line). Panel **b** 6- phase mean response curve resulting from a TSK (0.17) fit on simulated replicates of points placing over the McLPLA trajectory (black line) (see text for explanation). In both plots red dots represent said simulated replicates produced by assuming a uniform distribution of zero mean and variance = 0.1

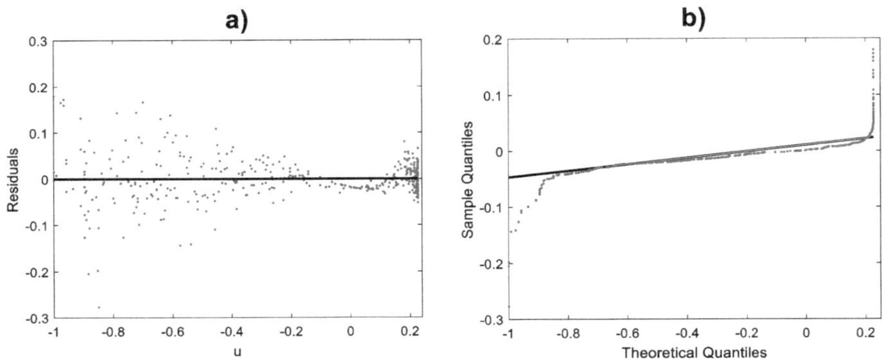

Fig. 9 Panel **a** residual diagram resulting from a TSK (0.37) fit on simulated replicates of points placing over the McLPLD. Panel **b** Normal-QQ plot for the TSK (0.37) fit

McLWD proxies the plots of Fig. 9 show arrangements of both residual and Normal QQ plots more consistent with underlying error distribution.

4 Discussion

In many instances, the linear model in log scales deriving from Huxley's formula of simple allometry (cf. Eq. (1)) fails to provide a consistent fit (Packard 2017a, b, 2013, 2009, 2012; Packard and Birchard 2008; Strauss and Huxley 1993). Facing related fitting intricacies extended Huxley's original perspective into a paradigm of multiple parameter complex allometry. This brought into consideration of all varieties of nonlinear of discontinuous formulations for modeling the allometric response-covariate linkage (e.g. Packard 2013; Macleod 2014; Bervian et al. 2006; Frankino et al. 2010; Lovett and Felder 1989). Complexity extension in geometrical space allometry endorses sigmoidal models (e.g. Nijhout and Wheeler 1996; Palestrini et al. 2000; Shingleton 2010; Rasmussen and Tan 1992; Emlen 1996; Mahmood 2013). The general amniote testes mass allometry examined in Macleod (2014), and reviewed here readily exemplifies said splitting. Envisioned allometry composes three phases, which brings about the problem of estimation of transition thresholds. For that aim Macleod (2014) approach centered in the inflection points of an additive modeling smoother for the analyzed general amniote testes mass data. Thresholds draw off from the resulting partition of the covariate domain into three size intervals. A first one for species <30 g, another for species between 30 and 10,000 g, and the third for species >10,000. Electronic scanning of envisioned piecewise linear arrangement presented in Fig. 1 in Macleod (2014) facilitated proxies for incumbent thresholds that we distinguish here utilizing the u_{rb1} and u_{rb2} symbols and that we recognized through present analysis as reference breakpoints. Echavarría-Heras et al. (2019, 2020) offered estimates for breakpoints associating to a weighted polyphasic loglinear allometry pattern. Determining criterion based on identifying the cutting points of the firing strengths of a Takagi–Sugeno–Kang fuzzy model. Illustration of suitability of the Echavarría-Heras et al. (2019, 2020) procedure mainly focused on study cases of biphasic allometry. Our effort here aimed to an exploration of the capabilities of a TSK construct to deal with identification of breakpoints in allometries composing three phases. The necessary breakpoints were generically denoted by means of the u_{mb1} and u_{mb2} symbols. Presently analyzed McLWD proxy values show noticeable dispersion. As a reason, just barely agreement between reference and breakpoint estimates resulted from the BLRP o the HRPLP slants. As it can be ascertained from reproducibility index values in Table 1, on spite of a noticeable spreading in the addressed proxy data the TSK approach delivered a reasonable identification of the inherent sigmoidal pattern. Moreover, the values of reproducibility strength measures associating to the fitted three-phase TSK scheme closely match those corresponding to conventional additive modeling or broken stick regression slants. Regarding the consistency of breakpoint projection, we must emphasize that the celebrated approximation capabilities of the TSK allowed the adaptation of complexity as required to identify user knowledge-based priors. This feature of the TSK scheme when fitted in simulated surrogates offered suitably accurate projection of reference break point values. A flexible structure inherent in a TSK fuzzy model is missing in the aforesaid conventional approaches. Figure 8 provides a pictorial going over of this assertion by showing how the TSK managed to (1) interpolate data to suitable tolerance (2) detect break points properly while (3) keeping a sigmodal-like mean response. Simulation assay results unravel the relevance of data quality as a factor influencing accuracy of TSK projections (Fig. 9).

5 Conclusion

It is worth pointing out that this work's essential aim was exploring the adequacy of the TSK fuzzy model as a device producing a curve that fits an apparent pattern of sigmoidal allometry. Thus, the biological interpretation of results is beyond the scope of the present research design. Being this vital point clarified, we may firstly emphasize that despite high dispersion observed in presently analyzed data (McLWD), the addressed TSK protocol displayed competence at performing both identifications of previously envisioned inherent sigmoidal allometry, and elucidation of possible breakpoints. However, even though we contributed here by elucidating how the TSK performs at dealing with the identification of sigmoidal allometry descriptors such as the topology of mean response and the number of required breakpoints in a direct way from available data, it deems also necessary to add that results could not support at this stage a clear advantage of this method over conventional approaches. Indeed, performed simulation assay reveals that further examination on the extent of uncertainty propagation due to high dispersion in data judges necessary. None less to say that a lack of adequacy of assumed error structure also bears relevance in setting the overall analytical approach's suitability. This sets data quality as a relevant issue in overall dependability of the TSK approach. Error propagation of breakpoint estimates on retransformed projections of the TSK output could sensibly affect the precision of predictions and biological interpretation. This issue could conform an interesting and important research endeavor.

References

Akaike, H. 1974. A new look at the statistical model identification. *IEEE Transactions on Automatic Control*, 19(6).

Azeem, M.F., M. Hanmandlu, and N. Ahmad. 2000. Generalization of adaptive neuro-fuzzy inference systems. *IEEE Transactions Neural Networks* 11 (6): 1332–1346.

Beckman, R.J., and R.D. Cook. 1979. Testing for two-phase regressions. *Technometrics* 21: 65–69.

Bervian, G., N. Fontoura, and M. Haimovici. 2006. Statistical model of variable allometric growth: Otolith growth in Micropogonias furnieri (Actinopterygii, Sciaenidae). *Journal of Fish Biology* 68: 196–208.

Bezdek, J.C., and S.K. Pal. 1992. *Fuzzy models for pattern recognition*. New York: IEEE Press.

Bull, W.B. 1975. Allometric Change of Landforms. *Geological Society of America Bulletin* 86: 1489.

Castro, J.R., O. Castillo, M.A. Sanchez, O. Mendoza, A. Rodríguez-Díaz, and P. Melin. 2016. Method for higher order polynomial sugeno fuzzy inference systems. *Information Science* 351: 76–89.

Champernowne, D.G. 1953. A model of income distribution. *Economic Journal* 63 (250): 318–351.

Chiu, S.L. 1994. Fuzzy model identification based on cluster estimation. *Journal of Intelligent & Fuzzy Systems* 2 (3): 267–278.

Coccia, M. 2018. New directions in measurement of economic growth, development and under development. *Journal of Economics and Political Economy - JEPE* 4 (4): 382–395.

Cohn, D., S. Ghahramani, and M. Jordan. 1997. Active learning with mixture models. In *Multiple Model Approaches to Modeling and Control*, ed. R. Murray-Smith, and T. Johansen. London: Taylor and Francis.

De Robertis, A., and K. Williams. 2008. Weight-length relationships in fisheries studies. The standard allometric model should be applied with caution. *Transactions of the American Fisheries Society*, 137(3): 707–719.

Dechnik-Vázquez, Y.A., L. García-Barrios, N. Ramirez-Marcial, M. van Noordwijk, and A. Alayon-Gamboa. 2019. Assessment of browsed plants in a sub-tropical forest frontier by means of fuzzy inference. *Journal of Environmental Management* 236: 163–181.

Dreyer, O. 2001. Allometric scaling and central source systems. *Physical Review Letters*, 87(3)

Echavarria-Heeras, H., C. Leal-Ramirez, E. Villa-Diharce, and A. Montesinos-Lopez. 2019a. Examination of the effects of curvature in geometrical space on accuracy of scaling derived projectinos of plant biomass units: Applicationsto the assessment of average leaf biomass in Elgrass Shoots. *BioMed Research International*, 3613679: 1–23.

Echavarría-Heras, H., C. Leal-Ramírez, J.R. Castro-Rodríguez, E. Villa-Diharce, and O. Castillo. 2018. A takagi-sugeno-kang fuzzy model formalization of eelgrass leaf biomass allometry with application to the estimation of average biomass of leaves in shoots: Comparing the reproducibility strength of the present fuzzy and related crisp proxies. In *Fuzzy logic augmentation of neural and optimization algorithms*, ed. O. Castillo, P. Melin, and J. Kacprzyk, 329–362. Switzerland: Springer.

Echavarria-Heras, H., C. Leal-Ramirez, E. Villa-Diharce, and N. Cazarez-Castro. 2018. On the suitability of an allometric proxy for nondestructive estimation of average leaf dry weight in eelgrass shoots I: Sensitivity analysis and examination of the influences of data quality, analysis method, and sample size on precision. *Theoretical Biology and Medical Modelling* 15: 4.

Echavarria-Heras, H., C. Leal-Ramirez, E. Villa-Diharce, and J.R. Castro-Rodríguez. 2019. A generalized model of complex allometry I: Formal setup, identification procedures and applications to non-destructive estimation of plant biomass units. *Applied Science* 9: 1–42.

Echavarria-Heras, H.A., J.R. Castro-Rodriguez, C. Leal-Ramirez, and E. Villa-Diharce. 2020. Assessment of a Takagi–Sugeno-Kang fuzzy model assembly for examination of polyphasic loglinear allometry. *PeerJ* 8: e8173.

El Naschie, M.S. 2004. A review of E-infinity theory and the mass spectrum of high energy particle physics. *Chaos, Solitons & Fractals* 19: 209–236.

Eleveld, D.J., J.H. Proost, H. Vereecke, A.R. Absalom, E. Olofsen, J. Vuyk, and M.M.R.F. Struys. 2017. An allometric model of remifentanil pharmacokinetics and pharmacodynamics. *Anesthesiology* 126 (6): 1005–1018.

Emlen, D. 1996. Artificial selection on horn length-body size allometry in the horned beetle onthophagus acuminatus (Coleoptera: Scarabaeidae). *Evolution* 50 (3): 1219–1230.

Ertel, J.E., and E.B. Fowlkes. 1976. Some algorithms for linear spline and piecewise multiple linear regression. *Journal of the American Statistical Association* 71: 640–648.

Forbes, T.L., and G.R. López. 1989. Determination of critical periods in ontogenetic trajectories. *Functional Ecology* 3: 625–632.

Frankino, W.A., D.J. Emlen, and A.W. Shingleton. 2010. Experimental approaches to studying the evolution of animal form: The shape of things to come. In *Experimental evolution: Concepts, methods, and applications of selection experiments*, ed. T. Garland and M.R. Rose, 419–478. Berkeley: University of California Press.

Gan, M.T., M. Hanmandlu, and A.H. Tan. 2005. From Gaussian mixture model to additive fuzzy systems. *IEEE Transactions on Fuzzy Systems* 13 (3): 303–316.

Glazier, D., M. Powell, and T. Deptola. 2013. Body-size scaling of metabolic rate in the trilobite Eldredgeops rana. *Paleobiology* 39 (1): 109–122.

Gould, S.J. 1966. Allometry and size in ontogeny and phylogeny. *Biological Reviews* 41: 587–640.

Huxley, J.S. 1932. *Problems of relative growth*. London: Methuen.

Jang, J.S., C.T. Sun, and E. Mizutani. 1997. Neuro-Fuzzy and soft computing: A computational approach to learning and machine intel.

Ji-Huan, H., and L. Jun-Fang. 2009. Allometric scaling laws in biology and physics. *Chaos Solitons & Fractals*, 41(4).

Kalbfleisch, J.G. 1985. *Probability and statistical inference, volume 2: Statistical inference*, 2nd ed. Berlin: Spinger.

Kolokotrones, T., V. Savage, E.J. Deeds, and W. Fontana. 2010. Curvature in metabolic scaling. *Nature* 464: 753–756.

Kwak, H.S., H.G. Im, and E.B. Shim. 2016. A model for allometric scaling of mammalian metabolism with ambient heat loss. *Integrative Medicine Research* 5 (1): 30–36.

Larsen, P.M. 1980. Industrial applications of fuzzy logic control. *International Journal of Man-Machine Studies* 12: 3–10.

Lemaître, J.F., C. Vanpé, F. Plard, and J.M Gaillard. 2014. The allometry between secondary sexual traits and body size is nonlinear among cervids. *Biology Letters*, 10.

Lin, L.I.K. 1989. A concordance correlation coefficient to evaluate reproducibility. *Biometrics* 45: 255–268.

Liu, G., R. Li, J. He, W. Li, J. Lu, W. Long, P. Gao, G. Cai, and M. Tang. 2018. Scaling relation of earthquake seismic data. *Physica A: Statistical Mechanics and Its Applications* 492: 2092–2102.

Lovett, D., and D.L. Felder. 1989. Application of regression techniques to studies of relative growth in crustaceans. *Journal of Crustacean Biology* 9 (4): 529–539.

Lu, M., J.S. Caplan, J.D. Bakker, J.A. Langley, T.J. Ozdzer, B.G. Drake, and J.P. Megonigal. 2016. Allometry data and equations for coastal marsh plants. *Ecology* 97 (12): 3554.

MacLeod, C.D. 2010. Assessing the shape and topology of allometric relationships with body mass: A case study using testes mass allometry. *Methods in Ecology and Evolution* 1: 359–370.

MacLeod, C.D., and R.C. MacLeod. 2009. The relationship between body mass and relative investment in testes mass in amniotes and other vertebrates. *Oikos* 118: 903–916.

Macleod, C. 2014. Exploring and explaining complex allometric relationships: A case study on amniote testes mass allometry. *Systems* 2: 379–392.
Mahmood, I. 2013. Evaluation of sigmoidal maturation and allometric models: prediction of propofol clearance in neonates and infants. *American Journal of Therapeutics* 20 (1): 21–28.
Mamdani, E.H. 1977. Application of fuzzy logic to approximate reasoning using linguistic systems. *IEEE Transactions on Computers*, C-26(12): 1182–1191.
Maritan, A., R. Rigon, J.R. Banavar, and A. Rinaldo. 2002. Network allometry. *Geophysical Research Letters*, 29(11).
Mayhew, T.M. 2009. A stereological perspective on placental morphology in normal and complicated pregnancies. *Journal of Anatomy* 215: 77–90.
McBride, G.B. 2005. A proposal for strength-of-agreement criteria for Lin's concordance correlation coefficient. *NIWA client report: HAM2005–062*. Hamilton, New Zeeland: National Institute of Water & Atmospheric Research.
Montesinos-López, A., E. Villa-Diharce, H. Echavarría-Heras, and C. Leal-Ramirez. 2018. *Journal of Coastal Conservation* 23: 71–91.
Moore, B.R., M. Page-Sharp, J.R. Stoney, K. Ilett, J.D. Jago, and K.T. Batty. 2011. Pharmacokinetics, pharmacodynamics, and allometric scaling of chloroquine in a murine malaria model. *Antimicrobial Agents and Chemotherapy* 55 (8): 3899–3907.
Muggeo, V.M. 2003. Estimating regression models with unknown break-points. *Statistics in Medicine* 22: 3055–3071.
Muggeo, V.M. 2008. Segmented: An R package to fit regression models with broken-line relationships. *R News* 8 (1): 20–25.
Myhrvold, N.P. 2016. Dinosaur metabolism and the allometry of maximum growth rate. *PLoS ONE* 11 (11): e0163205.
Neukum, G., and B.A. Ivanov. 1994. Crater size distributions and impact probabilities on Earth from lunar, terrestrial planet, and asteroid cratering data. In *Hazards due to comets and asteroids*, ed. T. Gehrels, 359–416. Tucson, AZ: University of Arizona Press.
Newman, M.E.J. 2007. Power laws, Pareto distributions and Zipf's law. *Contemporary Physics* 46: 323–351.
Nijhout, H.F., D.E. Wheeler. 1996. Growth models of complex allometries in holometabolous insects. *The American Naturalist*, 148: 40–56.
Ofstad, E.G., I. Herfindal, E.J. Solberg, and B.E. Sæther. 2016. Home ranges, habitat and body mass: Simple correlates of home range size in ungulates. *Proceedings of the Royal Society B*, 283.
Packard, G.C. 2009. On the use of logarithmic transformations in allometric analyses. *Journal of Theoretical Biology* 257: 515–518.
Packard, G.C. 2012. Is non-loglinear allometry a statistical artifact? *Biological Journal of the Linnaean Society* 107 (4): 764–773.
Packard, G.C. 2013. Is logarithmic transformation necessary in allometry? *Biological Journal of the Linnean Society* 109: 476–486.
Packard, G.C. 2016. Relative growth by the elongated jaws of gars: A perspective on polyphasic loglinear allometry. *Journal of Experimental Zoology (Molecular and Developmental Evolution)* 326B: 168–175.
Packard, G.C. 2017a. Misconceptions about logarithmic transformation and the traditional allometric method. *Zoology* 123: 115–120.
Packard, G.C. 2017b. The essential role for graphs in allometric analysis. *Biological Journal of the Linnaean Society* 120: 468–473.
Packard, G.C., and G.F. Birchard. 2008. Traditional allometric analysis fails to provide a valid predictive model for mammalian metabolic rates. *Journal of Experimental Biology* 211: 3581–3587.
Packard, G.C, and J. Huxley. 2012. Uca pugnax and the allometric method. *Journal of Experimental Biology*, 215.

Palestrini, C., A. Rolando, and P. Laiolo. 2000. Allometric relationships and character evolution in Onthophagus taurus (Coleoptera: Scarabaeidae). *Canadian Journal of Zoology* 78: 1199–1206.

Paul, R.A., C.D. Smyser, C.E. Rogers, I. English, M. Wallendorf, D. Alexopoulos, and T.E. Inder. 2014. An allometric scaling relationship in the brain of preterm infants. *Annals of Clinical and Translational Neurology* 1 (11): 933–937.

Pouliquen, O. 1999. Scaling laws in granular flows down rough inclined planes. *Physics of Fluids* 11 (3): 542–548.

Ramírez-Ramírez, G., L. Ramírez-Avilés, F.J. Solorio-Sánchez, J.A. Navarro-Alberto, and J.M. Dupuy-Rada. 2019. Shifts in tree allometry in a tropical dry forest: Implications for above-ground biomass estimation. *Botanical Sciences* 97 (2): 167–179.

Rasmussen, T.D., and C.L. Tan. 1992. The allometry of behavioral development: Fitting sigmoid curves to ontogenetic data for use in interspecific allometric analyses. *Journal of Human Evolution* 23 (2): 159–181.

Rodríguez, J.M., E. Angón, MA. González, J. Perea, C. Barba, A. García. 2017. Allometric relationship and growth models of juveniles of Cichlasoma festae (Perciforme: Cichlidae), a freshwater species native in Ecuador. *Revista de Biología Tropical*, 65(3): 1185–1193.

Samaniego, H., and M.E. Moses. 2008. Cities as organisms: Allometric scaling of urban road networks. *Journal of Transport and Land Use* 1: 21–39.

Savage, V.M., J.F. Gillooly, W.H. Woodruff, G.B. West, and A.P. Allen. 2004. The predominance of quarter-power scaling in biology. *Functional Ecology* 18: 257–282.

Shingleton, A. 2010. Allometry: The study of biological scaling. *Nature Education Knowledge* 3 (10): 2.

Snell, O. 1892. Die Abhängigkeit des Hirngewichts von dem Körpergewicht und den geistigen Fähigkeiten. *Arch Psychiatr* 23 (2): 436–446.

Solana-Arellano, M.E., H.A. Echavarría-Heras, C. Leal-Ramírez, and K.S. Lee. 2014. The effect of parameter variability in the allometric projection of leaf growth rates for eelgrass (Zostera marina L.). *Latin American Journal of Aquatic Research*, 42(5): 1099–1108.

Sommerton, D.A. 1980. A computer technique for estimating the size at sexual maturity in crabs. *Canadian Journal of Fisheries and Aquatic Sciences* 37: 1488–1494.

Strauss, R.E., and J.S. Huxley. 1993. The study of allometry since Huxley. In: *Problems of relative growth, new edition*, ed. D.H. Thompson. Baltimore: Johns Hopkins University Press.

Sugeno, M., and G.T. Kang. 1988. Structure identification of fuzzy model. *Fuzzy Sets and Systems* 28: 15–33.

Sutherland, G.D., A.S. Harestad, K. Price, and K. Lertzman. 2000. Scaling of natal dispersal distances in terrestrial birds and mammals. *Conservation Ecology* 4: 16.

Takagi, T., and M. Sugeno. 1985. Fuzzy identifications of systems and its applications to modeling and control. *IEE Transactions on Systems, MAN and Cybernetics* 15 (1): 116–132.

Thompson, D'Arcy W. 1992. *On growth and form* (Canto ed.). Cambridge University Press.

Tidière, M., J.F. Lemaître, C. Pélabon, O. Gimenez, and J.M. Gaillard, 2017. Evolutionary allometry reveals a shift in selection pressure on male horn size. *Journal of Evolutionary Biology*, 30.

Tsuboi, M.W., B. Van Der, B.T. Kopperud, J. Erritzøe, K.L. Voje, A. Kotrschal, K.E. Yopak, S.P. Collin, A.N. Iwaniuk, and N. Kolm. 2018. Breakdown of brain–body allometry and the encephalization of birds and mammals. *Nature Ecology and Evolution* 2: 1492–1500.

Wang, L.-X., and J.M. Mendel. 1992. Fuzzy basis functions, universal approximation, and orthogonal least-squares learning. *IEEE Transactions Neural Networks*, 3(5): 807–814.

Wang, C., K. Allegaert, M.Y.M. Peeters, D. Tibboel, M. Danhof, and C.A.J. Knibbe. 2014. The allometric exponent for scaling clearance varies with age: a study on seven propofol datasets ranging from preterm neonates to adults. *British Journal of Clinical Pharmacology*, **77**(1).

West, G.B., and J.H. Brown. 2005. The origin of allometric scaling laws in biology from genomes to ecosystems: Towards a quantitative unifying theory of biological structure and organization. *Journal of Experimental Biology* 208: 1575–1592.

William, J. 1979. Coffey allometric growth in urban and regional social-economic systems. *Canadian Journal of Regional Science* II (1): 50–65.

Wolinsky, M.A., D.A. Edmonds, J. Martin, and C. Paola, 2010. Delta allometry: Growth laws for river deltas. Geophysical Research Letters, 37.

Ying, H. 1998. General SISO Takagi-Sugeno fuzzy systems with linear rule consequent are universal approximators. *IEEE Transactions on Fuzzy Systems* 6 (4): 582–587.

Zadeh, L.A. 1965. Fuzzy sets. *Information and Control* 8: 338–353.

Zeng, W.S., and S.Z. Tang. 2011a. Goodness evaluation and precision analysis of tree biomass equations. *Scientia Silvae Sinicae* 47: 106–113.

Zeng, W.S., and S.Z. Tang. 2011b. Bias correction in logarithmic regression and comparison with weighted regression for nonlinear models. *Nature Precedings* 24: 137–143.

Zeng, W.S., L.J. Zhang, X.Y. Chen, Z.C. Cheng, K.X. Ma, and L. ZhiHua. 2017. Construction of compatible and additive individual-tree biomass models for Pinus tabulaeformis in China. *Canadian Journal of Forest Research* 47: 467–475.

Zeng, W.S., H.R. Duo, X.D. Lei, X.Y. Chen, X.J. Wang, Y. Pu, and W.T. Zou. 2017. Individual tree biomass equations and growth models sensitive to climate variables for Larix spp China. *European Journal of Forest Research* 136 (2): 233–249.

Zimmerman, H.J. 1991. *Fuzzy set theory and its applications*, 2nd ed. Boston MA: Kluwer.

A New Hybrid Method Based on ACO and PSO with Fuzzy Dynamic Parameter Adaptation for Modular Neural Networks Optimization

Fevrier Valdez, Juan Carlos Vazquez, and Patricia Melin

Abstract Bio-inspired algorithms are metaheuristics that simulate insects or animal behavior for solving optimization problems. Ant colony Optimization (ACO) and Particle Swarm Optimization (PSO) are two of the main bio-inspired methods used to solve different kind of optimization problems. On the other hand, there are hybrid optimization methods consisting of the combination of two or more metaheuristics to solve a common set of problems. In this paper a new hy-brid method combining ACO and PSO is presented. This method is tested with the optimization of modular neural networks (MNNs) architectures. The MNN learns to classify images of human faces from the ORL face database. Another interesting contribution of this work is the implementation of fuzzy inference sys-tems to dynamically adjust the values of some parameters for the metaheuristics used in this paper. Therefore, in this paper, the main idea is to design a hybrid method to optimize the modular neural networks architectures. The achieved re-sults using the proposed hybrid method outperform the results with the individual optimization methods. The main contribution is the proposed new hybrid meta-heuristic that can solve complex problems more efficiently than ACO or PSO individually.

Keywords Particle swarm optimization · Ant colony optimization · Hybrid optimization · Fuzzy logic · Modular neuronal network · Face recognition

F. Valdez (✉) · J. C. Vazquez · P. Melin
Computer Science in the Graduate Division, Tijuana Institute of Technology, Tijuana, BC, Mexico
e-mail: fevrier@tectijuana.mx

J. C. Vazquez
e-mail: juancarlos_jcs27@hotmail.com

P. Melin
e-mail: pmelin@tectijuana.mx

O. Castillo and P. Melin (eds.), *Fuzzy Logic Hybrid Extensions of Neural and Optimization Algorithms: Theory and Applications*, Studies in Computational Intelligence 940,
https://doi.org/10.1007/978-3-030-68776-2_20

1 Introduction

Ant colony Optimization (ACO) (Dorigo and Gambardella 1997), Particle Swarm Optimization (PSO) (Eberhart and Kennedy 1995), Bee Colony Algorithm (BCA) (Karaboga and Basturk 2007), Firefly Algorithm (FA) (Yang 2009), Cuckoo Search (CS) (Deb 2010), Bat Algorithm (BA) (Yang 2010), Grey Wolf Optimization (GWO) (Rodríguez et al. 2017) among others are bio-inspired optimization methods, whose robustness and ability of providing multiple solutions to a problem make them a powerful tool. This paper discusses how bio-inspired techniques are applied to find the optimal values for modular neural networks (MNNs) parame-ters. An Artificial Neural Network (ANN) is a biologically inspired computer program that replicates the nervous system of the human brain through mathematical models. A MNN similar to an ANN has the ability to learn interactively from the environment and can also show the ability to remember (Agatonovic-Kustrin and Beresford 2000).

The MNNs are based on the principle of divide and conquer, where a complex computational problem can be subdivided into smaller sub problems. Modular neural networks can be considered as a competitor of a conventional artificial neural network, but with more advantages because they can be s as experts in each sub problem called module (Azam 2000).

Ensure that your abstract reads well and is grammatically correct (Dai and Cao 2015; Melin et al. 2012; Vázquez et al.; Sánchez and Melin 2014; Valdezv et al. 2014).

ACO was proposed by Dorigo et al. (1999), Dorigo and Stützle (2004) and it is a population-based swarm intel-ligence algorithm. This algorithm has been inspired by the real ant colonies behavior and was originally designed for the traveling salesman problem (TSP) (Gambardella et al. 1995; Escario and Giron-Sierra 2015) and have since been extended to solve many types of optimization problems such as op-timal parameters for fuzzy control systems (Chen et al. 2017; Arnay et al. 2017; Wang et al. 2015). For this paper a version of ACO called Ant Colony System (ACS) is applied.

PSO was proposed by Kennedy, Eberhart and Shi, bird flocking or fish schooling was the inspiration as a social behavior of animals for this algorithm. The PSO has been successfully applied to different optimization problems where optimal architectures and parameters of artificial neural networks (Pulido and Melin 2016; Gaxiola et al. 2016) and Fuzzy Control System (Zadeh 1988; Vázquez and Valdez 2013; Hamada and Hassan 2018; Wang et al. 2017) are found.

A hybrid algorithm combines two or more algorithms or techniques that solve the same problem. This is generally performed to combine the advantages of each one, so that the hybrid algorithm is better than the individual algorithms. So, this paper de-scribes a hybrid method for the optimization of MNNs, which is based on the ACO algorithm and is hybridized with the PSO algorithm. In this case, the algorithms are switching between them over the course of our proposed method. In this paper, fuzzy logic is applied on these bio-inspired algorithms (Valdez et al. 2014; Vázquez and Valdez 2013; Bououden et al. 2015; Melin et al. 2013) for a dynamic ad-justing of some parameters on both algorithms (ACO and PSO); these parameters are the inertia weight (w) on PSO and evaporation local and global pheromone on ACO. A comparison of results between algorithms is performed.

This paper is organized as follows: In Sect. 2 the proposed hybrid optimization is described, Sect. 3 describes the neural networks, application and database to test the proposed method, Sect. 4 presents simulation results of different neural networks architectures, and conclusions are summarized in Sect. 5.

2 Proposed Method

In this paper two bio-inspired optimization algorithms (ACO and PSO) are used. The ACO algorithm was originally designed to solve route optimization problems, unlike of the PSO algorithm was designed for different optimization problems. For this work, the ACO algorithm is adapted to estimate parameters of modular neural networks. Fuzzy inference systems are designed to adjust their parameters for these algorithms.

2.1 Ant System and ACO Algorithm

The real behavior of ants has allowed to create some optimization algorithms based on it. These algorithms have been applied to find optimize solutions to mathematical problems, architectures and parameters. ACO was mainly created to solve discrete op-timization problems.

One of the most important variations of ACO is the Ant System (AS). In this kind of algorithm, the tour construction starts when the ants (m) are established on random nodes. At each step, each ant (k) uses a random proportional rule to calculate a prob-abilistic action to decide the next node to visit. An ant who currently is in the node i, calculate the probability to visit the node j using

$$P_{ij}^{k} = \frac{[\tau_{ij}]^{\alpha}[n_{ij}]^{\beta}}{\sum_{l \in N_i^k}[\tau_{ij}]^{\alpha}[n_{ij}]^{\beta}} \text{ if j } \in N_i^k \quad (1)$$

where $\eta_{ij}= 1_{dij}$ is a priori value, the relative influence of the pheromone trail and the heuristic information is represented by α and β respectively. N^k is the set of nodes that the ant κ has not visited. M^k represents the memory of an ant κ, this memory contains the order of nodes already visited allowing to define the feasible neighborhood in the construction rule given by Eq. (1). This memory also allows the ant κ to know the length of the graph T^k to return the route to deposit pheromone.

When all the ants have their tours, the next step is to update the pheromone trail. First a constant factor is used to decrease the pheromone value on all the arcs, and then in the arcs that have been traversed by the ants the pheromone is added. Pheromone evaporation is calculated by Eq. (2):

$$\tau_{ij} \leftarrow (1-\rho)\tau_{ij,}\forall(i,j)\in L \tag{2}$$

where ρ is the pheromone evaporation rate with value between [0, 1]. This parame-ter allows preventing accumulation of the pheromone trails, and this allows forgetting bad decisions previously taken. After evaporation, in all the arcs traversed by the ants deposit pheromone calculated by Eq. (3):

$$\tau_{ij} \leftarrow \tau_{ij} + \sum\nolimits_{k=1}^{m} \tau_{ij}, \forall(i,j)\in L \tag{3}$$

where C^k is the length of the graph T^k of the ant κ. When the arcs visited belong to the best tour of an ant, more pheromone is deposited on them. This allows to have more probability of being visited in future iterations.

As mentioned above, there are different variants of ACO, and for this work Ant Colony System (ACS) has been considered.

In ACS, when the global pheromone must be updated, the best ant in the current iteration is the only one able to add pheromone. This update is calculated by Eq. (4):

$$\tau_{ij} \leftarrow (1-\rho)\tau_{ij} + \rho\, O\, \tau_{ij}^{bs} \qquad \forall(i,j)\in T^{bs}, \tag{4}$$

The pheromone trail update in ACS only applies this update to the arcs of T*bs*, unlike of AS that applies this to all the arcs. Also in ACS, for the local pheromone trail update the ants when cross an arc (i, j) use a local pheromone update rule given by Eq. (5):

$$\tau_{ij} \leftarrow (1-\xi)\tau_{ij} + \xi\,\tau_0 \tag{5}$$

where ξ is a value between [0, 1]. The value of τ_0 is set to be the same as the initial value for the pheromone trails.

Fuzzy adaptation for ACO. For the fuzzy dynamic parameter adaptation of global and local pheromone evaporation, two fuzzy inference systems were developed. Both fuzzy inference systems have 3 triangular membership functions respectively called LOW, MEDIUM and HIGH in all their linguistic variables are shown in Fig. 1.

Fuzzy inference system for global pheromone evaporation. For the global pheromone trail update Eq. (6), a fuzzy inference system for pheromone evaporation is de-fined, this FIS consists the following components:

Two inputs, the first one represents the increment of pheromone trail, and the second one for pheromone trail. One output to represent the change in pheromone evaporation. Figure 2 illustrates the Fuzzy System to obtain the change in pheromone evapora-tion.

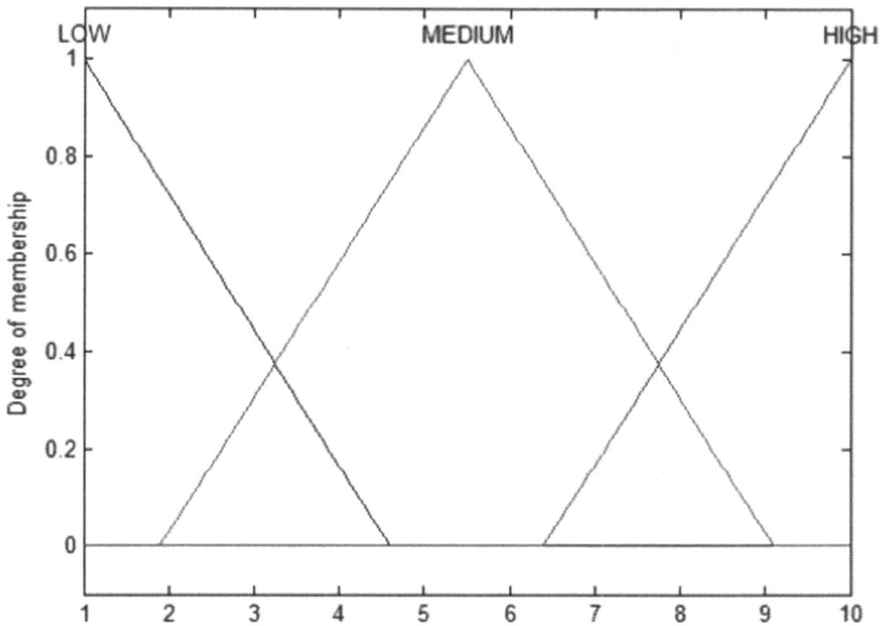

Fig. 1 Membership functions of the fuzzy inference system

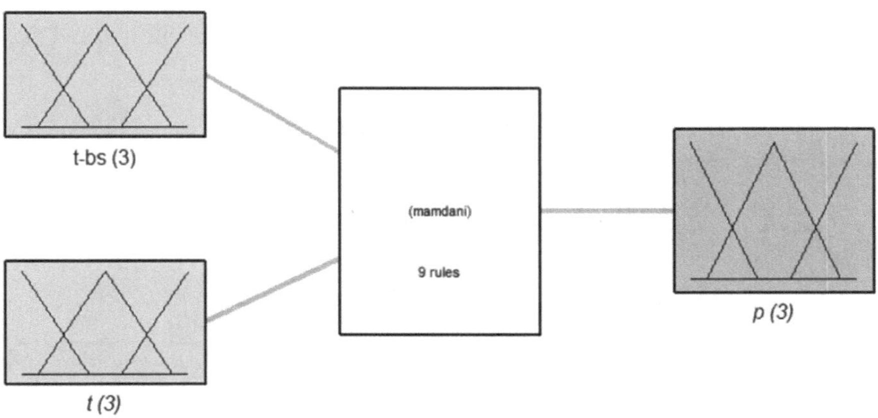

Fig. 2 Fuzzy system to adjust the pheromone ρ

Nine fuzzy if-then rules are considered to establish a new ρ value. The fuzzy if-then rules have been designed by experimental knowledge, and they are shown in Table 1.

The range of P and τ are normalized into [0, 1].

Table 1 Fuzzy if-then rules of the FIS (ACO)

Rules	Input		Output
	$\Delta\tau^{bs}$	τ	ρ
1	Low	Low	Low
2	Low	Medium	Medium
3	Low	High	High
4	Medium	Low	Low
5	Medium	Medium	Medium
6	Medium	High	High
7	High	Low	Low
8	High	Medium	Medium
9	High	High	High

Fuzzy inference system for local pheromone evaporation. For the local pheromone trail update Eq. (7), a fuzzy inference system is also proposed for pheromone evaporation, and this FIS has the following characteristics.

Two inputs, the first one represents the maximum probability of choosing the next node, and the second one represents pheromone trail. The output represents the change in pheromone evaporation.

Figure 3 illustrates the Fuzzy Inference System to obtain the change in pheromone evaporation.

Nine fuzzy if-then rules are considered from which the changes are calculated. The fuzzy if-then rules have been elaborated by experimental knowledge, and they are shown in Table 2.

The range of P and τ are normalized into [0, 1].

The flowchart of the ACO algorithm is shown in Fig. 4, where the implementation of both fuzzy inference systems is presented.

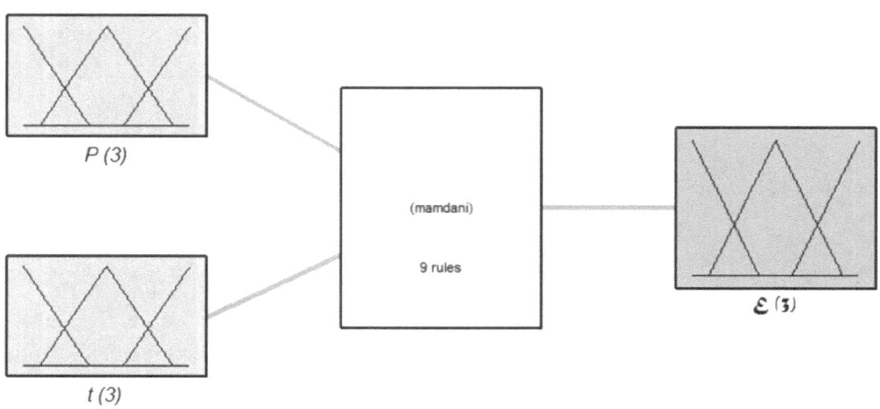

Fig. 3 Fuzzy system to adjust the pheromone evaporation ξ

Table 2 Fuzzy if-then rules of FIS #2 (ACO)

Rules	Input		Output
	P	τ	ξ
1	Low	Low	High
2	Low	Medium	Medium
3	Low	High	High
4	Medium	Low	Medium
5	Medium	Medium	Medium
6	Medium	High	Medium
7	High	Low	Low
8	High	Medium	Medium
9	High	High	High

2.2 *Particle Swarm Optimization*

PSO was proposed in Eberhart and Kennedy (1995), Kennedy and Eberhart (1995) and in this algorithm also each particle is a potential solution. The particles fly through a multidimensional search space, and the position of each particle is determined according to its own experience and its neighboring parti-cles. The position of a particle i at time step t is denoted by $x_i(t)$. This position changes by adding a velocity $v_i(t+1)$ to the current position, calculated by Eq. (6):

$$x(t+1) = x_i(t) + v_i(t+1) \tag{6}$$

This is the velocity vector and gives information about the experimental knowledge and socially exchange information from the neighboring particles. Originally, there are two PSO algorithms developed, the main difference is their neighborhoods size; the gbest (global best) and lbest (local best) PSO.

In this work, the gbest PSO variant is used. In this PSO variant the whole swarm is the neighborhood of a particle (Vázquez and Valdez 2013). The best position found by the swarm defines the social information and it is defined as $y(t)$. In PSO, each particle can move in a j dimensional space. In gbest PSO, the velocity of particle $i$is defined by Eq. (7):

$$\begin{aligned} v_{ij}(t+1) = v_{ij}(t) + c_1 r_{1j}(t)[y_{ij}(t) - x_{ij}(t)] \\ + c_2 r_{2j}(t)[\hat{y}_j(t) - x_{ij}(t)] \end{aligned} \tag{7}$$

where $v_{ij}(t)$ is $\hat{y}_j$ the velocity of particle i in dimension $j = 1,\ldots,n_x$ at time step t, is the position of particle i in dimension j at time step t, the personal best position, $y_{ij}(t)$, associated with particle i in dimension j is the best position the particle has visited since the first time step, y $(t)$is the global best position obtained so far b y any

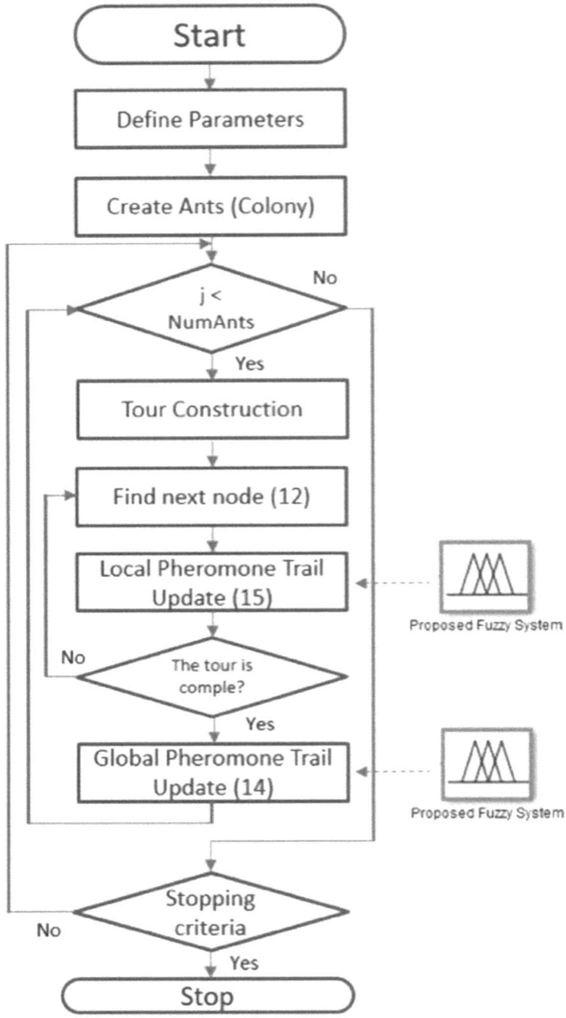

Fig. 4 The flowchart of the ACO algorithm

particle, c_1 and c_2 are respectively cognitive and social components with constant values, and $r_{1j}(t)$, $r_{2j}(t)$ ~ U (0,1) are random values between [0, 1].

In Eberhart and Shi (2000), a method to restrict the exploration and exploitation abilities was proposed. In this case, *inertia weight* is used to control the speed and direction of particles, the velocity equation changes to Eq. (8):

$$v_{ij}(t+1) = w \times v_{ij}(t) + c_1 r_{1j}(t)[y_{ij}(t) - x_{ij}(t)] \\ + c_2 r_{2j}(t)[\hat{y}_j(t) - x_{ij}(t)] \quad (8)$$

Fuzzy inference system for c_1, c_2 and w The fuzzy inference system is implemented over the velocity Eq. (10) from the PSO algorithm, where the inertia weight w, and learning factors c_1 and c_2 are adjusted.

The learning factors c_1 and c_2 have an important role in the velocity of a particle (Kennedy and Eberhart 1995), for this reason their values are very important, usually, their values are fixed when the algorithm starts. To establish inertia and learning factors values a fuzzy inference system was developed with the following characteristics.

Two inputs, one input represents the number of iterations *(NU)* that the best fitness has without changing its value, and the other input represents the current value of the inertia weight *(w)*. Three outputs, the first one represents the change in inertial weight *(chw)*, and the others the change in learning factors (c_1 and c_2). Each linguistic variable has 3 triangular membership function respectively called LOW, MEDIUM and HIGH.

Figure 5 illustrates the 3 membership functions for each linguistic variable of the proposed fuzzy inference system.

Figure 6 shows the fuzzy inference system with two inputs: the number of iterations that the best fitness has without changing its value *(NU)* and current inertia weight w and three outputs: the change in inertia weight *(chw)* and learning factors *(c_1 and c_2)*.

— Nine fuzzy if-then rules are developed. These fuzzy if-then rules have been ob-tained by experimental knowledge, and they are summarized in Table 3.

The range of *NU* is in [1, 10], the value for w is [0.2, 1.2] and the values of c_1 and c_2 are in [1, 2]. The mathematical description for c_1 and c_2 is defined in Eqs. 9 and 10:

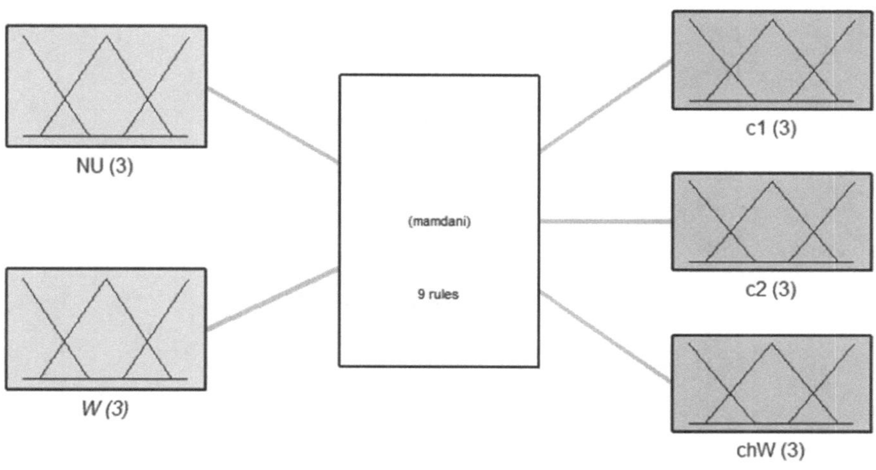

Fig. 5 Membership functions of the fuzzy system

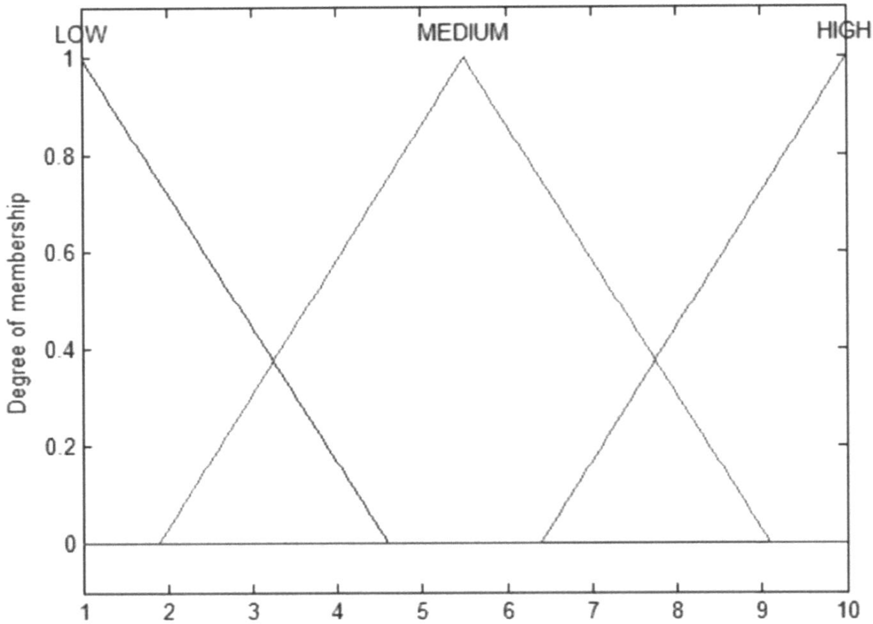

Fig. 6 Fuzzy system to adjust the *chw* and *c*1 and c2 parameters

Table 3 Fuzzy rules of the FIS for PSO

Rules	Input		Output		
	NU	*w*	*chw*	c_1	c_2
1	Low	Low	High	Low	Low
2	Low	Medium	Low	High	High
3	Low	High	Low	High	High
4	Medium	Low	High	Low	Low
5	Medium	Medium	Medium	Low	High
6	Medium	High	Low	High	High
7	High	Low	High	Low	Low
8	High	Medium	Low	High	High
9	High	High	Low	High	High

$$c_1 = \frac{\sum_{i=1}^{r_{c_1}} \mu_i^{c_1}(c_{1i})}{\sum_{i=1}^{r_{c_1}} \mu_i^{c_1}} \quad (9)$$

where c_1 represents the cognitive acceleration of the particle *i*, r_{c1} represents the num-ber of fuzzy if-then rules corresponding to c_1, c_{1i} represents the output result for the rule *i* corresponding to c_1 and $\mu_i^{c_1}$ represents the membership function of the rule *i* corresponding to c_2:

$$c_2 = \frac{\sum_{i=1}^{r_{c2}} \mu_i^{c_2}(c_{2i})}{\sum_{i=1}^{r_{c2}} \mu_i^{c_2}} \quad (10)$$

where c_2 represents the cognitive acceleration of the particle *i*, r_{c2} represents the number of fuzzy if-then rules corresponding to c_2. c_{2i} represents the output result for the rule *i* corresponding to c_2 and $\mu_i^{c_2}$ represents the membership function of rule corresponding to c_2.

The flowchart of the PSO algorithm is shown in Fig. 7, where the implementation of the fuzzy inference system is presented.

2.3 Hybrid Proposed Method

In this paper, both algorithms (ACO and PSO) are being switch according to which is having better results over the course of the iterations. The flowchart of the hybrid proposed method is shown in Fig. 8.

The process is summarized as follows:

Parameter are defined as follows: For the PSO algorithm, create a particle swarm. For the ACO algorithm, create an ant colony in each iteration: For the PSO and ACO algorithms, the evaluation fitness is performed. Dynamic adjustment of parameters: For the PSO algorithm from Eq. 10, if the fitness is higher with ACO algorithm. For the ACO algorithm from Eqs. 6 and 7, if the fitness is higher with PSO algorithm. Repeat steps 2–3 until find the objective function or when a maximum number of iterations is achieved.

3 The Neural Network Architectures and Representation

In this section the face database, the feature extraction applied to each image, the neural networks architectures and their representation for the optimization are described.

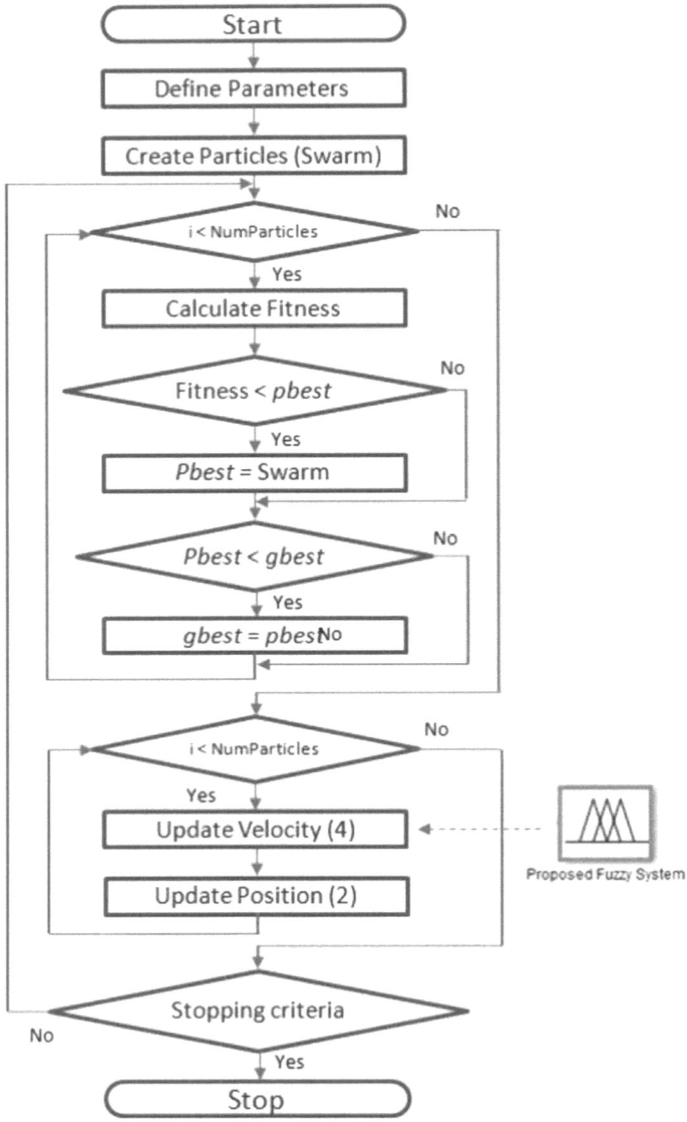

Fig. 7 The flowchart of the PSO algorithm

3.1 Face Database

The ORL Database (AT&T Laboratories Cambridge 2017; Samaria and Harter 1994) of faces was used to test efficiency of the proposed method. This database consists of 10 images of 40 distinct persons. Each image has a size of 92 * 112 pixels. This

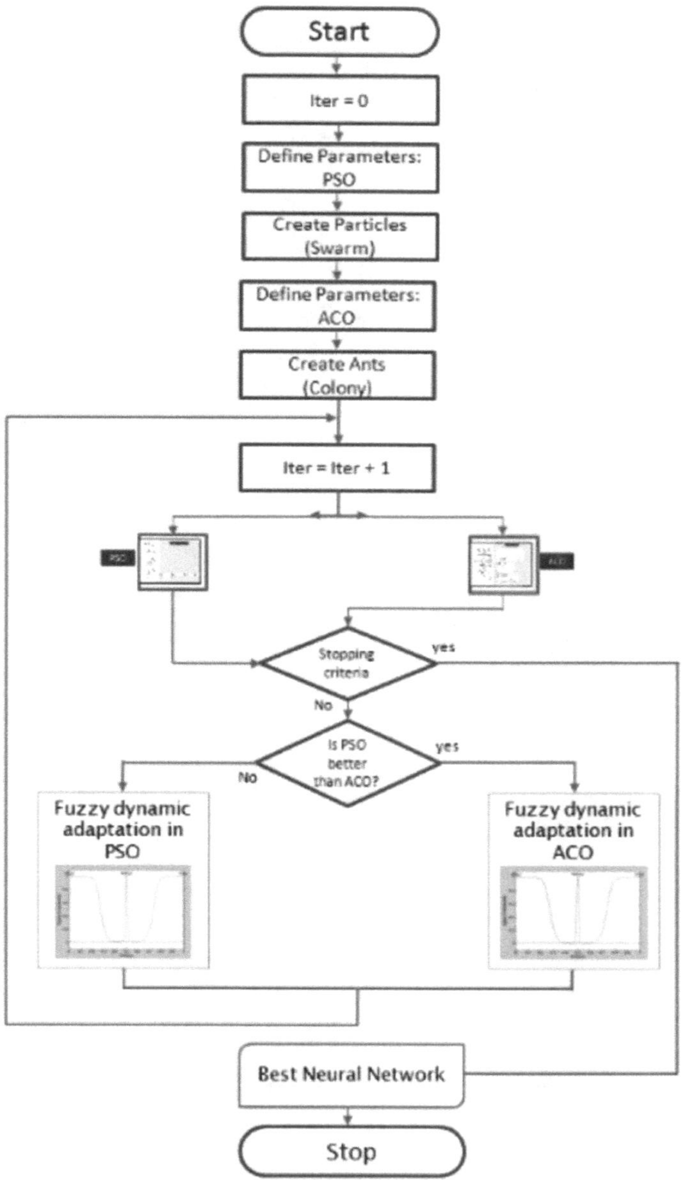

Fig. 8 The flowchart of the hybrid proposed method

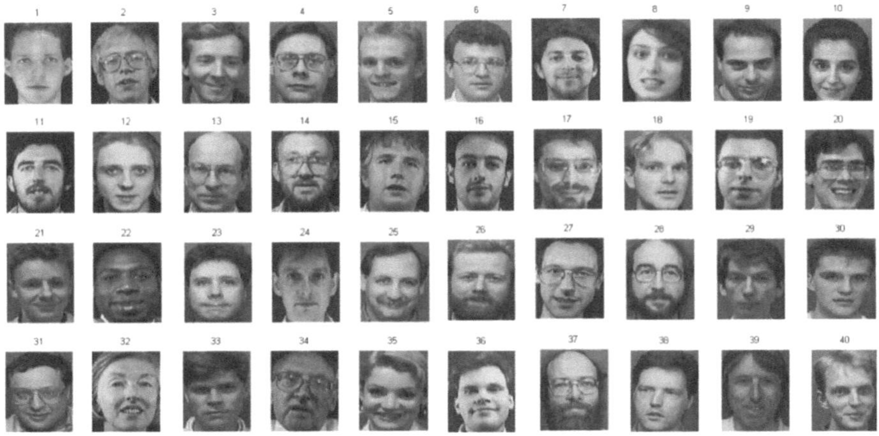

Fig. 9 Original images of persons

Fig. 10 Original images of one person

database contains images taken with variations of lighting, different facial expressions and facial details (with glasses or without glasses). Figure 9 shows examples of 40 images of people's faces before preprocessing.

Figure 10 shows examples of 10 images of one person. Four variations of the face database were created for each neural network architecture (described in Sect. 3.4).

3.2 *Local Binary Pattern*

In this work a feature extraction method called local binary pattern (LBP) (Swiderska-Chadaj et al. 2016; Ojala and Harwood 1996) is implemented for the face database. This feature extraction transforms an image into an image of integer labels describing small scale appearance of the image. The histogram is used for image analysis. Usually this feature extraction has been used for monochromatic images, but also can be used for color images and videos. In Fig. 11, the original image and its corresponding LBP image are presented.

Four neural networks were designed, one conventional neural network and three MNNs, which are described below.

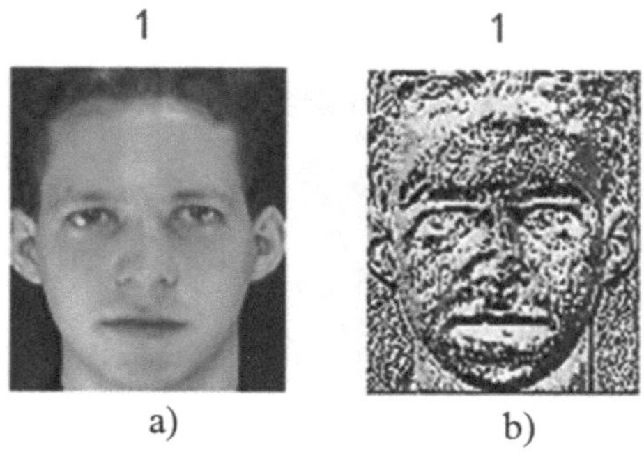

Fig. 11 Processing, **a** Original image to **b** LBP image

3.3 Neural Network Representation for Optimization

The representation of the neural network optimization using an ACO algorithm can be found in Fig. 12, where in the pheromone matrix all the possible solutions are represented. The first node plays the role of the nest and the food source is represented by the last node. Each ant constructs its route/tour from first to last node, and this route is used to set the parameters of the neural network. In this paper, the neural network optimization i s epresented by 7 nodes per n modules.

The representation of the neural network optimization using a PSO algorithm is shown in Fig. 13, where each particle is divided into sections. These sections are obtained multiplying the number of modules and the number of parameters used in each module.

Table 4 shows the search space for the optimization of neural networks architectures. For each ant and particle, its fitness is evaluated (recognition rate) according to the next expression:

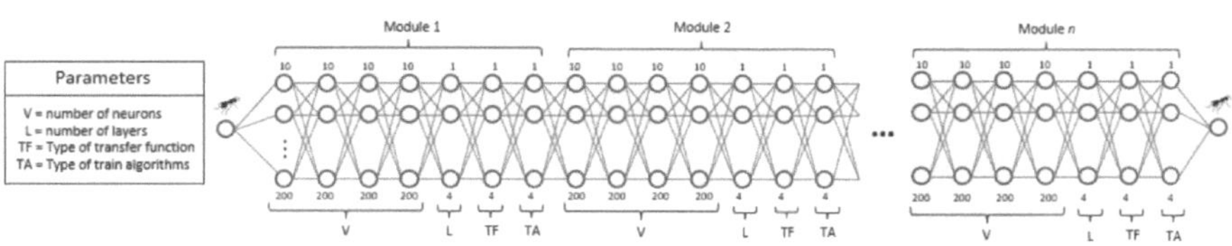

Fig. 12 Representation of ACO for the neural network optimization

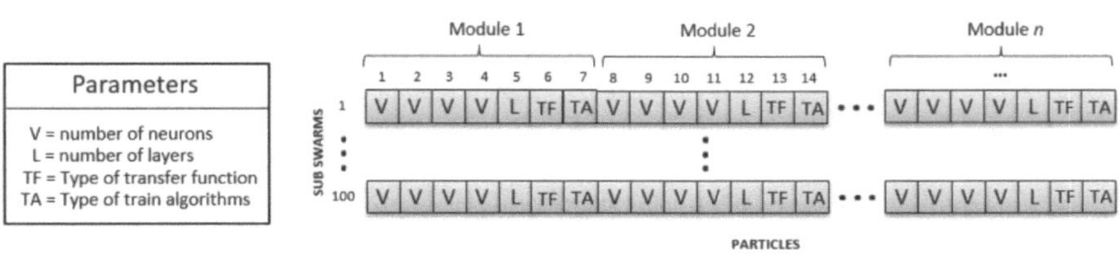

Fig. 13 Representation of PSO for the neural network optimization

Table 4 Optimization parameters

Parameters	Search space
Number of neurons from each layer	10–200
Number of layers	1–4
Type of transfer function	Logsig, tansig, purelin, satlins
Type of train algorithm	Trainscg, traingdx, traingda, trainlm

$$\text{Recognition\_rate} = (\text{number\_ of\_ persons\_ recognized} * 100)/ (\text{number\_ of\_ persons} * \text{imagen\_ to\_ recognotion})$$

The ACO algorithm uses 100 ants and the PSO algorithm 100 particles, and both algorithms use 100 iterations to minimize the fitness function.

3.4 Neural Network Architectures

After the LBP operator is applied, the database is used as input data to the neural net-work. In general, 7 images of each person of the ORL Database are used to training phase (280 images), 3 images of each person are used to recognition process (120 im-ages), and the method called "the winner takes all" is applied to integrate modules responses. The 4 different neural networks architectures used to optimize are described below.

Conventional artificial neural network Figure 14 shows the conventional artificial neural network architecture. In this architecture of the neural network the complete images are used as input data to a conventional artificial neural network.

Modular neural network (2 modules) Figure 15 shows the MNN architecture, where each image is divided into 2 parts and each part is respectively learned by a module.

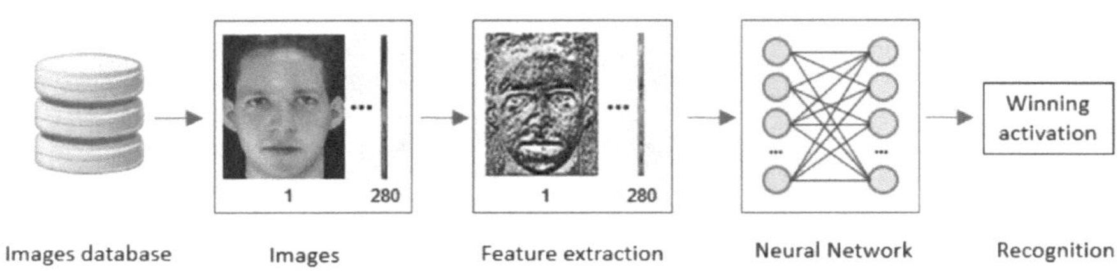

Fig. 14 Process of a conventional artificial neural network

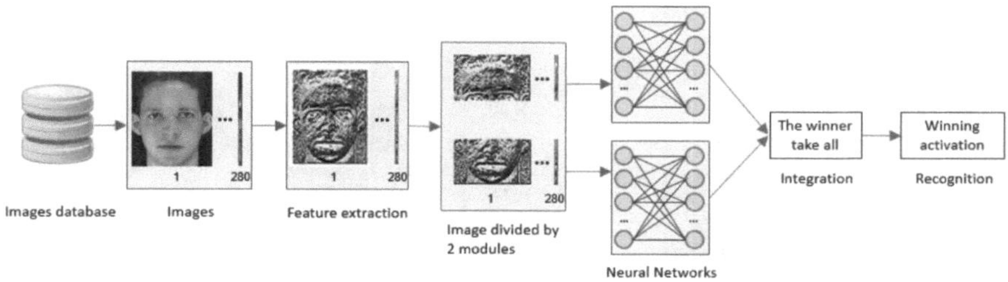

Fig. 15 Process of a modular neural network of two modules

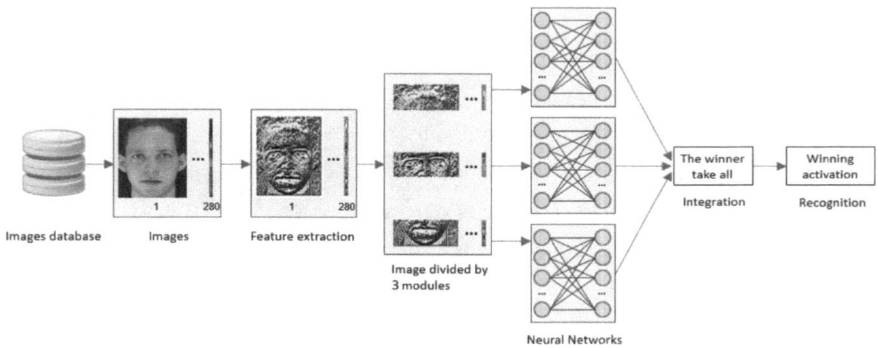

Fig. 16 Process of a modular neural network of 3 modules

Modular neural network (3 modules) Figure 16 shows the MNN architecture, where each image is divided into 3 parts (front, eyes and mouth) and each part is respectively learned by a module.

Modular neural network Figure 17 shows the MNN architecture, where each image is divided into 4 parts and each part is respectively learned by a module.

4 Simulation Results

In this section, the results of the comparison among ACO, PSO and the hybrid proposed method for the optimizations of different neural networks architectures (conventional and modular neural network) are presented. Cross validation is applied to estimate the accurately of each algorithm.

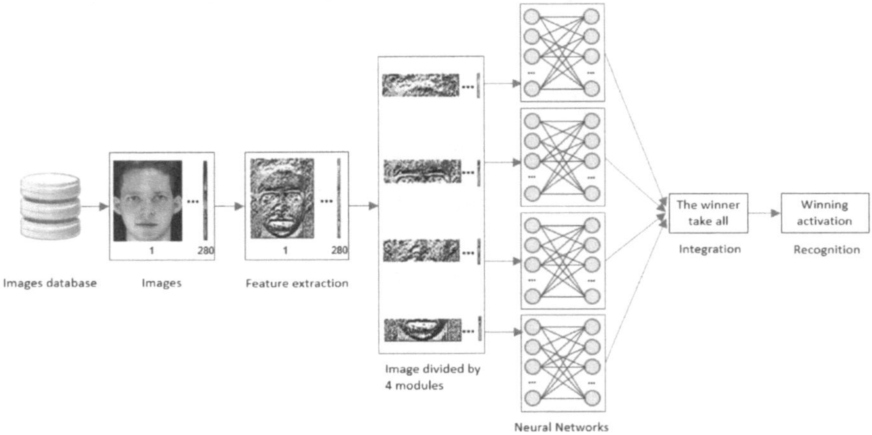

Fig. 17 Example of a modular neural network with 4 modules

4.1 Simulation Results for an Artificial Neural Network

The results of the optimization of ACO, PSO and the proposed hybrid method for a conventional artificial neural network are shown in Table 5. As it can be observed, the proposed hybrid method is better than the single algorithms.

In Fig. 18, these results are graphically represented.

Table 5 Simulation results (artificial neural network)

Recognition rate			
Cross validation	PSO	ACO	Hybrid optimization
1	75.2	65.7	82.3
2	74.9	71.4	81.6
3	79.3	73.4	83.6
4	75.9	63.6	78.8
5	76.8	69.6	84.2
6	76.3	72.4	80.1
7	75.3	70.8	78.8
8	78.5	70.1	84.3
9	75.6	66.4	80.3
10	79.2	68.6	80.7
Average	76.8	69.2	81.5

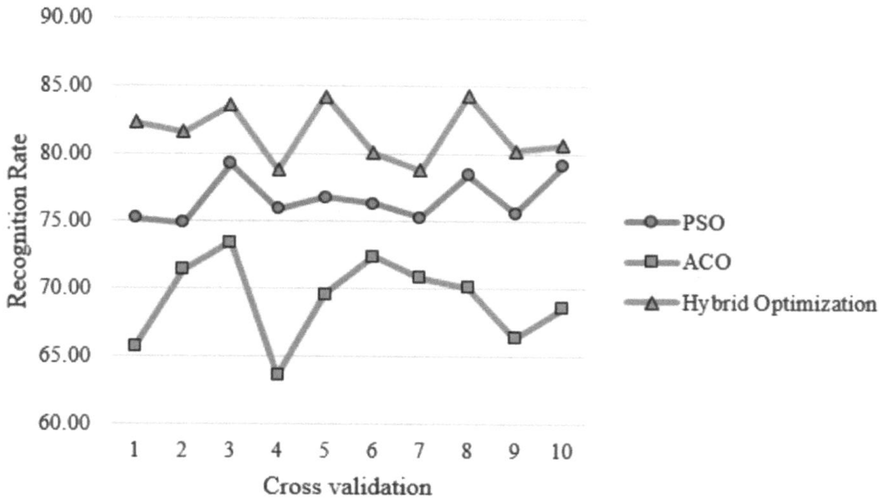

Fig. 18 Artificial neural network results

4.2 *Simulation Results for a MNN of 2 Modules*

Table 6 shows the results of the optimization of ACO, PSO and the proposed hybrid method for a modular neural network with 2 modules. In Table 6, can be noted that the proposed hybrid method is better than the single algorithms. In Fig. 19, these results are graphically represented.

Table 6 Simulation results (modular neural network, 2 modules)

Recognition rate			
Cross validation	PSO	ACO	Hybrid optimization
1	76.9	70.6	83.9
2	75.2	71.9	83.9
3	78.2	74.4	84.6
4	77.6	65.3	77.9
5	78.3	69.5	84.9
6	76.5	73.6	79.3
7	75.4	72.4	81.1
8	76.3	63.3	83.2
9	77.5	62.4	78.1
10	78.7	72.9	84.3
Average	77.06	69.7	82.1

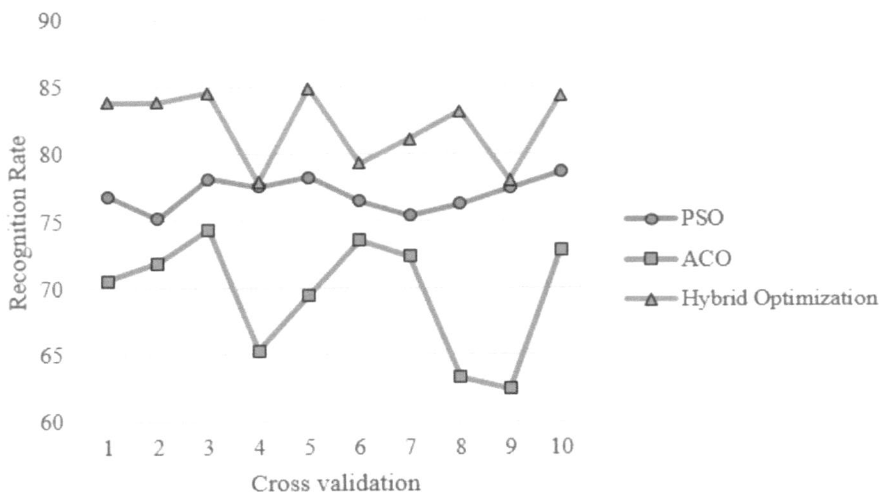

Fig. 19 Modular neural network with 2 modules

4.3 *Simulation Results for a MNN of 3 Modules*

Table 7 shows the results of the optimization of ACO, PSO and the proposed hybrid method for a modular neural network. It can be observed that the proposed hybrid method is better than the single algorithms. In Fig. 20, these results are graphically represented.

Table 7 Simulation results (modular neural network, 3 modules)

Recognition rate			
Cross validation	PSO	ACO	Hybrid optimization
1	78.4	70.3	85.1
2	76.3	68.4	79.4
3	80.6	72.7	85.7
4	81.3	76.4	78.7
5	79.2	73.9	86.2
6	80.8	70.1	78.4
7	78.5	66.9	83.9
8	81.6	71.9	85.8
9	78.4	65.4	82.4
10	80.2	69.3	82.9
Average	79.53	70.53	82.85

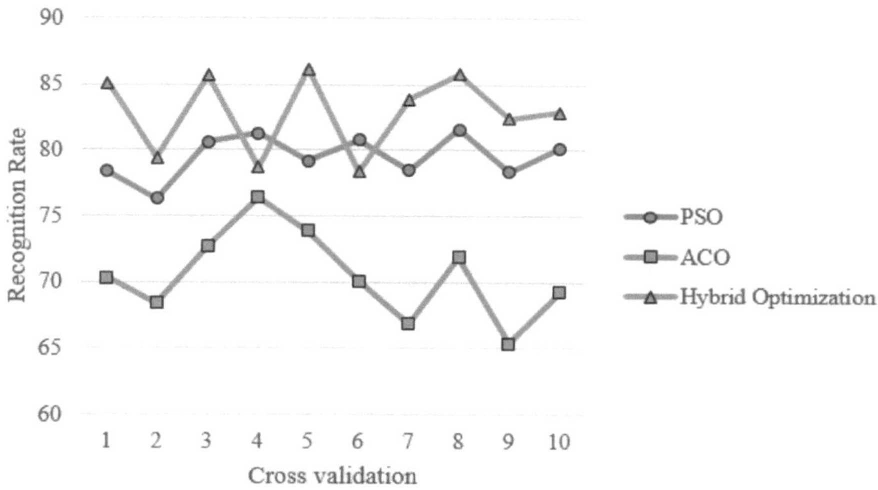

Fig. 20 Modular neural network with 3 modules

4.4 Simulation Results for a MNN with 4 Modules

Table 8 shows the results of the optimization of ACO, PSO and proposed hybrid method for a modular neural network. It can be observed that the proposed hybrid method is better than the single algorithms. In Fig. 21, these results are graphically represented.

Table 8 Simulation results (modular neural network, 4 modules

Recognition rate			
Cross validation	PSO	ACO	Hybrid optimization
1	80.7	75.3	85.6
2	78.2	70.2	84.4
3	79.4	76.2	86.8
4	82.3	80.3	85.3
5	83.4	72.9	86.4
6	82.5	74.4	85.7
7	79.8	65.6	83.6
8	82.1	73.3	85.1
9	76.3	66.4	83.3
10	83.2	71.6	86.2
Average	80.79	72.62	85.24

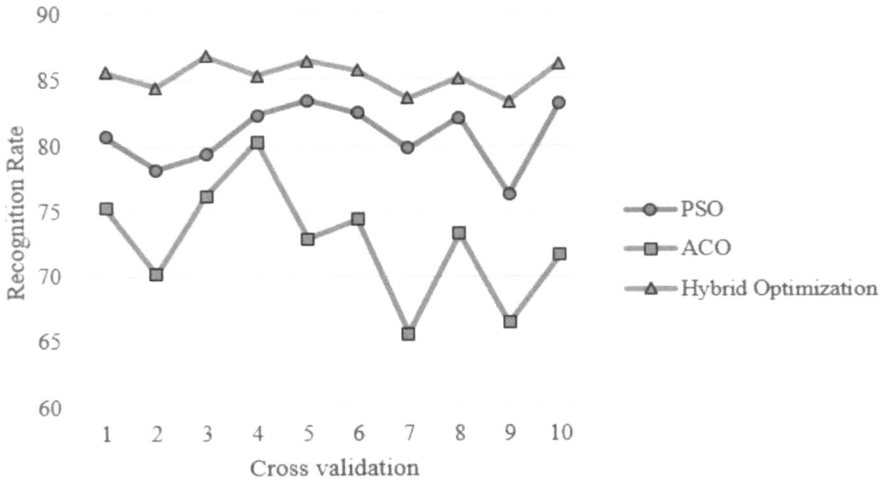

Fig. 21 Modular neural network with 4 modules

Table 9 Statistical test. T-values

Method	Architecture		
	ANN MNN (2 modules)	MNN (3 modules)	MNN (4 modules)
PSO versus hybrid optimization	5.63 5.29	3.03	5.38
ACO versus hybrid optimization	10.3 7.60	8.67	8.69

4.5 *Statistical Comparison*

As results and averages show, the proposed optimization allows obtaining better results. With the results achieved in this work, statistical comparisons were performed. In Table 9, the t-values of these comparisons (among ACO, PSO and the proposed hybrid opti-mization) are presented. In all the tests, the proposed optimization is better, especially more with respect to the ACO algorithm.

5 Conclusions

In this paper, Ant Colony Optimization (ACO), Particle Swarm Optimization (PSO) and a new hybrid method (ACO/PSO) are applied to optimize four neural network ar-chitectures, where each architecture is respectively divided from 1 to 4 modules (there are modular neural networks).

For the parameter adjustment, three fuzzy inference systems were designed. One fuzzy inference system was developed for the PSO algorithm to update the inertia weight (w), C_1 and C_2. For the ACO algorithm, two fuzzy inference system were designed to adjust local pheromone trail and global pheromone trail parameters. These algorithms (ACO and PSO) are used to create a hybrid optimization method. To test the proposed method, neural networks optimization is applied for face recognition, where each algorithm is implemented (ACO, PSO and proposed hybrid method). The hybrid method applies both algorithms where in each iteration depending on their fitness func-tion, the parameters of the worst algorithm are adjusted. The parameters to optimize of the neural network are the number of neurons for each layer, the number of layers, type of transfer function, and type of train algorithm.

To measure the performance (training and simulation) of neural networks, the im-ages of the of ORL database were used. The results of the optimization of ACO, PSO and the hybrid method are compared. The hybrid method shows better results in com-parison with the ACO and PSO algorithms. As future work, a different hybrid method can be considered.

Acknowledgements We would like to express our gratitude to CONACYT, and Tecnológico Nacional de México/Tijuana Institute of Technology for the facilities and resources granted for the development of this research.

References

Agatonovic-Kustrin, S., and R. Beresford. 2000. Basic concepts of artificial neural network (ANN) modeling and its application in pharmaceutical research. *Journal of Pharmaceutical and Biomedical Analysis* 22 (5): 717–727. https://doi.org/10.1016/S0731-7085(99)00272-1.

Arnay, R., F. Fumero, and J. Sigut. 2017. Ant Colony Optimization-based method for optic cup seg-mentation in retinal images. *Applied Soft Computing* 52: 409–417.

AT&T Laboratories Cambridge. 2017. The Database of Faces. https://www.cl.cam.ac.uk/research/dtg/attarchive/facedatabase.html. Accessed 2017.

Azam, F. 2000. Biologically inspired modular neural networks. Ph.D. dissertation, Virginia Polytechnic Institute and State University, Blacksburg, Virginia, May 2000.

Bououden, S., M. Chadli, and H.R. Karimi. 2015. An ant colony optimization-based fuzzy predictive control approach for nonlinear processes. *Information Sciences* 299: 143–158.

Chen, Z., S. Zhou, J. Luo, J. 2017. A robust ant colony optimization for continuous functions. *Expert Systems with Applications* 81: 309–320.

Dai, K., and J.Z.F. Cao. 2015. A novel decorrelated neural network ensemble algorithm for face recog-nition. *Knowledge-Based Systems* 89: 541–552.

Deb, X.S.Y.S. 2010. Engineering Optimisation by Cuckoo Search. *International Journal of Mathematical Modelling and Numerical Optimisation* 1 (4): 330–343.

Dorigo, M., and L.M. Gambardella. 1997. Ant colony system: a cooperative learning approach to the traveling salesman problem. *IEEE Transactions on Evolutionary Computation* 1(1): 53–66. https://doi.org/10.1109/4235.585892.

Dorigo, M., and T. Stützle. 2004. *Ant Colony Optimization*. Scituate, MA, USA: Bradford Company.

Dorigo, M., G. Caro, and L.M. Gambardella. 1999. Ant algorithms for discrete optimization. *Artificial Life* 5 (2): 137–172.

Eberhart, R., and J. Kennedy. 1995. A new optimizer using particle swarm theory. In *MHS'95. Proceedings of the sixth international symposium on micro machine and human science*, 39–43, Oct 1995.
Eberhart, R., and Y. Shi. 2000. Comparing inertia weights and constriction factors in particle swarm optimization. In *Congress on evolutionary computation.*
Escario, J.B., and J.F.J.J.M.Giron-Sierra. 2015. Ant colony extended: Experiments on the travelling salesman problem. *Expert Systems with Applications.*
Gambardella, L., M. Dorigo. 1995. Ant-Q: A reinforcement learning approach to the traveling salesman problem. In *Twelfth international conference on machine learning.*
Gaxiola, F., P. Melin, F. Valdez, J.R. Castro, and O. Castillo. 2016. Optimization of type-2 fuzzy weights in backpropagation learning for neural networks using GAs and PSO. *Applied Soft Computing* 38: 860–871.
Hamada, M., and M. Hassan. 2018. Artificial neural networks and particle swarm optimization algo-rithms for preference prediction in multi-criteria recommender systems. *Informatics* 5(2): 1–16.
Karaboga, D., and B. Basturk. 2007. A powerful and efficient algorithm for numerical function opti-mization: artificial bee colony (ABC) algorithm. *Global Optimization* 39: 459–471. https://doi.org/10.1007/s10898-007-9149-x.
Kennedy, J., and R.C. Eberhart. 1995. Particle swarm optimization. In *Proceedings of the 1995 IEEE international conference on neural networks*, 1942–1948.
Melin, P., F. Olivas, O. Castillo, F. Valdez, and J.S.M. Valdez. 2013. Optimal design of fuzzy classification systems using PSO with dynamic parameter adaptation through fuzzy logic. *Expert Systems with Applications.*
Melin, P., D. Sánchez, and O. Castillo. 2012. Genetic optimization of modular neural networks with fuzzy response integration for human recognition. *Information Sciences* 197: 1–19.
Ojala, T., and M.P.D. Harwood. 1996. A comparative study of texture measures with classification based on feature distributions. *Pattern Recognition* 29 (1): 51–59.
Pulido, M. and P. Melin (2016). Genetic algorithm and particle swarm optimization of ensemble neu-ral networks with type-1 and type-2 fuzzy integration for prediction of the Taiwan Stock Exchange. In *IEEE international conference on intelligent systems.*
Rodríguez, L., O. Castillo, J. Soria, P. Melin, F. Valdez, C.I. Gonzalez, G.E. Martinez, and J. Soto. 2017. A fuzzy hierarchical operator in the grey wolf optimizer algorithm. *Applied Soft Computing* 57: 315–328. https://doi.org/10.1016/j.asoc.2017.03.048.
Samaria, F.S., and A.C. Harter. 1994. Parameterisation of a stochastic model for human face identification. In *Proceedings of 1994 IEEE workshop on applications of computer vision*, 138–142, Dec 1994. https://doi.org/10.1109/acv.1994.341300.
Sánchez, D. and P. Melin. 2014. Optimization of modular granular neural networks using hierarchical genetic algorithms for human recognition using the ear biometric measure. *Engineering Applications of Artificial Intelligence* 27, 41–56.
Swiderska-Chadaj, Z., T. Markiewicz, B. Grala, and J. Slodkowska. 2016. Local binary patterns and unser texture descriptions to the fold detection on the whole slide images of menin-giomas and oligodendrogliomas. In *XIV mediterranean conference on medical and bio-logical engineering and computing 2016*, 388–392. Cham: Springer International Publishing.
Valdez, F., P. Melin, and O. Castillo. 2014. Modular Neural Networks architecture optimization with a new nature inspired method using a fuzzy combination of Particle Swarm Optimization and Genetic Algorithms. *Information Sciences* 270: 143–153.
Vázquez, J., and F. Valdez. 2013. Fuzzy logic for dynamic adaptation in PSO with multiple topologies. In *IFSA/NAFIPS.*
Wang, S., S. Yuan, M. Ma, and R.C. Luo. 2015. Wavelet phase estimation using ant colony optimization algorithm. *Journal of Applied Geophysics* 122: 159–166.
Wang, C., Z. Shi, and F. Wu. 2017. An improved particle swarm optimization-based feed-forward neural network combined with RFID sensors to indoor localization. *Information* 8 (9): 1–18.

Vázquez, J.C., M. Lopez and P. Melin. Real time face identification using neural networks approach. In *Soft computing for recognition based on biometrics*, vol. 312, 155–169.
Yang, X. 2009. Firefly algorithms for multimodal optimization. In O., W., T, Z. (eds.) *Lecture notes in computer science*, vol. 5792. Berlin, Heidelberg: Springer.
Yang X.S. 2010. A new metaheuristic bat-inspired algorithm. In *Studies in computational intelligence*, ed. J.R., G., D.A., P., C., C., G., T., N, K. vol. 284. Berlin, Heidelberg: Springer.
Zadeh, L. 1988. Fuzzy logic. *Computer* 21 (4): 83–93.

Knowledge Discovery Using an Evolutionary Algorithm and Compensatory Fuzzy Logic

Carlos Eric Llorente-Peralta, Laura Cruz-Reyes, and Rafael Alejandro Espín-Andrade

Abstract Database Knowledge Discovery attention has been growing last decades using different approaches as part of a new era when information is multiplied in proportion and importance. Fuzzy Logic predicates approach is one of them, fundamental because of their interpretability properties. A new concept of transdisciplinary interpretability has been introduced by using a new axiomatic approach: Compensatory Fuzzy Logic. Several ways have been used as fuzzy predicates searching techniques, notably a Genetic Algorithm, part of a Data Analysis Platform called Eureka Universe. This paper presents two Genetic Programming Algorithm Approaches, with outstanding results and illustrated by a case study.

Keywords Knowledge discovery · Data mining · Genetic programming · Compensatory fuzzy logic · Genetic algorithm · Generalized membership function

1 Introduction

At present, the automated discovery of knowledge in databases is becoming increasingly important; the accelerated growth of data generated by the users and services has contributed to the interest in its development. Among the various literature methods developed for this purpose, the proposed algorithm is among those that make use of logic to discover patterns.

C. E. Llorente-Peralta
Tecnológico Nacional de México, Instituto Tecnológico de Tijuana, Tijuana, Mexico
e-mail: ceric.lop@hotmail.com

L. Cruz-Reyes (✉)
Tecnológico Nacional de México, Instituto Tecnológico de Ciudad Madero, Ciudad Madero, Mexico
e-mail: lauracruzreyes@itcm.edu.mx

R. A. Espín-Andrade
Universidad Autónoma de Coahuila, Saltillo, Mexico
e-mail: espinr@uadec.edu.mx

O. Castillo and P. Melin (eds.), *Fuzzy Logic Hybrid Extensions of Neural and Optimization Algorithms: Theory and Applications*, Studies in Computational Intelligence 940,
https://doi.org/10.1007/978-3-030-68776-2_21

Thus, the compensatory fuzzy logic (CFL) has been developed by various researchers in the last decade. CFL is a multivalued logic that uses fuzzy sets and predicate logic to evaluate a given fuzzy logical predicate and its interference in a database (Ceruto Cordovés et al. 2013, 2014; Espin-Andrade et al. 2016; Marin Ortega et al. 2013; Martinez Alonso and Espin-Andrade 2013).

In related works, it is mentioned how incorporating the CFL in the use of ontologies allows enhancing formal representation and its use in knowledge management (Racet-Valdéz et al. 2010). In agreement with the results of other authors, they indicate that this approach has facilities to obtain logical predicates that reflect reality (Ceruto Cordovés et al. 2013, 2014; Martinez Alonso et al. 2014). Some other works have focused on using the CFL as a system that allows us to carry out inference processes through Klenne's logical axioms and Reichenbach's implication (Espin-Andrade et al. 2012).

It is also important to mention that in this article, it is used a novelty membership function, which is called a generalized continuous linguistic variable (GCLV) (González et al. 2021), in which, unlike what is found in the literature, the knowledge of experts is not used to carry out the vagueness modeling process, if not that this is carried out by adapting the parameters to a data set through the same function.

Using the information obtained in the literature, an evolutionary algorithm is proposed that uses the CFL to make knowledge discovery in the form of fuzzy logical predicates, which are treated from a genetic programming method (GP). Furthermore, a second algorithm is used to optimize the GCLV for each set that is part of the optimized predicate. The use of these two simultaneous concepts has not been shown in the literature.

Besides, previous publications have not made a description of the algorithms used to carry out the tests presented, but rather that algorithms that are part of commercial software have been used, thus focusing on the results of the CFL tests in heuristic algorithms.

In Sect. 2, the concepts that support the proposal are detailed. Section 3 presents the proposed nested algorithms and their components; in section four, a case study is presented. Finally, in section five, the conclusions reached, and future work are presented

2 Background and Definitions

In this section, the relevant concepts for the development of this document are presented.

2.1 Knowledge Discovery in Databases

In the literature, it is mentioned that the knowledge discovery in databases (KDD) is an automated process where discovery and analysis operations are combined in such a way it is defined as "The non-trivial process of identification of valid, novel, potentially useful patterns and fundamentally understandable to the user from of the data" (Fayyad et al. 1996). To achieve this objective, algorithms are used to produce a group of patterns and/or models on the data (Fayyad et al. 1996).

The KDD is divided into different stages, one of which is data mining (DM). At this stage, the search and discovery of patterns are carried out that are not describable at first glance and have a high degree of interest due to their usability in the company's processes and objectives (Quinlan 1986). Some works use predicate logic for DM with promising results (Atanassov 2015; Weitschek et al. 2013).

2.2 Genetic Programming

Traditionally, genetic programming (GP) carries out the evolution of programs. In this generation, after generation, the populations of programs are stochastically transformed into new populations. The results are expected to be better; however, the results cannot be guaranteed (Poli et al. 2008; Rokach and Maimon 2005).

In GP, programs are represented by syntax trees instead of lines of code; for example, in the expression $Var_1 \wedge Var_2 \vee Var_3$ the variables (Var_1, Var_2, Var_3) represent the leaves of the tree, also called terminals. On the other hand, the logical operators ($\wedge$, $\vee$) representing functions are internal nodes; these, in turn, are expressed in prefix notation (Janikow 1996; Murthy 1998; Poli et al. 2008).

Through the construction of these trees, the process of creating the initial random population is carried out. In this case, the growth tree method and the full tree method are used to exemplify.

In a complete tree, all leaves are the same depth. Nodes are randomly taken from the set of functions until the maximum depth of the tree is reached (see Fig. 1). A growth tree allows choosing from the entire set of variables and functions at any point in the construction except for the initial node and the leaves; this is done while the depth limit of the tree is not reached (see Fig. 2) (Espejo et al. 2010; Olaru and Wehenkel 2003; Poli et al. 2008).

In the evolutionary algorithms used in genetic programming, two individuals are received for the crossing, e.g. crossing points are randomly selected, and the subtrees are replaced at the selected points (see Fig. 3). In a mutation example, a mutation point is selected and replaced with a new sub-tree (see Fig. 4) (McKay et al. 2010; Murthy 1998; Poli et al. 2008).

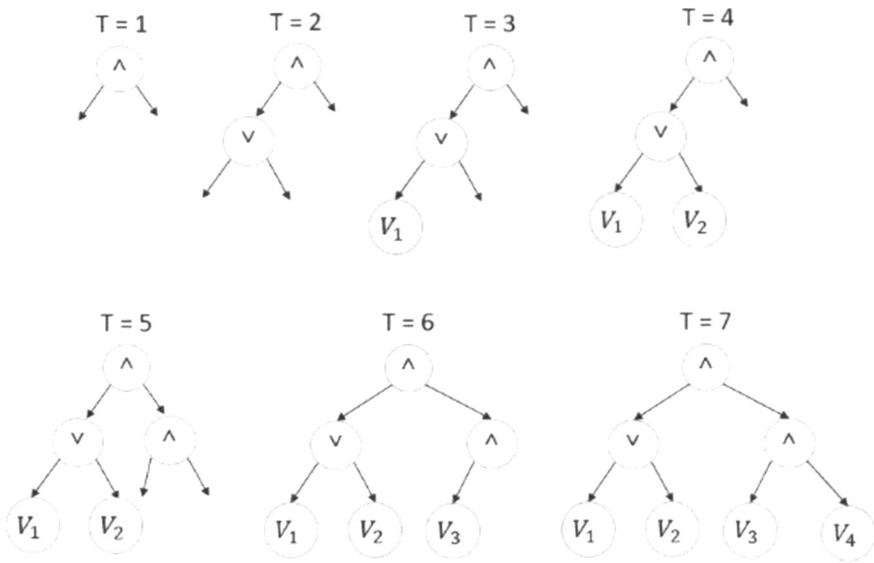

Fig. 1 Construction of the complete tree seven times

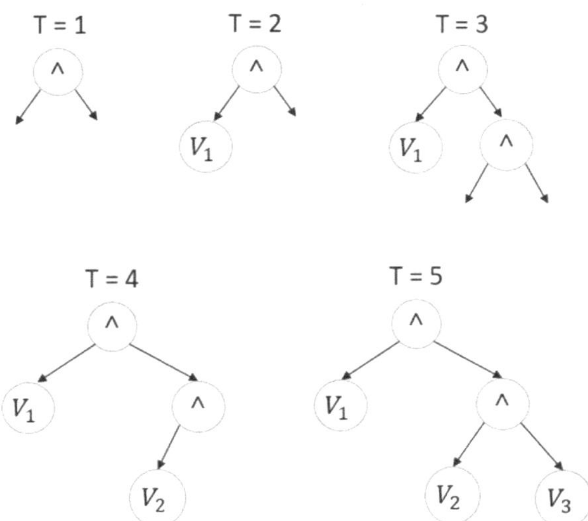

Fig. 2 Construction of a growing tree in five stages

2.3 *Compensatory Fuzzy Logic*

Compensatory fuzzy logic (CFL) is a new multivalent logic system that breaks with the traditional axiomatics of logic systems to achieve semantically better behavior than classical systems (Espin-Andrade et al. 2014a). since in the CFL, the idea is

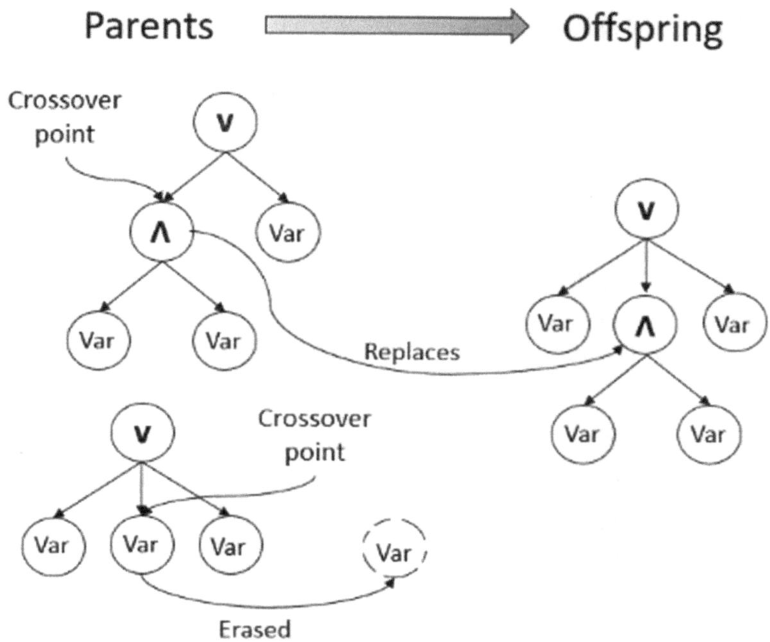

Fig. 3 Graphic representation of the subtree crossing method

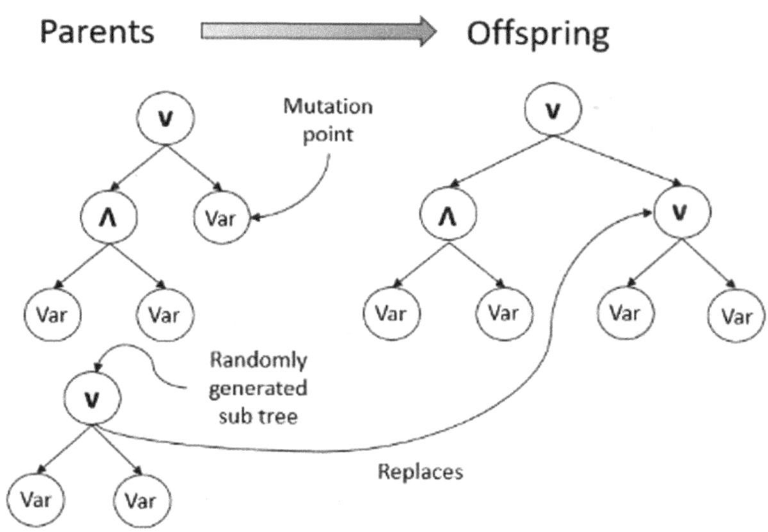

Fig. 4 Graphic representation of the headless chicken method

proposed that the increase or decrease of one set's true value can be compensated with the corresponding decrease or increase of the other (Espin-Andrade et al. 2014a).

The CFL is formed by a quartet of continuous operators that satisfy the group of following axioms: conjunction (c), disjunction (d), fuzzy strict order (o) and negation (n) (Espin-Andrade et al. 2014a, 2015; Rosete Suarez et al. 2011).

Compensation Axiom: $min(x_1, x_2, \ldots, x_n) \leq c(x_1, x_2, \ldots, x_n) \leq max(x_1, x_2, \ldots, x_n)$

Commutativity or symmetry axiom: $c(x_1, x_2, \ldots, x_n) = c(x_1, x_2, \ldots, x_n)$

Strict growth axiom: if $x_1 = y_1, x_2 = y_2, \ldots, x_{i-1} = y_{i-1} = x_{i+1} = y_{i+1}, \ldots, x_n = y_n$ they are different from zero, and $x_i > y_j$ then $c(x_1, x_2, \ldots, x_n) > c(x_1, x_2, \ldots, x_n)$

Veto axiom: If $x_i = 0$ for some i then $c(x) = 0$

Fuzzy reciprocity axiom: $o(x, y) = n[o(y, x)]$

Fuzzy transitivity axiom: If $o(x, y) \geq 0.5$ and $o(y, z) \geq 0.5$ then $o(x, y) \geq \max(o(x, y), o(y, z))$

De Morgan's Laws: $n(c(x_1, x_2, \ldots, x_n)) = d(n(x_1), n(x_2), \ldots, n(x_n))$

$$n(d(x_1, x_2, \ldots, x_n)) = c(n(x_1), n(x_2), \ldots, n(x_n))$$

where:
- x: is a linguistic variable existing in a set.
- i: is the i_th variable.
- j: It is the j_th variable.
- n: is the last variable in the set.

In this case, the compensatory fuzzy logic based on the geometric mean is used, for which the operations of conjunction, disjunction, and negation are calculated through the following formulas:

$$\text{Conjunction}: \quad c(x_1, x_2, \ldots.., x_n) = (x_1, x_2, \ldots.., x_n)^{\frac{1}{n}} \tag{1}$$

$$\text{Disjunction}: \quad d(x_1, x_2, \ldots.., x_n) = 1 - (1 - x_1, 1 - x_2, \ldots.., 1 - x_n)^{\frac{1}{n}} \tag{2}$$

$$\text{Negation}: \quad 1 - x \tag{3}$$

An implication operator should preferably start from definitions that use combinations of the conjunction, disjunction, and negation operators. Therefore the formula $i_1(x, y) = d(n(x), y)$ or $i_2(x, y) = d(n(x), c(x, y))$ can be used in a general way. The equivalence defined from the implication operator would be the following

$e(x, y) = c(i(x, y), i(y, x))$ (Espin-Andrade et al. 2016; Martinez Alonso et al. 2014).

For the universal and existential quantifiers, whatever a fuzzy predicate p on the universe U is, these are defined respectively as (Martinez Alonso et al. 2014):

$$\forall_{x \in U} p(x) = \begin{Bmatrix} e^{\frac{1}{n}\sum\limits_{x \in U} \ln(p(x))} if\ p(x) \neq 0 \\ 0\, if\ any\ x p(x) = 0 \end{Bmatrix} \quad (4)$$

$$\exists_{x \in U} p(x) = \begin{Bmatrix} 1 - e^{\frac{1}{n}\sum\limits_{x \in U} \ln(1-p(x))} if\ p(x) \neq 0 \\ 0\, if\ any\ x p(x) = 0 \end{Bmatrix} \quad (5)$$

In CFL, it is necessary to represent the fuzzy terms that exist in natural language. This process is called vagueness modeling that uses linguistic labels; with these labels, experts' knowledge is taken and applied to the problem of interest. Said linguistic variables obtain their value in scales provided by the decision-maker, such as the one defined in Table 1 (Espin-Andrade, González et al. 2014b).

An important concept within the CFL is the fuzzy sets. As an example, Zadeh proposed the set of "tall men". For example, he expresses the limitation of classical logic to classify a man as tall or short, since this classification can vary with only one centimeter of difference (Zadeh 1965).

In the CFL, a function is used that defines the transition from high to low, and for this, it assigns a value between 0 and 1 to the different heights. Depending on this value, it is considered that it belongs to the set or not.

In definition, a fuzzy set is a qualitative or quantitative characteristic of some object, which is given a membership value in the interval [0,1] employing a function that determines the degree of truth of the attribute (Espin-Andrade et al. 2015; Zadeh 1965). These membership functions can have different structures: straight, triangular, sigmoidal, in Z, Gaussian, among others (Ceruto Cordovés et al. 2013).

Table 1 Example table of vagueness modeling

Truth value	Category
0	False
0,1	Almost false
0,2	Quite false
0,3	Something false
0,4	More false than true
0,5	As true as false
0,6	More true than false
0,7	Something real
0,8	Quite true
0,9	Almost true
1	True

In this work, use is made of a function that can be optimized by varying its parameters, thus seeking its best configuration. In the literature, some works make use of the parameter optimization methodology, one of which is parameterized optimization in the particle swarm optimization algorithm to avoid being trapped in local optimum and improving performance (Valdez et al. 2017). In another work, the use of the Shadowed Type-2 Fuzzy Systems algorithm for the adaptation of dynamic parameters is proposed, obtaining favorable results in the proposed study cases (Castillo et al. 2019).

This function is called generalized linguistic continuous variable function (GCLV) is a function that adapts to the data of the input set based on the parameters that make it up, which are α, γ, and m allowing in this way to obtain a function sensitive to the data and with significant variability in its construction (González et al. 2021).

Some of the advantages of GCLV are (González et al. 2021):

(a) It contains functions of at least three types of shapes: increasing, decreasing, and convex.
(b) Linguistic labels modify the members of the GCLV.
(c) The GCLV exploits the possibilities as a universal approximator of the sigmoidal function.
(d) Its parameters have a meaning, as presented in Dombi's theory.
(e) The GCLV s the conditions suggested by Valente de Oliveira to guarantee semantics may be met.

The GCLV is calculated from the next formula (González et al. 2021):

$$GCLV(x : \alpha, \gamma, m) = \frac{Sigm(x : \alpha, \gamma)^m * (1 - Sigm(x : \alpha, \gamma)^{1-m})}{M}, \tag{6}$$

where:

$$M = m^m * (1 - m)^{1-m}$$

In Fig. 5, it can be seen how the fuzzy set named height is shown in a high degree in the dotted line (positive sigmoidal), medium in the star line (convex), and low in the solid line (negative sigmoidal).

Predicate evaluation function

Using an optimization algorithm, like the proposed in this work, it is possible to obtain compensatory fuzzy logic predicates that satisfy a minimum evaluation requirement. This value is obtained by using the universality quantifier, and in turn, this result is expressed in degrees of truth value (TV).

For this case, the compensatory fuzzy logic based on the geometric mean is used, which, according to Eq. 1, would be expressed as follows:

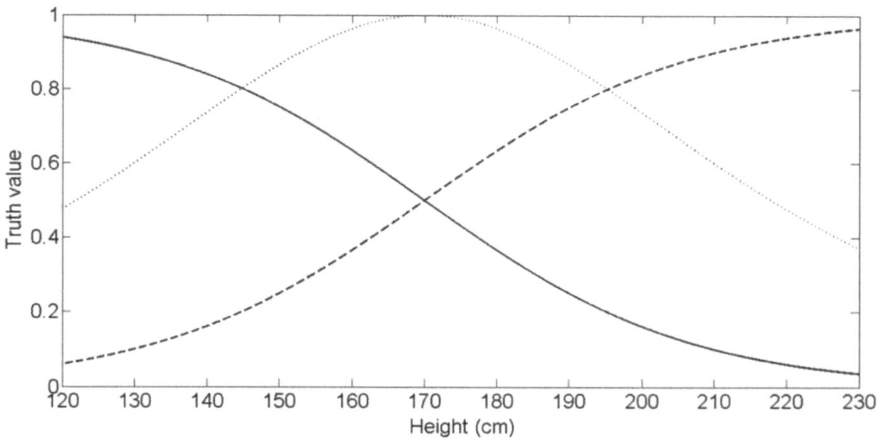

Fig. 5 Family of membership functions formed from a GCLV

$$\forall_{x \in U} p(x) = \left\{ \begin{array}{c} e^{\frac{1}{n}\sum\limits_{x \in U} \ln(p(x))} if\ p(x) \neq 0 \\ 0\, if\ any\ x p(x) = 0 \end{array} \right\}, \tag{7}$$

where:

$P =$ universe of predicates that can be generated by using operators and variables,
$p =$ predicate selected from the universe of predicates,
$U =$ Set of objects that make up the instance,
$x =$ Object that is part of the set of objects.

In order to illustrate the Eq. 7, an example is presented. For this example, the compensatory fuzzy logic based on the geometric mean is applied to a data set with three attributes: fixed acidity, volatile acidity, and citric acid attributes. Using a logical conjunction operator, the following predicate is built $p =$ fixed_acidity $\wedge$ volatile_acidity $\wedge$ citric_acid. Each attribute of the predicate is treated by the GCLV (see Eq. 3); the result of this treatment is observed in Table 2. In Table 3, the result of the conjunction (see Eq. 1) of these linguistic variables is presented.

The final predicate evaluation is obtained using the for all quantifier (Eq. 7); the result is presented in Table 4.

3 Solution Methodology

In a previous work, the first version of this algorithm was developed, which was used to obtain fuzzy logic predicates that allowed to carry out a classification process (Cruz-Reyes et al. 2019). In the present work, that algorithm will be called GA-GPv1, and the new proposed algorithm as GA-GPv2.

Table 2 Original data and its GCLV truth value

Fixed_acidity	Volatile_acidity	Citric_acid
11.6 (0.999)	0.580 (0.937)	0.66 (0.001)
10.4 (0.994)	0.610 (0.908)	0.49 (0.013)
7.4 (0.542)	1.185 (0.004)	0.00 (0.099)
10.4 (0.994)	0.440 (0.990)	0.42 (0.034)
8.3 (0.845)	1.020 (0.037)	0.02 (0.129)
7.6 (0.624)	1.580 (2.01E-5)	0.00 (0.099)
6.8 (0.299)	0.815 (0.383)	0.00 (0.099)
7.3 (0.500)	0.980 (0.062)	0.05 (0.190)
7.1 (0.415)	0.875 (0.216)	0.05 (0.190)

Table 3 Partial evaluation of a predicate using conjunction

$p(x)$
0.107603808
0.228712974
0.060869588
0.322487062
0.160208276
0.010781477
0.225259041
0.181403127
0.257840743

Table 4 Quantifier calculation for all

$ln\ (p(x))$	For all
−2.2292992	**0.1273201**
−1.4752874	
−2.7990216	
−1.1316923	
−1.8312806	
−4.5299257	
−1.4905042	
−1.7070335	
−1.3554132	

The difference between GA-GPv1 and GA-GPv2 is that in the discovery process, the former only carries out genetic mutation operations, and in the GA-GP_{v2} are added the cross-deck method and the optimization of the GCLV function (OGCLV).

3.1 Knowledge Discovery Algorithm Using Compensatory Fuzzy Logic

This section describes the operation of the GA-GP algorithm, which is a genetic algorithm that discovers rules in data instances using CFL. In Algorithm 1, the operation of GA-GP is observed.

Algorithm 1: GA-GP

Input: *generations, max_population, N_Best.*
Output: set of *N_Best* CFL predicates and their *truth value.*
1: *population* = random_population( ); *// Initialize a predicates population*
2: evaluate(*population*); *// truth value is calculated through eq 7*
3: **for** *i* = 0 **to** *generations*
4: sort(*population*); *// sort from highest to lowest truth value*
5: **for** *j* = 0 **to** (*max_population* * 0.95), with increments of two *//cross 95% of the population*
6: **if** *j* ≤ (*max_population* * 0.1) *//10% of the best individuals are crossed*
7: *kids* = cross($individual_j$, $individual_{j+1}$)
8: **else** *//crosses 85% of the population randomly*
9: *kids* =*kids* ∪ cross($individual_{rand}$, $individual_{rand}$)
10: **for** *j* = 0 **to** (*max_population* * 0.05) *//mutates 5% of the population*
11: *kids* = *kids* ∪ mutation($individual_{rand}$)
12: evaluate(*kids*);
13: replace(*population, kids*) *//replace the worst individuals with kids*
14: **for** *j* = 0 **to** max_population * 0.1 *// optimizes 10% of the population*
15: OGCLV($individuo_{rand}$); *// Optimize GCLVs of each selected predicate and obtain its fitness using Eq. 7*
16: **return** *N_Best* individuals whit their *truth value*

The Algorithm 1 receives some variables, such as the maximum number of generations (generations), the maximum number of individuals (compensatory fuzzy logical predicates) per generation, and the number of best predicates that will be returned.

In line 1, the population is initialized, for which the genetic programming trees seen in Sect. 2 are used. These trees represent compensatory fuzzy logical predicates. The method used for this construction is the half-and-a-half ramp, which uses half the growing trees and half the full trees (see Sect. 2).

In the decision trees constructed, the leaf nodes are linguistic states, and the internal nodes are CFL operators (Espejo et al. 2010; Olaru and Wehenkel 2003; Poli et al. 2008). The CFL operators used are conjunction ($\wedge$), disjunction ($\vee$), implication ($\rightarrow$), equivalence ($\leftrightarrow$), and negation ($\neg$), which are taken randomly when they are not specified.

To represent the decision trees, we opted for what was recommended in the literature, where each individual is represented by a string in prefix notation (Espejo et al. 2010; Olaru and Wehenkel 2003; Poli et al. 2008) as well as the use of auxiliaries for logical operators of CFL whose arity can be 2 to *n* possible variables. Figure 6 shows the representation of the trees presented in Figs. 1 and 2.

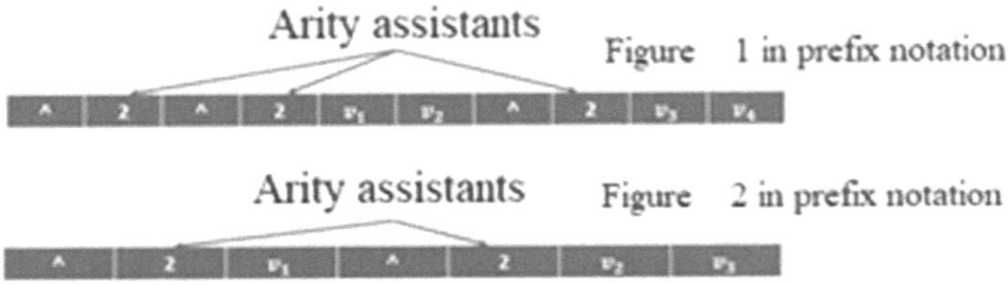

Fig. 6 Representation in prefix notation of trees in Figs. 1 and 2

Line 2 corresponds to the evaluating of each of the individuals in the population. This evaluation is achieved using Eq. 7. An example of this procedure can be seen in Sect. 2.

The genetic code's primary functioning is in lines 3–15. Within this process of evolution of generations, the genetic operations of crossing and mutation are carried out.

From lines 4–13, the selection process called uniform selection is carried out. Whitley proposed and is used in non-generational GA; uniform selection replaces a few individuals in each generation. It is useful when members of the population collectively (and not individually) solve a problem (Whitley 1989). The main processes of this method are in lines 4, 12, and 13.

From lines 5–9, we observe the crossing process for which 95% of the population is used, 10% of the best individuals are crossed in order to obtain new descendants with high truth value (lines 6–7). Later, individuals are randomly crossed with 85% of the population (lines 8–9).

In the crossing process, the sub-tree crossing method is used, seen in Sect. 2 of the document. New individuals are created by replacing the subtree rooted at the crossover point in a copy of the first parent with a copy of the subtree rooted at the second parent's crossover point (Espejo et al. 2010; Olaru and Wehenkel 2003; Poli et al. 2008).

In lines 10 and 11, the mutation process is observed, where 5% of the randomly selected population is used. In this case, the headless chick method is used in which a mutation point is selected, a new sub-tree is generated, and this is integrated into the selected crossing point (Espejo et al. 2010; Olaru and Wehenkel 2003; Poli et al. 2008).

In lines 15 and 16, the optimization of 10% of the population is performed using the OGCLV genetic algorithm, which optimizes the parameters of the GCLVs belonging to the current predicate. The number of predicates to optimized is selected in order to maintain computational efficiency. The OGCLV algorithm returns the best parameter settings discovered, as well as the truth value reached.

In line 16, a group of the n best-discovered predicates is returned with their corresponding truth value.

3.2 Generalized Continuous Linguistic Variable Algorithm

This section presents the algorithm called optimization for generalized continuous linguistic variable (OGCLV), which utilizes vagueness modeling of GCLV (González et al. 2021). Generally, this process is carried out using experts' knowledge; however, the GCLV generates a series of families of functions that, by optimizing parameters, are adapted to the data of the set. In Algorithm 2, the operation of this is presented.

For the development of the processes, the algorithm receives a series of variables, which are the maximum number of generations (generations), the maximum number of individuals per generation (max_population), and the predicate to optimize (predicate).

Algorithm 2: OGCLV

Input: *generations, max_population, predicate.*
Output: set of *optimized parameters* of each GCLV of the predicate and the *truth value*

1: *population* = random_population( ); *// initialize sets of parameters for the predicate*
2: evaluate(*population*); *// truth value is calculated through eq 7*
3: **for** $i = 1$ **to** *generations*
4:     sort(*population*); *// sort from highest to lowest truth value*
5:     **for** $j = 1$ **to** (*max_population* * 0.95) *//cross 95% of the population*
6:         **if** $j \leq$ (*max_population* * 0.1) *//10% of the best individuals are crossed*
7:             *kids* = cross_deck($individual_j$, $individual_{j+1}$);
8:         **else** *//crosses 85% of the population randomly*
9:             *kids* = *kids* ∪ cross_deck($individual_{rand}$, $individual_{rand}$);
10:     **for** $j = 1$ **to** (*max_population* * 0.05) *//mutates 5% of the population*
11:         *kids* = *kids* ∪ mutation($individual_{rand}$);
12:     evaluate(*kids*);
13:     replace(*population, kids*) *//replace the worst individuals with kids*
14: **return** set of *optimized parameters* and the *truth value of the received predicate*

For the population generation process in line 1, a numeric vector is used for the parameters of each GCLV of a given predicate. The parameters β represent the minimum value of a dataset attribute that corresponding to a truth value near zero (0.01 for example). The parameter γ defines the diffuse value 0.5 and *m* shows the tendency of the membership function to be a sigmoidal, a negative sigmoidal or a convex function. For all *n* attributes that are part of the predicate to optimize, the individual's size will be 3 * *n*. In Fig. 7, an example is showed for $n = 3$.

To explore values for each parameter during the population initialization, the Eqs. 8 to 10 are used to genera them randomly with a uniform distribution.

β_1	γ_1	m_1	β_2	γ_2	m_2	β_3	γ_3	m_3

Fig. 7 Chromosome to optimize the GCLVs of a predicate

$$\beta = Rand(Min, \ldots., Max) < \gamma \tag{8}$$

$$\gamma = Rand(Min, \ldots., Max) > \beta \tag{9}$$

$$m = Rand[0, 1] \tag{10}$$

This means that the parameter β is selected randomly from the data and must be less than γ. The parameter γ is also selected randomly from the data and must be greater than β. For the case of m, it is assigned a random value between 0 and 1.

For the selection method, the uniform selection is used again (lines 4–13) (Whitley 1989). This method is also used in the main algorithm; this method's main processes are observed in lines four, twelve, and thirteen.

From lines 5–9, we observe the crossing process for which 95% of the population is used, 10% of the best individuals are crossed in order to obtain a new descendent with high truth value (lines 6–7), later, individuals are randomly crossed with 85% of the population (lines 8–9).

In lines 10 and 11, the mutation process is executed, where 5% of the randomly selected population is used. The limit crossing method is used in which a mutation point is selected, and the current value is exchanged for a random value generated between a minimum and maximum value.

Line 14 returns the set of optimized parameters for each GCLV of the predicate. Also it returns its truth value.

For the crossing process, the cross-deck method is proposed, which is inspired by a card game's shuffling. In Algorithm 3, its design is presented.

Algorithm 3: Cross deck

Input: *parent1, parent2 // sets of parameters that are part of parents predicate*
Output: *kid // set of parameters that are part of a descendent predicate*
1. **for** $i = 1$ **to** length(*parent1*) // *from 1 to the maximum size of the individual*
2.   *cards* = random [0,0.3] * length(*parent1*); *//chooses up to 30% alleles of the individual*
3.   **for** $j = 1$ **to** ($j \leq cards$ and $i \leq$ length(*parent1*)), with global increments of i
4. $kid_i = parent1_i$;
5.   *cards* = random [0,0.3] * length(*parent2*) *//chooses up to 30% alleles of the individual*
6.   **for** $j = 1$ **to** ($j \leq cards$ and $i \leq$ length(*parent2*)), with global increments of i
7. $kid_i = parent2_i$;
8.   **if** $i <$ length (*parents*);
9.   decrement i
10. **return** *kid*

---

4 Experimentation

This section presents the experiments carried out to compare the algorithms GA-GP$_{v1}$ and GA-GP$_{v2}$, which carry out knowledge discovery and optimization of predicates in a very similar way. However, some differences may impact on the final performance. It will be analyzed the capability to build predicates, the variety of construction, and the quality of parameter optimization. It is also pertinent to mention that 30 tests were carried out for each experiment so that the results obtained can be observed objectively.

The equipment used in each of the experiments is a Dell machine with an Intel Core i7-7500u processor with 2.7 GHz * 2, a 16 Gb RAM, and a Windows 10 operating system. The language used to compile the program in java by using NetBeans. Table 5 describes the settings used in the algorithm in both versions.

The instance used is a sample of the red wine dataset (Cortez et al. 2009), which consists of the registry of 1599 wines from which 12 attributes are taken, and these are rated with different degrees of quality.

Experiment 1: Analyzing the capability to build predicates

A series of CFL predicates are selected to test the GA-GP$_{v1}$ and GA-GP$_{v2}$ algorithms in their predicate optimization process. This experiment has the objective to analyze the capability to build predicate with high truth value in equal condition of building. Table 6 shows the group of selected predicates; each of the predicates will be optimized finding the GCLVs most adequate to a given dataset.

For the KDD process, in Table 7, we can see the average truth value obtained in the experimentation carried out using the predicates of Table 4.2 for different minimum truth values defined for the optimization process. We can see how the use of the crossover method in GA-GP$_{v2}$ affects positively its performance.

Experiment 2: Analyzing the variety of predicate construction

Both algorithms were executed establishing a minimum truth value for the discovery of predicates. In Table 8, the variability in predicates construction can be observed because, in the GA-GP$_{v1}$ algorithm, only complete trees are built. In GA-GP$_{v2}$, the ramp method of half and half is used to also obtain incomplete trees.

Table 5 Configuration of the algorithms used for the KDD process

GA-GP		OGCLV
Poblaciốn	100	50
Tree depth	3	
Percentage of cross	95%	95%
Percentage of mutation	5%	5%
number of generations.	100	50
Minimum truth values	0.99, 0.98, 0.97, 0.96, 0.95	

Table 6 Predicates used in the GCLV optimization

Predicated to optimize
(IMP (AND "new citric_acid" "new free_sulfur_dioxide" "new alcohol" "new fixed_acidity" "new pH" "new residual_sugar" "new volatile_acidity" "new sulphates" "new density" "new chlorides" "new total_sulfur_dioxide") "new quality")
(IMP (NOT "new chlorides") "new quality")
(IMP (OR "new pH" "new density" "new fixed_acidity" "new chlorides") "new quality")
(IMP (OR "new volatile_acidity" "new density") "new quality")
(IMP (NOT "new sulphates") "new quality")
(IMP (OR "new chlorides" "new total_sulfur_dioxide" "new sulphates" "new citric_acid" "new fixed_acidity" "new free_sulfur_dioxide" "new pH" "new density" "new residual_sugar") "new quality")
(IMP (OR "new sulphates" "new volatile_acidity" "new residual_sugar" "new pH") "new quality")
(IMP (OR "new alcohol" "new residual_sugar" "new volatile_acidity" "new sulphates" "new pH" "new fixed_acidity" "new citric_acid" "new density" "new total_sulfur_dioxide") "new quality")
(IMP (OR "new fixed_acidity" "new pH") "new quality")
(IMP (NOT "new volatile_acidity") "new quality")
(IMP (OR "new sulphates" "new residual_sugar" "new free_sulfur_dioxide" "new chlorides") "new quality")
(IMP (OR "new residual_sugar" "new total_sulfur_dioxide" "new pH" "new citric_acid" "new density" "new free_sulfur_dioxide" "new alcohol") "new quality")
(IMP (IMP "new chlorides" "new volatile_acidity") "new quality")
(IMP (IMP "new total_sulfur_dioxide" "new density") "new quality")
(IMP (OR "new citric_acid" "new free_sulfur_dioxide" "new alcohol" "new fixed_acidity" "new pH" "new residual_sugar" "new volatile_acidity" "new sulphates" "new density" "new chlorides" "new total_sulfur_dioxide") "new quality")

Table 7 Truth values reached by the algorithms through experimentation

Results			
Minimum truth value	Truth values found by GA-GP$_{v1}$	Truth values found by GA-GP$_{v2}$	Improvement fraction
0.99	0.973650	**0.99185**	0.018200
0.98	0.937841	**0.99364**	0.055799
0.97	0.926196	**0.99129**	0.065094
0.96	0.944852	**0.96223**	0.017378
0.95	0.964967	**0.97610**	0.011133
General average	0.949501	**0.98302**	0.033521
standard deviation	0.019513	0.01359	

Table 8 Predicates constructed by the algorithms in experimentation

Discovered predicates		
Minimum truth value	Predicates built by GA-GP$_{v1}$	Predicates built by GA-GP$_{v2}$
0.99	NOT (IMP "mid residual_sugar" "low citric_acid")	(IMP(NOT(IMP"midchlorides" "lowresidual_sugar"))"lowsulphates")
0.98	NOT (IMP "high residual_sugar" "low alcohol")	(OR(IMP"midchlorides" "lowchlorides")"lowresidual_sugar")
0.97	AND (AND "low residual_sugar" "mid free_sulfur_dioxide" "mid residual_sugar" "high alcohol" "high density" "mid alcohol" "low volatile_acidity" "low fixed_acidity" "low total_sulfur_dioxide" "high residual_sugar" "mid chlorides" "high pH" "low chlorides") (AND "low pH" "low residual_sugar" "mid sulphates" "high total_sulfur_dioxide" "mid chlorides" "mid density" "low chlorides" "mid citric_acid")	(IMP"midsulphates" (NOT"midresidual_sugar"))
0.96	NOT (IMP "high residual_sugar" "low free_sulfur_dioxide")	(IMP(AND"highchlorides" (AND"middensity" "highchlorides" "midresidual_sugar" "highpH" "midchlorides" "midtotal_sulfur_dioxide"))"lowalcohol")
0.95	NOT (IMP "mid sulphates" "low total_sulfur_dioxide")	(OR"lowalcohol" (IMP"midvolatile_acidity" "midcitric_acid")(NOT"midchlorides")(EQV"midchlorides" "lowresidual_sugar"))

Experiment 3: analyzing the quality of parameter optimization

For this analysis, Table 9 shows the average results obtained for each predicate of Table 8. The highest truth value obtained for the optimized predicate was marked with bold. Note that in many cases there is an important difference between the results obtained by GA-GP$_{v1}$ and those obtained by GA-GP$_{v2}$.

Through the results obtained, the Wilcoxon test was used to verify statistically, if there is a difference between one and the other. To carry out this test, the STAC

Table 9 Truth values achieved by GA-GP$_{v1}$ and GA-GP$_{v2}$

Predicate	GA-GP$_{v2}$	GA-GP$_{v1}$
1	**0.976430137**	0.811614770
2	0.982110969	**0.988766950**
3	**0.903058920**	0.795068110
4	**0.916270852**	0.841222380
5	0.973208652	**0.973600450**
6	**0.866627112**	0.732014310
7	**0.909053824**	0.819276530
8	**0.874235399**	0.725561380
9	**0.918507379**	**0.844661268**
10	0.966643202	**0.967475837**
11	**0.916384717**	0.870367931
12	**0.885982721**	0.748333851
13	**0.933014655**	0.902903347
14	**0.927128220**	0.863925856
15	**0.861977535**	0.707722381
Averange	0.920708000	0.839501020
Stándard deviation	0.039789090	0.090873919

platform (Rodríguez-Fdez et al. 2015) was used, which determines statistically that, there is evidence to conclude that the GA-GP$_{v2}$ algorithm is superior GA-GP$_{v1}$, by obtaining a p-Value of 0.0035. Besides, GA-GP$_{v2}$ has shown less variability in this experiment.

5 Conclusions

Knowledge discovery is a problem that increases daily due to the large amount of data generated worldwide; it is not a simple problem from this perspective. That is why it is essential to develop methods that allow extracting accurate, usable, and relevant information.

In the last decade, a group of researchers has used compensatory fuzzy logic (CFL). However, these works use commercial software to discover CFL predicates, leaving the process as a black box. For this reason, it is proposed an evolutionary algorithm called GA-GP. This algorithm allows realizing the knowledge discovery process through the use of genetic programming methods.

This work proposes a new version of the GA-GP algorithm, in which is added a crossing method to provide further flexibility to the solutions construction process. We also make variations to the evolutionary optimization process of a generalized linguistic continuous variable (GLCV), a new membership function adaptable to a given dataset.

We present a case study to compare the two versions of GA-GP. In this comparison, given a set of predicates structures, the new version offers greater flexibility in building better quality solutions. Besides, letting the algorithms discover predicates freely, a new version constructs more variety in the sense of balance and unbalance. When comparing the optimization results statistically, we demonstrate that there is significant evidence to affirm the new version improves the previous one.

Future works propose to continue improving the proposed algorithm through other genetic operators and evolutionary optimization methods.

References

Atanassov, K. 2015. Intuitionistic fuzzy logics as tools for evaluation of Data Mining processes. *Knowledge-Based Systems* 80:122–130. https://doi.org/10.1016/j.knosys.2015.01.015

Castillo, O., P. Melin, F. Valdez, J. Soria, E. Ontiveros-Robles, C. Peraza, and P. Ochoa. 2019. Shadowed Type-2 Fuzzy systems for dynamic parameter adaptation in harmony search and Differential Evolution Algorithms. *Algorithms* 12(1). https://doi.org/10.3390/a12010017

Ceruto Cordoves, T., O. Lapeira Mena, A. Rosete Suarez, and R.A. Espin-Andrade. 2013. Discovery of fuzzy predicates in database. *Fourth International Workshop on Knowledge Discovery, Knowledge Management and Decision Support* 45–54. https://doi.org/10.2991/.2013.6

Ceruto Cordovés, T., A. Rosete Suárez, and R.A. Espín-Andrade. 2014. Knowledge discovery by fuzzy predicates. *Studies in Computational Intelligence* 537:187–196. https://doi.org/10.1007/978-3-642-53737-0_13

Cortez P., A. Cerdeira, F. Almeida, T. Matos, and J. Reis. 2009. Modeling wine preferences by data mining from physicochemical properties. In *Decision Support* Systems, Elsevier, 47(4):547–553. https://archive.ics.uci.edu/ml/datasets/wine+quality, https://doi.org/10.1016/j.dss.2009.05.016

Cruz-Reyes, L., R.A. Espin-Andrade, F.L. Irrarragorri, C. Medina-Trejo, J.F. Padrón Tristán, D.A. Martinez-Vega, and C.E. Llorente Peralta. 2019. Use of compensatory fuzzy logic for knowledge discovery applied to the warehouse order picking problem for real-time order batching. In: *Handbook of Research on Metaheuristics for Order Picking Optimization in Warehouses to Smart Cities,* 62–88. IGI Global. https://doi.org/10.4018/978-1-5225-8131-4.ch004

Espejo, P. G., S. Ventura, and F. Herrera. 2010. A survey on the application of genetic programming to classification. *IEEE Transactions on Systems, Man and Cybernetics* Part C: Applications and Reviews, 40(2):121–144. https://doi.org/10.1109/TSMCC.2009.2033566

Espin-Andrade, R.A., E. González, and E. Fernandez. 2012. A compensatory inference system. *CLAIO* 4404–4415. http://citeseerx.ist.psu.edu/viewdoc/download?doi=10.1.1.1052.2991&rep=rep1&type=pdf

Espin-Andrade, R.A., E. Fernández, and E. González. 2014a. Compensatory fuzzy logic: A frame for reasoning and modeling preference knowledge in intelligent systems. *Soft Computing for Business Intelligence. Studies in Computational Intelligence.* Springer, Berlin, Heidelberg., 537:3–23. https://doi.org/10.1007/978-3-642-53737-0_1

Espin-Andrade, R.A., E. González, E. Fernández, and M. Martinez Alonso. 2014b. Compensatory fuzzy logic inference. *Soft Computing for Business Intelligence* 25–43. https://doi.org/10.1007/978-3-642-53737-0_2

Espin-Andrade, R.A., E.G. Caballero, W. Pedrycz, and E. R. Fernández González. 2015. Archimedean-compensatory fuzzy logic systems. *International Journal of Computational Intelligence Systems* 8(2):54–62. https://doi.org/10.1080/18756891.2015.1129591

Espin-Andrade, R.A., E. Gonzalez, W. Pedrycz, and E. Fernandez. 2016. An interpretable logical theory: The case of compensatory fuzzy logic. *International Journal of Computational Intelligence Systems* 9(4):612–626. https://doi.org/10.1080/18756891.2016.1204111

Fayyad, U., G. Piatetsky-Shapiro, and P. Smyth. 1996. The KDD process for extracting useful knowledge from volumes of data. *Communications of the ACM,* 39(11):27–34. https://doi.org/10.1145/240455.240464

González, E., R.A. Espin-Andrade, L. Martinez, and L.A. Guerrero-ramos. 2021. Continuous linguistic variables and their applications to data mining and time series prediction. *International Journal of Fuzzy Systems* 1–22. https://doi.org/10.1007/s40815-020-00968-w

Janikow, C.Z. 1996. A genetic algorithm method for optimizing fuzzy decision trees. *Information Sciences* 89(3–4):275–296. https://doi.org/10.1201/9780203713402

Marin Ortega, P.M., R.A. Espin-Andrade, and J. Marx Gomez. 2013. Multivalued fuzzy logics: A sensitive analysis. *Fourth International Workshop on Knowledge Discovery, Knowledge Management and Decision Support* 1–7. https://doi.org/10.2991/.2013.1

Martinez Alonso, M., and R.A. Espin-Andrade. 2013. Knowledge discovery by Compensatory Fuzzy Logic predicates using a metaheuristic approach. *Fourth International Workshop on Knowledge Discovery, Knowledge Management and Decision Support* 17–26. https://doi.org/10.2991/.2013.3

Martinez Alonso, M., R.A. Espín-Andrade, V.L. Batista, and A. Rosete Suárez. 2014. Discovering knowledge by fuzzy predicates in compensatory fuzzy logic using metaheuristic algorithms. *Soft Computing for Business Intelligence* 537:161–174. https://doi.org/10.1007/978-3-642-53737-0_11

McKay, R.I., N.X. Hoai, P.A. Whigham, Y. Shan, and M. O'neill. 2010. Grammar-based genetic programming: A survey. *Genetic Programming and Evolvable Machines* 11(3):365–396. https://doi.org/10.1007/s10710-010-9109-y

Murthy, S.K. (1998). Automatic construction of decision trees from data: A multi-disciplinary survey. *Data Mining and Knowledge Discovery* 345–389. https://doi.org/10.1023/A:1009744630224

Olaru, C., and L. Wehenkel. 2003. A complete fuzzy decision tree technique. *Fuzzy Sets and Systems* 138(2):221–254. https://doi.org/10.1016/S0165-0114(03)00089-7

Poli, R., W.B. Langdon, and N.F. McPhee. 2008. A field guide to genetic programing. *Lulu.Com* 3:233. http://www.essex.ac.uk/wyvern/2008-04/WyvernApril087126.pdf

Quinlan, J. R. 1986. Induction of decision trees. *Machine Learning* 1(1):81–106. https://doi.org/10.1023/A:1022643204877

Racet-Valdéz, A., R.A. Espin-Andrade, and J. Marx-Gómez. 2010. Compensatory fuzzy ontology. *International Conference in ICT Innovations* 2009. Springer, Berlin, Heidelberg., 35–44. https://doi.org/10.1007/978-3-642-10781-8_5

Rodríguez-Fdez I., A. Canosa, M. Mucientes, A. Bugarín. 2015. STAC: a web platform for the comparison of algorithms using statistical tests. In Proceedings of the 2015 IEEE International Conference on Fuzzy Systems (FUZZ-IEEE)

Rokach, L., and O. Maimon. 2005. Decision trees. *Data Mining and Knowledge Discovery Handbook*. Springer, Boston, MA. https://doi.org/10.1007/0-387-25465-x_9

Rosete Suarez, A., Ceruto Cordovés, T., & Espin-Andrade, R.A. 2011. A General Method for Knowledge Discovery using Compensatory Fuzzy Logic and Metaheuristics. *Gathering Knowledge Discovery, Knowledge Management and Decision Making* 240–271

Valdez, F., J.C. Vazquez, P. Melin, and O. Castillo. 2017. Comparative study of the use of fuzzy logic in improving particle swarm optimization variants for mathematical functions using co-evolution. *Applied Soft Computing Journal*, 52, 1070–1083. https://doi.org/10.1016/j.asoc.2016.09.024

Weitschek, E., G. Felici, and P. Bertolazzi. 2013. Clinical data mining: Problems, pitfalls and solutions. *24th International Workshop on Database and Expert Systems Applications*, 90–94. https://doi.org/10.1109/DEXA.2013.42

Whitley, D. 1989. The GENITOR algorithm and selection pressure: Why rank-based allocation of reproductive trials is best. *Proceedings of the Third International Conference on Genetic Algorithms*, 116–123.

Zadeh, L.A. 1965. Fuzzy sets. *Information and Control* 8(3):338–353. https://doi.org/10.1061/9780784413616.194